Nutrients and Nutraceuticals for Active & Healthy Ageing

Seyed Mohammad Nabavi •
Grazia D'Onofrio • Seyed Fazel Nabavi
Editors

Nutrients and Nutraceuticals for Active & Healthy Ageing

Springer

Editors
Seyed Mohammad Nabavi
Applied Biotechnology Research Center
Baqiyatallah University of Medical Sciences
Tehran, Iran

Grazia D'Onofrio
Clinical Psychology Service, Health Department, Fondazione IRCCS Casa Sollievo della Sofferenza
San Giovanni Rotondo, Italy

Seyed Fazel Nabavi
Applied Biotechnology Research Center
Baqiyatallah University of Medical Sciences
Tehran, Iran

ISBN 978-981-15-3551-2 ISBN 978-981-15-3552-9 (eBook)
https://doi.org/10.1007/978-981-15-3552-9

This Springer imprint is published by the registered company Springer Nature Singapore Pte Ltd.
The registered company address is: 152 Beach Road, #21-01/04 Gateway East, Singapore 189721, Singapore

Contents

About the Editors

Seyed Mohammad Nabavi is a Biotechnologist and Senior Scientist in the Applied Biotechnology Research Center, Baqiyatallah University of Medical Sciences, Tehran, Iran. His research interests are focused on natural products. Till now, he has authored/co-authored over 250 publications in international journals indexed in ISI and has more than 7000 citations to published articles and also more than 10 books.

Grazia D'Onofrio is working as Psychologist and Researcher in the Clinical Psychology Service, Health Department, Fondazione IRCCS Casa Sollievo della Sofferenza, San Giovanni Rotondo, Italy. In 2009, she was awarded the title of Expert in Integrated Psychodynamic Psychotherapy. Her research interests are focused on various aspects of dementia, psychological and behavioral symptoms, and multidimensional impairment. She has authored numerous scientific articles which are published in international journals of high repute and book chapters. Currently, she is Editorial Board Member for some of the international journals and Guest Editor for special issue on the Alzheimer's disease. She is participating in various international projects, namely European and Japan-European projects (Horizon 2020) on the development of technological systems to promote healthy and active aging.

Seyed Fazel Nabavi is a Biotechnologist and Senior Scientist in the Applied Biotechnology Research Center, Baqiyatallah University of Medical Sciences, Tehran, Iran. His research interests are focused on natural products. Till now, he has authored/co-authored over 200 publications in international journals indexed in ISI and has more than 6000 citations to published articles and also more than 5 books.

Demographic and Epidemiological Aspects of Aging 1

Della Grace Thomas Parambi, M. K. Unnikrishnan, Akash Marathakam, and Bijo Mathew

Abstract

With the unprecedented surge in geriatric populations, public health policies should prioritize plans for ensuring independence and dignity of elders. Demographic transformations have social, political, and economic implications that influence funding and provisioning geriatric care. While maximum life span is genetically determined, East Asia has shown the fastest improvements in life expectancy at birth, increasing from 45 years (1950) to 74 years (2005). 65-year olds would increase from 12.5% to 20% in the USA by 2030, while China and India would encounter larger numbers. Health promotion schemes and health care management strategies have propelled an abrupt rise in the survival rates of elders, accompanied by the increasing need for trained personnel, specialized care, and budgetary policies to earmark funds from young taxpayers to pay for geriatric care. "Compression of morbidity" hypothesis optimistically proposes that age-related disease and disability can be postponed to terminal years of life. Education, supportive technologies, and treatments improve the quality of life of elderly citizens in developed industrialized societies, but the obesity epidemic is a looming threat. Bioethicists warn that rising geriatric populations demand detailed plans to avert bankruptcy of insurance companies. It is important to understand that the infirmity of old age is, at least partly, the inevitable outcome of

D. G. T. Parambi
College of Pharmacy, Department of Pharmaceutical Chemistry, Jouf University, Sakaka, Al Jouf, Saudi Arabia

M. K. Unnikrishnan · A. Marathakam
Department of Pharmaceutical Chemistry, National College of Pharmacy, Calicut, Kerala, India

B. Mathew (✉)
Division of Drug Design and Medicinal Chemistry Research Lab, Department of Pharmaceutical Chemistry, Ahalia School of Pharmacy, Palakkad, Kerala, India

S. M. Nabavi et al. (eds.), *Nutrients and Nutraceuticals for Active & Healthy Ageing*,
https://doi.org/10.1007/978-981-15-3552-9_1

deleterious mutations accumulating in post-reproductive life. Reproductive success being key to survival, harmful mutations that express in early reproductive period of life are selectively deleted, while deleterious mutations expressing in post-reproductive life would accumulate. Antagonistic pleiotropy proposes that genes serving a favorable outcome in early life would accumulate, even if harmful in later life, leading to senescence. Genes promoting inflammatory repair in early life have been implicated in chronic inflammatory diseases of old age. Public health policies, especially in lifestyle disorders, can learn from evolutionary biology, but such research is scanty. While global mortality rates have plummeted towards the end of the twentieth century, death rates from Alzheimer's disease, heart disease, cerebrovascular disease, influenza, and pneumonia registered exponential increases with age. Most common chronic conditions among elderly Americans are hypertension, coronary heart diseases, chronic joint symptoms, and stroke. Patterns in comorbidity-co-occurrence of chronic conditions help assess disease burden and prevalence. Hypertension, cholesterol, sedentary life, obesity, and smoking have been associated with mortality from CHD, cancer, and stroke. Lung, colon, prostate, bladder, and rectum cancers show the highest incidence in men, whereas breast, uterine, colon, rectum, and lung cancers are the highest in women. An OECD report suggests that dementia affects 30% of those aged 85–89. Epidemiological data on "functional impairment" or "disability," reveal that greater understanding of the occurrence, causes, and consequences of disability help formulate preventive therapies for minimizing disability. Metrics of disability can be a powerful tool for estimating adverse reactions, presence and severity of multiple pathogeneses, including physical, cognitive, and psychological, as well as their potential synergistic effects. Activities of daily living (ADL), instrumental activities of daily living (IADL), and objective assessment of physical performance help assess capacity for independent community life, an essential aspect of long-term geriatric care.

Keywords

Aging · Demography · Epidemiology · Life expectancy · Disability · Mortality

Aging is an inescapable outcome of a long life that each of us values highly. While the ability to live long is, at least partly, the gift of modern healthcare, the absolute conquest of aging per se may not yet be a priority in public health. On the other hand, ensuring a life of independence, respectability, and dignity is indeed the foremost public health concern in a world that is progressively getting older every day, with a significant proportion of the geriatric population suffering from chronic ailments without any definite cure.

Today, the geriatric community comprises a greater portion of humanity than at any other time in history. Yesteryears teach us that demographic transformations have influenced and will continue to have a far-reaching impact on the social, political, and economic status of all sections of the society (US Census Bureau n.d.; Wang et al.

2012). According to WHO, in about 5 years, the number of people 65 years or older would outnumber the children under 5 years.

It is estimated that the geriatric population that touched 524 million in 2010 would soar to 1.5 billion in 2050. This dramatic increase in the population of old people has won considerable attention and has been a subject of extensive debate and discussion. The current and future of geriatric health profiles constitute a boundless concern of the world today (United Nations n.d.-a; Kinsella and Phillips n.d.; Population Division. U.S. Census Bureau 2000). United Nations population projection study (Fig. 1.1) also reveals that the population of 85 years and older is expected to rise by 351% between 2010 and 2050 (Fig. 1.2).

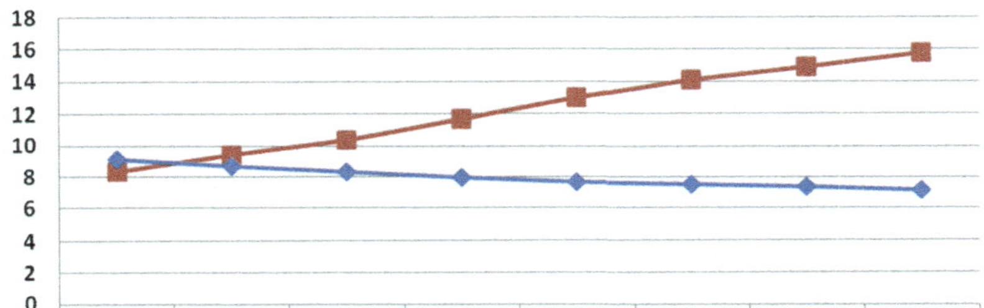

	2015	2020	2025	2030	2035	2040	2045	2050
AGE 65 +	8	9	10.4	11.7	13	14.1	14.9	15.8
age<5	9.1	8.7	8.3	8	8	7.5	7	7

Fig. 1.1 Young children and old population as a percentage of world population (Source: United Nations: Population Division; Department of Economic and Social Affairs: POP/DB/WPP/Rev.2017/POP/F09-1)

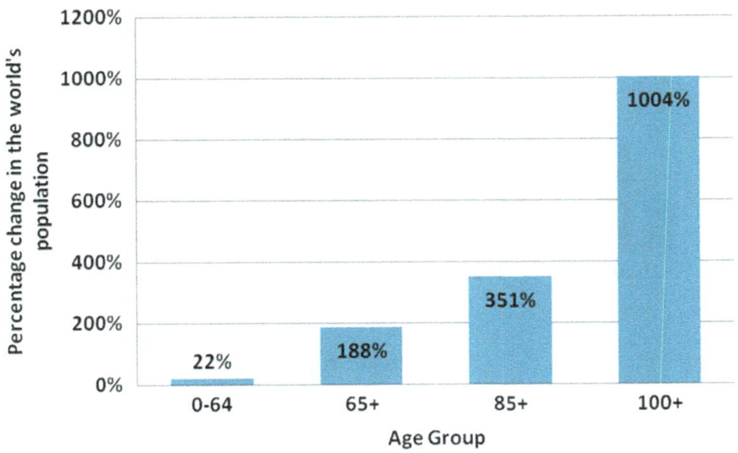

Fig. 1.2 Population projection study from 2010 to 2050 between ages

1.1 The Average Life Expectancy Is Increasing

The bioscientific, medical model of disease and longevity suggests that particular genotypes might live much longer than actually predicted (Yashin et al. 1998; Toupance et al. 1998; World Health Organization. National Institute on Aging, National Institutes of Health, U.S. Department of Health and Human Services 2011). An anthropological study suggests that the average length of life in Japan has risen to 83 years today and at least 81 years in several other countries. Most dramatic progress was observed in East Asia, where life expectancy at birth increased from 45 years in 1950 to 74 years in 2005. Such data are in tune with life expectancy improvements, which constitute a major part of the socio-politico-structural changes spreading across the world at different scales (Oeppen and Vaupel 2002; Christensen et al. 2009).

It was only in the early 1980s that the demographers started analyzing the increase in life expectancy, driven by decline in fertility and mortality (Linda 2004; Simmons 1945; Treas and Logue 1986; Kinsella and Taueber 1993; Martin 1987). Adults 65 years or older now constitute 12.5% of the population of the USA and are predicted to grow by 20% by 2030. Highly populated countries like China and India would witness a greater numerical surge than developed countries like the USA (Fig. 1.3) (National Centre for Health Statistics 1978; United Nations n.d.-b).

Health promotion schemes and health care management strategies in many countries have propelled an abrupt rise in the survival rates of persons 85 years and older. There has been a steadily rising academic interest on account of the looming public health crisis from an aging population, accompanied by the

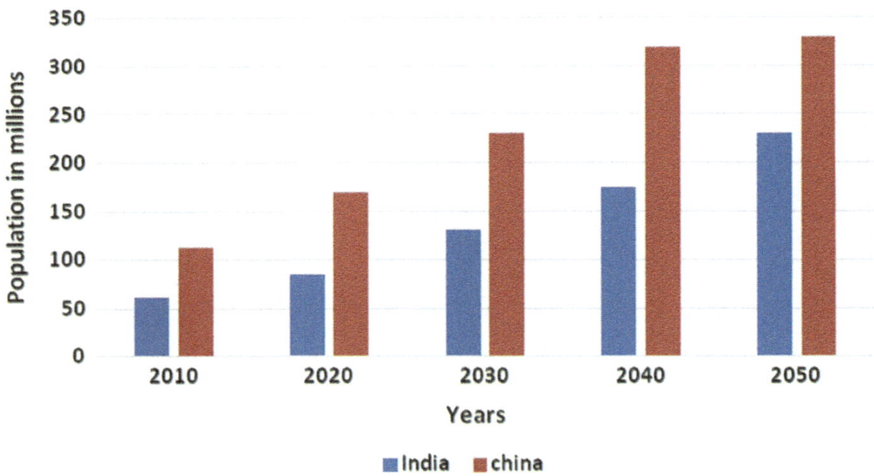

Fig. 1.3 Projected population of aged 65 and older in India and China (source: United Nations. World population prospects)

increasing need for trained personnel, specialized care, and challenging budget constraints. Many developed countries have to earmark funds from young taxpayers to meet the rising public health costs of geriatric care.

1.2 Biodemography of Aging (James 2003)

The challenges of a booming life expectancy in the 1980s were a hot topic for every gerontologist. James Fries (1980, 1996) one among the most prominent experts, in his widely cited article, has asserted that there are two kinds of deaths: premature and senescent. Premature deaths occur as unanticipated events such as accidents or diseases. Senescent death, on the other hand, knocks down a person when he or she approaches his or her maximum life span. Every human being has a maximum life span, distributed around 85 years with a mean deviation of 7 years (Cheung et al. 2005). According to the Fries theory, nothing can be done to alter the maximum life span, which is innate, fixed, and beyond the influence of any possible natural, behavioral, or medical therapies. This reminds us of Aristotle who compared premature death to a fire extinguished by splashing water and senescent death to a fire burning itself out. He also proposed that every human is born with a certain amount of fuel and that no fuel could be added to surmount the maximum potential life span.

Fries, Prof of Medicine, Stanford University, in 1980, proposed that illness can be postponed to the few final years of terminal life. This is known as "compression of morbidity" hypothesis, a noteworthy idea with significant and optimistic implications of practical value in public health. In a study that followed up thousands of University of Pennsylvania alumni for more than two decades, researchers showed that the compression of morbidity hypothesis is valid. If the age of onset of the first chronic infirmity could indeed be postponed, the therapeutic burden from the geriatric populations could also be minimized, much against the notion that any aging society would continue to consume more and more resources. Evidence for the possibility of delaying the onset of infirmity is mixed, at best. A review on the compression of morbidity has reported cross-national evidence in its support. Education, availability of supportive technologies, and medical treatments have been found to improve the quality of life of older persons in developed industrialized societies. However, if this is to continue to the next generation, the obesity epidemic should be contained. Nutritional abundance and higher incomes, thanks to modern technology, have aggravated the obesity epidemic.

American bioethicists, in their eloquent arguments for rationing medical care, hypothesize that increasing longevity and the consequent rise in the population of the elderly, demands meticulous plans to avert the possible bankruptcy of insurance companies. Faria (2015), professor of neurosurgery, Mercer University School of

Medicine, has also added more resourceful approaches with the implementation of Obama care. Reforms should also focus on convincing Americans who are accustomed to receive medical care by third-party payers.

1.3 The Inevitability of Aging: Lessons from Evolution

Many theories explain the universality of the aging phenomenon (Fabian and Flatt 2011; Rose and Evolutionary 1991; Rose and Charlesworth 1980; Hughes et al. 2002). JBS Haldane and Medawar proposed the theory of deleterious mutations accumulating in the post-reproductive period of life. Medawar argued that reproductive success being key to survival, harmful mutations that express in early life, i.e. during the reproductive period, are selectively deleted over the course of multiple generations. On the other hand, deleterious mutations in genes that express in post-reproductive life would accumulate, because selection pressure eases after reproductive age. Selection pressure also favors higher fertility rates among the young because of high mortality rates from injury and disease. Thus, over a period of many generations, the cumulative load of deleterious mutations that express late in life would result in infirmity that generally accompanies aging.

Hamilton (1966, 1996) proposed that the accumulation of favorable mutations promoting early reproductive fitness would eventually increase mortality and decrease fertility later in life, even without harmful mutations or trade-off effects.

George Williams developed on the idea of accumulating deleterious mutations in later life to propose what is now called the theory of antagonistic pleiotropy (Williams 1957). Williams argues that genes often serve multiple functions at different time points in the lifetime of a living being. Such pleiotropic genes, which express a favorable outcome in early life and a harmful outcome in later life, are likely to accumulate over many generations. On the other hand, genes with harmful outcomes in early life would be lost. Thus, over a period of many generations, there would be a tendency to accumulate pleiotropic genes that result in senescence. Antagonistic pleiotropy has been demonstrated in humans, at least circumstantially. It has been shown that genes whose expression promotes inflammatory repair in early life are also implicated in many chronic inflammatory diseases that characterize geriatric populations. A higher pro-inflammatory ratio of cytokines TNFα/IL-10 was shown to be associated with death from cardiovascular ailments in geriatric populations. This is compatible with the antagonistic pleiotropy hypothesis because higher ratios of TNFα/IL-10 would protect individuals from infection during the reproductive stage of life (Van Den Biggelaar 2004).

Public health policies, especially in lifestyle disorders, can learn a great deal from the lessons in evolutionary biology. There is already an impressive array of researchers working in this direction, but the impact is yet to be felt in mainstream public health research.

1.4 Epidemiology of Aging

The advocacy of International Cooperation in National Research in Demography (CICRED) has endeavored to create demographic compilations summarizing multiple characteristics of elderly populations (rural/urban residence, marital status, and labor participation), in addition to the preparation of individual country reports (Minino et al. 2007). Such compilations have helped the international community in the development of relevant monographs for many developing and developed countries.

1.5 Mortality

A plethora of mortality reports (Bureau of the Census 1991) implicates heart disease, followed by cancer, as the most common causes of death, in younger individuals. On the other hand, the five major causes of death in the total populations are listed as heart disease (Hadley and Rossi 2005; Centers for Disease Control and Prevention, National Center for Health Statistics n.d.; Lolak et al. 2006), cancer (Blank and Bellizzi 2008; Extermann and Aapro 2000; Extermann and Hurria 2007; Balducci 2003; Wells and Balducci 1997; Balducci and Extermann 2000; Hayat et al. 2007), stroke (Lakatta 1985), lower respiratory tract infections, and Alzheimer's disease. An exponential increase in mortality rate with age is seen for heart disease, cerebrovascular disease, influenza, and pneumonia. An even steeper rise is observed for Alzheimer's disease (Launer et al. 1999; Obadia et al. 1997; Lobo et al. 2000; Suh and Shah 2001; Rice et al. 2001). On the other hand, cancer and lower respiratory tract infections do not show a steep rise with age, probably because victims under this category are likely to be cigarette smokers who die young. The case is similar to severe diabetics among who, a disproportionately large fraction die young. Mortality rates have witnessed a dramatic fall towards the second half of the twentieth century when compared to the first.

1.6 Disease Status

According to a study by the National Institute of Aging, the most commonly reported chronic conditions in the USA among the older population were hypertension, coronary heart diseases, and chronic joint symptoms and stroke.

Table 1.1 Diseases associated with aging

Age-related change	Diseases
High blood pressure and cholesterol	Heart attack and stroke
Low carbohydrate metabolism	Diabetes
Bone loss	Fractures and osteoporosis
Amyloid plaques in brain	Dementia, AD, Parkinson's disease
Damage to respiratory system	Bronchitis, emphysema
Damage to cartilage	Arthritis
Opacity of lens and increased ocular pressure	Cataract and glaucoma
Uncontrolled cell proliferation	Cancer

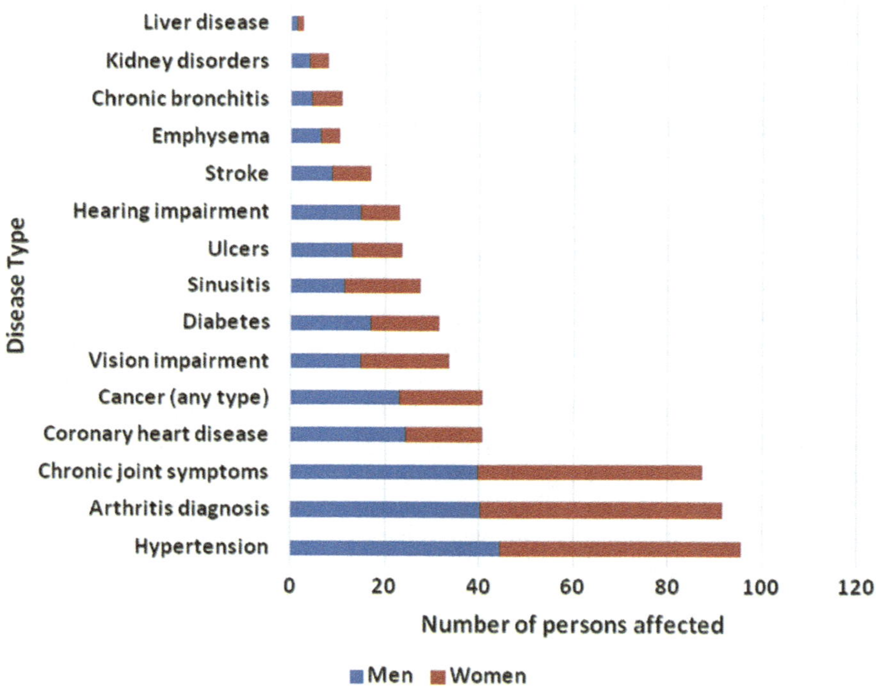

Prevalence and incidence of chronic conditions are also dependent on race and ethnicity (Fries 1980). The concept of comorbidity-co-occurrence of multiple chronic conditions (Neale and Kendler 1995; Fried et al. 2004; Barnett et al. 2012; Wolff et al. 2002; Freid et al. 2012; Guralnik 1996) is a useful indicator in assessing both disease burden and prevalence. Table 1.1 shows age-related degenerative changes in the body and the diseases that possibly arise as a consequence (Khaw 1997).

Widespread research in lifestyle disorders and functional impairment has provided immense insight into identifying multiple risk factors (B.P, cholesterol, sedentary life, obesity, and smoking) for fatal conditions such as CHD, cancer, and stroke.

Cancer mortality rates do not directly reflect the incidence rates of diagnosed cancers. Data generated by the SEER survey by "National Cancer Institute and Cancer Registries" have shown that highest incidence in men are seen for lung, colon, prostate, bladder, and rectum and in women for breast, uterine, colon, rectum, and lung cancers. However, incidence of prostrate, breast, and lung cancers drop at the oldest ages.

According to WHO, by 2030, non-communicable diseases are expected to constitute more than one half of the total disease burden, while infectious and parasitic diseases would account for 30% and 10%, respectively, in low-income countries (Frenk et al. 1991; Brandling-Bennett 1991; Oakes et al. 1991).

Prevalence estimates for Alzheimer's disease, the most common form of dementia, vary markedly between nations probably because diagnoses and reporting systems are not standardized.

A report by the Organization for Economic Cooperation and Development (OECD) suggests that dementia affected 30% of those aged 85–89. The projections by the "Alzheimer's Disease International" indicate that 115 million people worldwide are affected by AD/dementia.

1.7 Disability and Longer Lives

When resolving "functional impairment" or "disability," among population subgroups, we should particularly pay attention to geriatric populations, which constitute the largest numbers. Recent epidemiological data reveal that a greater understanding of the occurrence, determinants, and consequences of disability in the older community offers insight into various preventive therapies for minimizing disability. Disability status has also been identified as one of the most powerful tools for estimating adverse reactions, presence and severity of multiple pathogeneses, including physical, cognitive, and psychological, as well as their potential synergistic effects on health status (Manton et al. 1986; Manton and Myers 1987).

Tools used for assessing disability were gradually adapted from those employed in clinical research for assessing the effects of medical conditions such as stroke on physical and mental functioning, work ability, and the need for formal and informal care of older people. Instead of assessing basic tasks (such as standing, reaching, and gasping) as measures of disability; activities of daily living (ADL) and instrumental activities of daily living (IADL) are selected for assessment because they are more essential for independent living in the community. There is no single instrument which exactly fits into the measurement of disability. Lack of standardization of different techniques makes it difficult to compare disability ratings across various studies (Lauretani et al. 2008; McDermott et al. 2008; Semba et al. 2007; Bandeen-Roche et al. 2006; Onder et al. 2005; Chang et al. 2004; Valenti et al. 2004; Ferrucci et al. 2002; Guralnik et al. 2001; Simonsick et al. 2001).

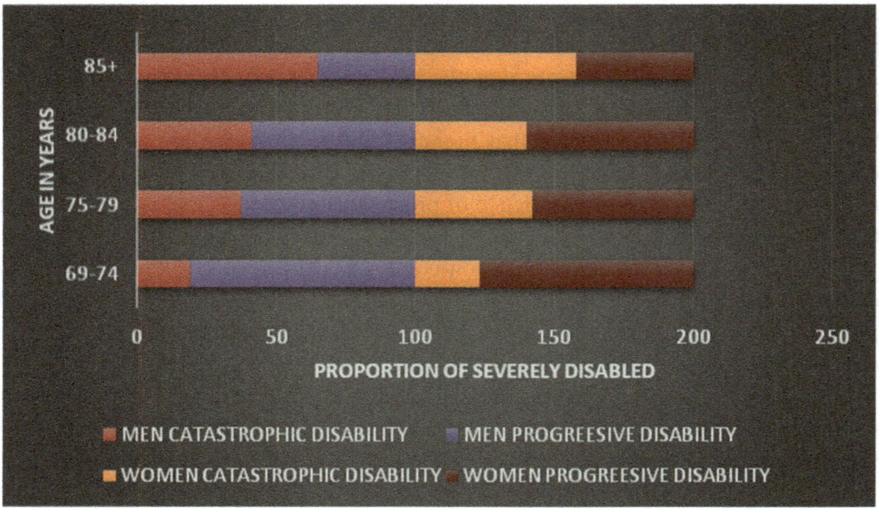

Fig. 1.4 Percentage of "catastrophic and progressive disability among old ages

There are two terms being constantly employed in understanding the dynamics of disability progression (Guralnik and Ferrucci 2003).

1. Progressive disability: Indicates passive decline of health that results from one or more ongoing chronic conditions.
2. Catastrophic disability: Occurs instantaneously in response to stroke or hip fracture.

   A survey by the National Institute on Aging under National Institute of Health, USA, shows that progressive disability is rising faster than catastrophic disability. This long-term survey among the older population with severe ADL disability observed that catastrophic disability is seen in younger people, while the proportion of victims with progressive ADL disability is much higher among older people (Fig. 1.4).

Many investigations towards the assessment of the dynamics of disability among old people have focused on the pathological changes that trigger the onset of disability. Theoretical pathway, one among the postulates proposed, takes away pragmatic support and states that impairment and functional limitations, follow disease and lead to disability. Functional limitations are fundamental to assessment and are defined as restrictions experienced while executing basic physical and mental activities.

Objective measures for assessment of physical performance in research and clinical settings have been widely appreciated and regarded as indicators of functional limitations and can be used to measure actual disability. Researchers claim that objective performance is also a measure of assessment of function and disability over a discrete period of time, across different countries and cultures.

Disability has been steadily gaining attention on account of more and more research data being generated on disability status with complex patterns of disease, epidemiological details, etc. Tackling disability is intimately connected to the long-term geriatric care.

1.8 Summary

We live in a steadily aging society. The proportion of elderly citizens in the world's population is rising alarmingly. Modern medicine has helped us to fight disease and prolong life, but we have been less successful in combating the debility of senescence. Despite all the breakthroughs in genomics, proteomics, and metabolomics, overcoming the disability, morbidity, and prostration in old age has been mostly ineffective. Paradoxically, geriatric care-needs are the highest among the rich developed industrial societies. The problem is more than what meets the eye. In a few decades from now, the developing economies, with much larger populations thriving on a much lower income, will have to grapple with a much greater burden. Institutionalized care of the old is not yet a priority for the developing countries. Without technological resources, skilled workforce with multiple competencies, the impending geriatric burden is likely to be the biggest drain on healthcare resources and money.

Institutionalized geriatric care should become the foremost public health priority in the very same societies where family is the exclusive care giver. Larger families and greater intimacy between individual family members make geriatric care a part of the accepted family norm in poorer societies. However, during the course of the demographic transition towards nuclear families with fewer children and better incomes, the tables have turned. The need for institutionalized care places a much greater budgetary burden upon the government. In this context, the government should formulate feasible strategies for training in specialized care, expanding insurance cover, generating supportive technologies, and creating a skilled workforce, in addition to legally empowering helpless elders who are vulnerable to exploitation.

Conquering senility does not seem to be, at least for now, a realistic scientific goal. Given the implications of the evolutionary baggage built into the dynamics of survival, public health policy makers should pay more attention to the biological compulsions along the path towards senescence. Focus should be on preventive strategies, the looming obesity epidemic, for instance. In this context, the only way forward appears to be a concerted effort by public health policy initiatives built on a healthy economic foundation.

References

Balducci L (2003) Geriatric oncology. Crit Rev Oncol Hematol 46(3):211–220
Balducci L, Extermann M (2000) Cancer and aging: an evolving panorama. Hematol Oncol Clin North Am 14(1):1–16
Bandeen-Roche K, Xue QL, Ferrucci L et al (2006) Phenotype of frailty: characterization in the women's health and aging studies. J Gerontol A Biol Sci Med Sci 61(3):262–266
Barnett K, Mercer SW, Norbury M et al (2012) Epidemiology of multimorbidity and implications for health care, research, and medical education: a cross-sectional study. Lancet 380:37–43
Blank TO, Bellizzi KM (2008) A gerontologic perspective on cancer and aging. Cancer 112(Suppl 11):2569–2576
Brandling-Bennett A (1991) Cholera in the Americas. Paper presented at the International Union for the Scientific Study of Population Seminar on Causes and Prevention of Adult Mortality in Developing Countries, Santiago, Chile
Bureau of the Census (1991) Global aging: comparative indicators and future trends. Wall chart. U.S. Center For International Research, Washington; Various years international data base on aging (data base). U.S. Center for International Research, Washington
Centers for Disease Control and Prevention, National Center for Health Statistics (n.d.) National Health Interview Survey, sample adult questionnaire: trends in health and aging. http://www.cdc.gov/nchs/agingact.htm. Accessed 10 July 2008
Chang M, Cohen-Mansfield J, Ferrucci L et al (2004) Incidence of loss of ability to walk 400 meters in a functionally limited older population. J Am Geriatr Soc 52(12):2094–2098
Cheung SLK, Robine JM, Jow-Ching Tu E, Caselli G (2005) Three dimensions of the survival curve: horizontalization, verticalization, and longevity extension. Demography 42:243–258
Christensen K, Doblhammer G, Rau R, Vaupel JW (2009) Aging populations: the challenges ahead. Lancet 374(9696):1196–1208
Extermann M, Aapro M (2000) Assessment of the older cancer patient. Hematol Oncol Clin North Am 14(1):63–77
Extermann M, Hurria A (2007) Comprehensive geriatric assessment for older patients with cancer. J Clin Oncol 25(14):1824–1831
Fabian D, Flatt T (2011) The evolution of aging. Nat Educ Knowl 3(10):9
Faria MA (2015) Bioethics and why I hope to live beyond age 75 attaining wisdom!: a rebuttal to Dr. Ezekiel Emanuel's 75 age limit. Surg Neurol Int 6:35
Ferrucci L, Penninx BW, Volpato S et al (2002) Change in muscle strength explains accelerated decline of physical function in older women with high interleukin-6 serum levels. J Am Geriatr Soc 50(12):1947–1954
Freid VM, Bernstein AB, Bush MA (2012) Multiple chronic conditions among adults aged 45 and over: trends over the past 10 years. NCHS Data Brief 100:1–8
Frenk J, Bobadilla JL, Lozano R (1991) The epidemiologic transition: the Latin American experience. Paper presented at the International Union for the Scientific Study of Population Seminar on Causes and Prevention of Adult Mortality in Developing Countries, Santiago, Chile
Fried LP, Ferrucci L, Darer J et al (2004) Untangling the concepts of disability, frailty, and comorbidity: implications for improved targeting and care. J Gerontol A Biol Sci Med Sci 59:255–263
Fries JF (1980) Ageing, natural death, and the compression of morbidity. N Engl J Med 313:407–428
Fries JF (1996) Physical activity, the compression of morbidity and the health of the elderly. J R Soc Med 89:64–68
Guralnik JM (1996) Assessing the impact of comorbidity in the older population. Ann Epidemiol 6 (5):376–380
Guralnik JM, Ferrucci L (2003) Assessing the building blocks of function: utilizing measures of functional limitation. Am J Prev Med 25(3 Suppl 2):112–121

Guralnik JM, Ferrucci L, Balfour JL et al (2001) Progressive versus catastrophic loss of the ability to walk: implications for the prevention of mobility loss. J Am Geriatr Soc 49(11):1463–1470
Hadley EC, Rossi WK (2005) Exceptional survival in human populations: National Institute on Aging perspectives and programs. Mech Ageing Dev 126(2):231–234
Hamilton WD (1966) The moulding of senescence by natural selection. J Theor Biol 12(1):12–45
Hamilton WD (1996) Narrow roads of gene land, vol 1. Freeman, New York
Hayat MJ, Howlader N, Reichman ME et al (2007) Cancer statistics, trends, and multiple primary cancer analyses from the surveillance, epidemiology, and end results (SEER) program. Oncologist 12(1):20–37
Hughes KA et al (2002) A test of evolutionary theories of aging. Proc Natl Acad Sci U S A 99:14286–14291
James WV (2003) The biodemography of aging. Popul Dev Rev 30:48
Khaw KT (1997) Epidemiological aspects of aging. Philos Trans R Soc Lond B 352:1833
Kinsella K, Phillips DR (n.d.) Global aging: the challenge of success. www.prb.org/pdf05/60.1globalaging.pdf. Accessed 27 Mar 2014. Updated Mar 2005
Kinsella K, Taueber CM (1993) An aging world II, International population report series. Bureau of the Census, Washington, p 95
Lakatta EG (1985) Health, disease, and cardiovascular aging. In: America's aging: health in an older society. Institute of Medicine, National Research Council, National Academy of Sciences, National Academy Press, Washington, pp 73–104
Launer LJ, Andersen K, Dewey ME et al (1999) Rates and risk factors for dementia and Alzheimer's disease: results from EURODEM pooled analyses. EURODEM Incidence Research Group and Work Groups, European studies of dementia. Neurology 52(1):78–84
Lauretani F, Semba RD, Bandinelli S et al (2008) Low plasma carotenoids and skeletal muscle strength decline over 6 years. J Gerontol A Biol Sci Med Sci 63(4):376–383
Linda JW (2004) Aging, health, and public policy-demographic and economic prospective, vol 30. Population Council, New York, pp 3–18
Lobo A, Launer LJ, Fratiglioni L et al (2000) Prevalence of dementia and major subtypes in Europe: a collaborative study of population-based cohorts. Neurologic diseases in the elderly research group. Neurology 54(11 Suppl 5):S4–S9
Lolak LJ, DeFrances CJ, Hall MJ (2006) National hospital discharge survey: 2004 annual summary with detailed diagnosis and procedure data. National center for health statistics. Vital Health Stat 13(162):1–218
Manton KG, Myers GC, Andrews GR (1987) Morbidity and disability patterns in four developing nations: their implications for social and economic integration of the elderly. J Cross Cult Gerontol 2(2):115–129
Manton KG, Dowd JE, Woodbury MA (1986) Conceptual and measurement issues in assessing disability cross-nationally: analysis of a WHO sponsored survey of the disablement process in Indonesia. J Cross Cult Gerontol 1(4):339–362
Martin LG (1987) Census data for studying elderly populations. Asia-Pac Popul J 2(2):69–82
McDermott MM, Tian L, Ferrucci L et al (2008) Associations between lower extremity ischemia, upper and lower extremity strength, and functional impairment with peripheral arterial disease. J Am Geriatr Soc 56(4):724–729
Minino AM, Heron M, Smith BL et al (2007) Deaths: final data for 2004. Natl Vital Stat Rep 55 (9):1–120
National Centre for Health Statistics (1978) Health United States (DHEW publication no. (PHS) 78-1232). National Center for Health Statics, Hyattsville
Neale MC, Kendler KS (1995) Models of comorbidity for multifactorial disorders. Am J Hum Genet 57:935–953
Oakes SC, Mitchell VS, Pearson GW, Carpenter CJ (eds) (1991) Malaria: obstacles and opportunities. A report of the Committee for the Study of Malaria Prevention and Control, Division of International Health, Institute of Medicine. National Academy Press, Washington

Obadia Y, Rotily M, Degrand-Guillaud A et al (1997) The PREMAP study: prevalence and risk factors of dementia and clinically diagnosed Alzheimer's disease in Provence, France. Prevalence of Alzheimer's disease in Provence. Eur J Epidemiol 13(3):247–253

Oeppen J, Vaupel JW (2002) Broken limits to life expectancy. Science 296:10291031

Onder G, Penninx BW, Ferrucci L et al (2005) Measures of physical performance and risk for progressive and catastrophic disability: results from the Women's Health and Aging Study. J Gerontol A Biol Sci Med Sci 60(1):74–79

Population Division. U.S. Census Bureau (2000) The Census Bureau on prospects for US population growth in the twenty-first century. Popul Dev Rev 26(1):197–200

Rice DP, Fillit HM, Max W et al (2001) Prevalence, costs, and treatment of Alzheimer's disease and related dementia: a managed care perspective. Am J Manag Care 7(8):809–818

Rose MR, Charlesworth B (1980) A test of evolutionary theories of senescence. Nature 287:141–142

Rose M, Evolutionary R (1991) Biology of aging. Oxford University Press, New York

Semba RD, Ferrucci L, Sun K et al (2007) Oxidative stress and severe walking disability among older women. Am J Med 120(12):1084–1089

Simmons LW (1945) The role of the aged in primitive society. Yale University Press, New Haven

Simonsick EM, Newman AB, Nevitt MC et al (2001) Measuring higher level physical function in well-functioning older adults: expanding familiar approaches in the Health ABC study. J Gerontol A Biol Sci Med Sci 56(10):M644–M649

Suh GH, Shah A (2001) A review of the epidemiological transition in dementia: cross-national comparisons of the indices related to Alzheimer's disease and vascular dementia. Acta Psychiatr Scand 104(1):4–11

Toupance B, Godelle B, Gouyon PH, Schachter F (1998) A model for antagonistic pleiotropic gene action for mortality and advanced age. Am J Hum Genet 62:1525–1534

Treas J, Logue B (1986) Economic development and the older population. Popul Dev Rev 12:645–673

United Nations (n.d.-a) World population ageing 2013. www.un.org/en/development/desa/population/publications/pdf/ageing/World Population Ageing Report 2013.pdf. Accessed 27 Mar 2014. Updated 2013

United Nations (n.d.-b) World population prospects: the 2010 revision. http://esa.un.org/unpd/wpp

US Census Bureau (n.d.) Global population at a glance: 2002 and beyond. www.census.gov/prod/2004pubs/wp02-1.pdf. Accessed 15 Apr 2014. Updated Mar 2004

Valenti G, Denti L, Maggio M et al (2004) Effect of DHEAS on skeletal muscle over the life span: the InCHIANTI study. J Gerontol A Biol Sci Med Sci 59(5):466–472

Van Den Biggelaar AH (2004) Inflammation underlying cardiovascular mortality is a late consequence of evolutionary programming. FASEB J 12(9):1022–1024. https://doi.org/10.1096/fj.03-1162fje

Wang H, Dwyer-Lindgren L, Lofgren KT et al (2012) Age-specific and sex-specific mortality in 187 countries, 1970–2010: a systematic analysis for the Global Burden of Disease Study 2010. Lancet 380:2071–2094

Wells NL, Balducci L (1997) Geriatric oncology: medical and psychosocial perspectives. Cancer Pract 5(2):87–91

Williams GC (1957) Pleiotropy, natural selection, and the evolution of senescence. Evolution 11:398–411

Wolff JL, Starfield B, Anderson G (2002) Prevalence, expenditures, and complications of multiple chronic conditions in the elderly. Arch Intern Med 162:2269–2276

World Health Organization. National Institute on Aging, National Institutes of Health, U.S. Department of Health and Human Services (2011) Global health and aging. NIH Publication no. 11-7737, pp 1–25

Yashin AI, Vaupel JW, Andreev KF, Tan Q, Iachine IA, Carotenuto L, De Benedictis G et al (1998) Combining genetic and demographic information in population studies of ageing and longevity. J Epidemiol Biostat 3:289–294

Ageing Phenotypes

2

Sethuraman Sathya and Kasi Pandima Devi

Abstract

People growing old face several problems like loss of body function and increased risk for diseases. Maintaining good health and body function is vital for healthy ageing. Biological complexity and diversity of the ageing process render no simple, reliable and single technique to measure someone's healthy ageing. Hence to know about the key features of healthy ageing, (1) the healthy ageing phenotypes (HAP) with five domains such as physiological and metabolic health, physical capability, cognitive function, psychological and subjective wellbeing and social wellbeing, and (2) the way to measuring the HAP are discussed in this chapter. This chapter covers very limited conceptual and practical problems in recognizing and measuring the healthy ageing phenotypes. In specific, the present chapter describes some aspects of HAP, yet which is not adequately detailed before. The conceptualization of HAP is at the early stage of research. The data discussed in the present study helps other researchers to assess the possible utility of the tools to measure the HAP and enable them to include still more tools in respective domains of HAP for their enhanced assessment.

Keywords

Healthy ageing phenotypes · Physiological and metabolic health · Social wellbeing

S. Sathya · K. Pandima Devi (✉)
Department of Biotechnology, Alagappa University, Karaikudi, Tamil Nadu, India

S. M. Nabavi et al. (eds.), *Nutrients and Nutraceuticals for Active & Healthy Ageing*,
https://doi.org/10.1007/978-981-15-3552-9_2

2.1 Introduction

Ageing is a common natural biological process in all organisms, leading to progressive and harmful changes in the body. The world's population continues to age, characterized by growth in both the number and proportion of older people. Currently, 10% of the world's population is 60 years or above, and it is predicted that this number will increase to 20% by 2050. The fundamental process of ageing is experienced as a lethargic deterioration resulting in infirmity, illness and death. However, the animal and human studies reveal that the process of ageing is malleable, which can be slowed and associated with disease progression. Hence ageing has been termed as the substrate on which age-associated diseases grow. The major determinants of ageing process is behavioural factors (diet, alcohol consumption, smoking and physical activity) and social factors (including relationships, roles and support), which could robustly be linked with health and wellbeing in older age (Khaw et al. 2008; Van Dam et al. 2008). However, scientific evidences reveal that the long-term efficiency of pragmatic and lifestyle-based interferences alters these two factors and thereby improves healthy and wellbeing in later life, which is still being researched. A way to overcome this problem is to identify different perspectives in the area of ageing research. We have accepted the conception of 'Healthy Ageing Phenotype' which is intended to encapsulate the potential to be engaged socially, productive and to function separately in both physical and cognitive levels (Franco et al. 2007). Ageing phenotype may also be defined as a state of being alive, while having extremely conserved neuro-endocrine control systems, metabolic and hormonal function at cellular and molecular levels (Franco et al. 2009). In addition, the spark workshops are tactical meetings in which many scientists from diverse disciplines and perspectives share informative knowledge, construct novel concepts and assess alternative idea for future research activities. In this meeting, the international expertises from complementary areas linked to ageing research assemble to: (1) find definitions of the 'Ageing Phenotype', (2) identify possible mechanisms to increase healthy life expectancy of the people and (3) highlight the areas of ageing research which are important for the future (Franco et al. 2007).

2.2 Ageing: The Human Frailty

The significance of measuring changes in ageing is underlined by the fact that the world population is ageing. Scientific literatures have reported that 29% of European people above 65 years will be affected from ageing by 2080. This is because of the combination of increased life expectancy and decreased birth rate (Franco et al. 2007). Indeed, many countries have elevated the age of retirement to sustain the workforce and to diminish the pension burden linked with greater longevity. The incessant development in the prevention of mortality as a result of advances in income, hygiene, sanitation and infection control, nutrition, education and medicine is generally believed to account for the increases in longevity. Moreover, ageing plays an important role in most common complex ailments and increases the burden of chronic disease (Franco et al. 2009). These pressures highlight the significance of finding ways to allow people to healthy ageing and to maintain good health and high level of welfare. However, the progression of targeted interventions may be an efficient additional resource and would be supportive to differentiate those who are ageing healthy from those who are ageing unwell. Therefore, we need to assess the healthy ageing phenotypes which help to maintain good health, improve quality of life and economy healthcare costs and enable aged people to participate productively in society and to offer their stability, improve the capacity for synthetic problem-solving, augmented ability to manage conflicts and ability to consider perspectives from other age groups (Murray et al. 2015; Rowe and Kahn 2015).

2.3 Healthy Ageing Phenotypes

Advancement in healthcare systems improves the longevity of the human life but not the fertility rates (Vaupel 2010). Globally people growing older with good health and functioning is considered as a prerequisite for successful ageing. It is not happening at most of the instances. Unavoidably, ageing is accompanied with social care, more years of ill health and a high rate of medical expenses at later years of life (Oeppen and Vaupel 2002; Yates et al. 2009). With the biological complexity, and the heterogeneity, of the ageing process, it is also not possible to measure the reliable healthy ageing of a particular individual with a single parameter. Also, there are no agreed definitions or phenotypes to measure healthy ageing (Franco et al. 2009). Still research on attaining healthy ageing associated with positive health and wellbeing in older people are under-researched. To recognize alternative side points in attaining healthy ageing, people from different countries, disciplines, and perspectives gathered to share their knowledge in finding clear definitions of Healthy Ageing Phenotypes (HAP), potential mechanisms and interventions and critical area to be focused on the future on ageing research (Franco et al. 2007).

HAP is defined as the condition of being alive with proper functioning of body systems. HAP is a multi-dimensional one and is determined by age, gender, and environmental factors. HAP is believed to measure in five different categories: (1) physiological and metabolic health, (2) physical capability, (3) cognitive function, (4) psychological wellbeing and (5) social wellbeing and which are assumed to modify or reduce the premature death, prevent morbidity and disability (Fig. 2.1) (Depp and Jeste 2006).

2.4 Physiological and Metabolic Health

Because of the changes in lifestyle and work pressure, stressors such as alcohol and smoking become temporary relief in due course. Chronic exposure to such stressors leads to the loss of homeostatic mechanism that is necessary to maintain the normal body functions (Caspersen et al. 1985). Loss of such homeostatic mechanisms eventually leads to the metabolic dysregulation, early signs of pre-disease symptoms, if not managed, ultimately leading to the functional loss, chronic disease

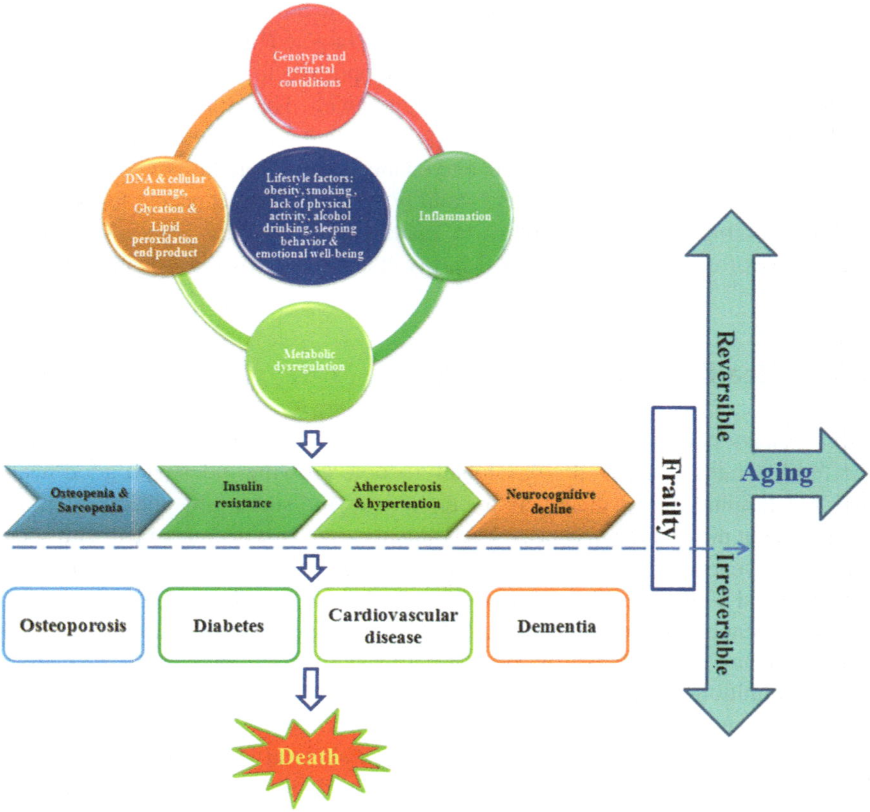

Fig. 2.1 Mechanisms and phases of divergence from the Healthy Ageing Phenotype

and finally death. Mortality is an ageing-related phenotype, which is often interlinked with the result of physiological and metabolic health (Table 2.1).

Physical activity refers to the outcome of skeletal muscles movement or exercise that consumes energy. Recent longitudinal cohorts review claimed that physical activity has been consistently interrelated with a decrease in weight gain, obesity, cardio metabolic diseases such as cardiovascular diseases and type II diabetes and age-related disorders like dementia and Alzheimer's disease (Reiner et al. 2013). There are literatures indicating that functioning of vital organs, inflammation, metabolic process and body mass index (BMI) are the crucial biomarkers of ageing, associated with physiological and metabolic health. Central adiposity plays a vital role in ageing and age-related risks. An increase in the BMI to a level of 5 kg/m^2 leads to an increase in the all-cause mortality rates to 30%, vascular mortality to 40%, diabetic, renal and hepatic mortality to 50–120%. Further, higher BMI results in impaired cognitive functions (de Hollander et al. 2012; Prospective Studies Collaboration 2009; Gallucci et al. 2013). In addition, rise in the level of blood pressure and glucose concentration are crucial predictors of greater ageing-related

Table 2.1 Tools to quantify elected domains and sub domains of the Ageing Phenotype

Domain	Subdomain	Tools
Physiological and metabolic health	Cardiovascular function	✓ Blood pressure ✓ Blood lipids
	Lung function	✓ Forced expiratory volume (FEV1) ✓ Blood glucose
	Glucose metabolism	✓ Glycated haemoglobin (HbA1C)
	Body composition	✓ Body mass index (BMI)
Physical capability strength	Strength	✓ Handgrip strength
	Locomotion	✓ Gait speed
	Endurance	✓ Walk endurance test
	Dexterity	✓ Pegboard dexterity test
	Dexterity	✓ Standing balance test
Cognitive function	Episodic memory	✓ Story recall ✓ Speed reaction time ✓ Symbol digit modalities test
	Executive function	✓ Word list recall ✓ Paired associate learning
	Life satisfaction	✓ Stroop ✓ Trail making tests A and B
	Positive and negative affect	✓ Satisfaction with life scale (SWLS) ✓ Positive and negative affect schedule (PANAS)
Psychological wellbeing	Quality of life	✓ Control, autonomy, pleasure and self-realization, quality of life scale (CASP-19) ✓ WHO quality of life-BREF (WHOQOL-BREF)
	Mental health	✓ Centre for epidemiological studies depression scale ✓ Warwick-Edinburgh mental wellbeing scale (WEMWBS)
	Resilience	✓ Psychological resilience scale
Social wellbeing	Social network	✓ NIH Toolbox: friendship ✓ PROMIS: satisfaction with social roles and activities
	Social functioning	✓ Revised UCLA loneliness scale ✓ Social support behaviour scale
	Perceived emotional/social support	✓ NIH Toolbox: emotional support ✓ Instrumental support ✓ Loneliness
	Sense of purpose	✓ Perceived rejection scale ✓ NIH Toolbox: psychological wellbeing ✓ Meaning and purpose

mortality, and is also connected with poorer cognitive function in later years of living (Einarson et al. 2011; Sarwar et al. 2010). In 2002, Lewington et al. (Prospective Studies Collaboration 2002) stated that difference of 20 mmHg or 10 mmHg in Systolic BP or Diastolic BP, respectively, is associated with twofold more mortality from several vascular causes. Lungs behavioural functions can be assessed with biomarkers like forced expiratory volume (FEV1). But the evidence of cognitive waning in an association of FEV1 is limited. Apart from these, there are other biomarkers, which are used to assess the cardiovascular functions connected with age-related disorders. However, they are still investigated on their beneficial use over identified ones (Weuve et al. 2011; Danesh et al. 2008; Emerging Risk Factors Collaboration 2010; Zittermann et al. 2011; Di Angelantonio et al. 2009).

2.5 Physical Capability

Cooper et al. (2010a, b) described the healthy ageing as a functional and physical ability of one to perform day-to-day life activity normally. Hence, seizing physical capability is an important measure to measure the HAP. Seizing physical capability deploys variety of measurement tools that capture different functions ranging from measuring hand strength to walking endurance. Although there are several simple and convenient to use test like balancing, gait speed, prediction of sit-up time longevity and so on, rise of considerable variation still question the testing protocols. To overcome these issues, standardized protocols for assessing the motor functions have been addressed by NIH Toolbox (Rantanen et al. 2012; Studenski et al. 2011; Reuben et al. 2013). This NIH Toolbox has been aimed and designed to answer standard questions from domains such as strength, locomotion, endurance, dexterity and balance to assess the HAP through motor functions. This NIH Toolbox test is applicable to the person above age 3. However, there are some test like timed up and go and timed sit-to stand tests, which are mobility-specific assessment criteria in older adults. The level of accurateness of the test largely depends on age of interest, incorporation of broad range of test and their complexity. Looking to the recent developments and the advancement in technology, these tests have now become more sensitive with increased difficulties. Such test may fail in the people ageing less healthy (Reuben et al. 2013; Hodes et al. 2013).

2.6 Cognitive Functions

Several aspects of cognitive functions decline on the onset of ageing process and hence offer a potential biomarker that can used to assess the healthy cognitive ageing. However, measure of these functions is highly interlinked (Salthouse 2010; Tucker-Drob and Salthouse 2011). Cognitive processing speed is an efficient measure that calculates the overall cognitive functions (Deary et al. 2010). Assessment of physical activity such as time taken for visual inspection and speeded reaction are responsive measures to measure the cognitive processing speed and also used to compare with the performance based measure of routine activities (Shipley et al. 2007; Langlois et al. 2013).Other cognitive functions that contribute unique pattern of change in response to age include episodic memory and executive function. Indices of these functions have significant part in assessing the dementia in later life like Alzheimer's disease (Verhaeghen 2011; Snowden et al. 2002).

Measuring the cognitive functions suggestion may vary among the researchers. However, several guidances can aid in selecting the cognitive function test. Subjective rating done by the participants shows relatively insignificant results, which is often influenced by the participant's mood and personal thinking (Stewart 2012; Gershon et al. 2012). Future tools measuring the cognitive functions also are able to maximize the participants with cognitive defects. In addition, Lezak et al. (2004) tabulated the recent advances and tools that measure the cognitive weakness.

2.7 Psychological and Subjective Wellbeing

The effects of psychological (PW) and subjective wellbeing (SW) are directly proportional to the age of the individuals. PW and SW significantly increase at the latter adulthood as of the cognitive and physiological dysregulation. This can be overcome by the individual themselves to accommodate the changes of ageing and to accept the real life situations. This phenomenon of correcting oneself in response to PW and SW is termed as wellbeing paradox (Baltes et al. 1990). Hence, PW and SW are potential markers of HAP, which includes the phenotypic changes and the indices of adaptive mechanism of wellbeing. PW and SW are complementary to each other but have separate assessment self-report. SW assesses the ability of a person in positive terms, which includes the level of high positive thoughts (PT) and life satisfaction and low negative thoughts (NT). PW assess the successful approaches and their solutions to overcome the obstacles of later life (Keyes et al. 2002).

There are more evidences that validate SW as potential marker for ageing phenotype. Meta-analyses done by Arent et al. (Arent et al. 2000) on PT and NT in older adults were strongly evidenced SW as the marker of HAP. Moreover, there are several tools including satisfaction with life scale and the life satisfaction index used in intervention studies which detect the changes in the SW components in older adults with good qualifications (Diener et al. 1985; Wood et al. 1969). However several intervention studies have included the life satisfaction measure, the results of those surveys ended up with lack of sensitivity and conceptual reliability. Subjective rating of health is a dominant predictor of SW (Idler and Benyamini 1997). Particularly, SF-36 (Short Form Health Survey), which assess the generic, coherent and easily administered quality of life, is more popular among subjective rating health, which is in intervention studies with healthier adults.

Similar to SW, the components of PW also gained considerable interest in terms of measuring the HAP (Ryff 1989). In PW, healthy ageing in older adults can be attained by successful commitment of age-related challenges, which include the aspects of challenged thriving such as autonomy, self-acceptance, personal growth and positive relation in communities. Meta-analysis also clearly indicated that, similar to SW, the scales of PW also respond to life style intervention studies and forecast the problems of healthy ageing (Ryff and Keyes 1995; Sin and Lyubomirsky 2009). Resilience is a key factor that is closely related to PW and provides a key pathway to attain successful ageing in the older adults (Windle et al. 2011). Resilience is a concept or process that aids the recovery of older adults from stress or trauma. This process helps the aged people to think, negotiate, manage and adapt to a particular situation. A study by Windle et al. (Windle et al. 2008) measures the responsiveness of the resilience process and appears to be the best suited measure for measuring healthy ageing.

2.8 Social Wellbeing

Apart from the issues disused earlier, engagement in social relationships and personal interest are also considered as serious issues of living with proper health in later years of life. Social wellbeing is a broad area with social relationship and personal development, which holds a number of models for ageing components (Rowe and Kahn 1987; World Health Organization 2015). However, though social wellbeing is considered as key outcome in ageing process, absence of proper definition and measures hold its ability to develop a HAP tool suitable for measuring healthy ageing. Yet factors that contribute to social wellbeing which include integration, engagement, participation, networks, ties, connections and connectedness suggest that active social participation is well interconnected with beneficial health outcomes and reduced mortality rates (Moen et al. 1992; Thomas 2011; Sirven and Debrand 2008; Kawachi and Berkman 2001; Kaplan et al. 1988; Cornwell et al. 2008). Nevertheless, some aspects of social wellbeing are crucially important and amenable to change. In the line of measuring the HAP, this diversity has further become a significant hindrance to the identification of appropriate measurement tools.

In conclusion, to quantify the HAP, there is an utmost need for the development of measurable markers that aids the intervention studies. This chapter covers very limited conceptual and practical problems in recognizing and measuring the healthy ageing phenotypes. In specific, the present chapter describes some aspect of HAP, yet which is not adequately detailed before. However, holistic assessment of HAP can be achieved by measuring the five domains of HAP such as physiological and metabolic health, physical capability, cognitive function, psychological and subjective wellbeing and social wellbeing. Unlike domains like cognitive function, selection of subdomain and their measurements are under-developed in other areas of HAP such as social wellbeing and have limitation in developing an appropriate tool to measure. This study covers the maximum possible assessment of HAP. Yet, some functions such as smell and vision have not been discussed and their reports on HAP is very meagre. Richer et al. (Richer et al. 2011) report the perception on relation between age and sense that modulate the sensing behaviour occurs during intervention studies. In addition, the NIH tool box also dictates some criteria for sense assessment. The conceptualization of HAP is at the early stage of research. The data provided in the present study helps other researchers to assess the possible utility of the tools to measure healthy ageing and enable them to include still some tools in the respective domains of HAP for their enhanced assessment.

Acknowledgement The authors wish to acknowledge the Computational and Bioinformatics facility provided by the Alagappa University Bioinformatics Infrastructure Facility (funded by DBT, GOI; File No. BT/BI/25/012/2012,BIF) and the grant provided by (1) DST-FIST (Grant No. SR/FST/LSI-639/2015(C)) (2) UGC-SAP (Grant No. F.5-1/2018/DRS-II(SAP-II)), (3) DST-PURSE (Grant No. SR/PURSE Phase 2/38 (G)), (4) RUSA 2.0 [F. 24-51/2014-U, Policy (TN Multi-Gen), Dept of Edn, GoI] and (5) the University Science Instrumentation Centre (USIC), Alagappa University.

References

Arent SM, Landers DM, Etnier JL (2000) The effects of exercise on mood in older adults: a meta-analytic review. J Aging Phys Act 8(4):407–430

Baltes PB, Baltes MM, European Network on Longitudinal Studies on Individual Development (1990) Successful aging: perspectives from the behavioral sciences. Oxford University Press, Oxford

Caspersen CJ, Powell KE, Christenson GM (1985) Physical activity, exercise, and physical fitness: definitions and distinctions for health-related research. Public Health Rep 100(2):126

Cooper R, Kuh D, Hardy R (2010a) Objectively measured physical capability levels and mortality: systematic review and meta-analysis. BMJ 341:c4467

Cooper R, Kuh D, Cooper C, Gale CR, Lawlor DA, Matthews F, Hardy R, FALCon and HALCyon Study Teams (2010b) Objective measures of physical capability and subsequent health: a systematic review. Age Ageing 40(1):14–23

Cornwell B, Laumann EO, Schumm LP (2008) The social connectedness of older adults: a national profile. Am Sociol Rev 73(2):185–203

Danesh J, Kaptoge S, Mann AG, Sarwar N, Wood A, Angleman SB, Wensley F, Higgins JP, Lennon L, Eiriksdottir G, Rumley A (2008) Long-term interleukin-6 levels and subsequent risk of coronary heart disease: two new prospective studies and a systematic review. PLoS Med 5(4): e78

de Hollander EL, Bemelmans WJ, Boshuizen HC, Friedrich N, Wallaschofski H, Guallar-Castillón P, Walter S, Zillikens MC, Rosengren A, Lissner L, Bassett JK (2012) The association between waist circumference and risk of mortality considering body mass index in 65-to 74-year-olds: a meta-analysis of 29 cohorts involving more than 58000 elderly persons. Int J Epidemiol 41(3):805–817

Deary IJ, Johnson W, Starr JM (2010) Are processing speed tasks biomarkers of cognitive aging? Psychol Aging 25(1):219

Depp CA, Jeste DV (2006) Definitions and predictors of successful aging: a comprehensive review of larger quantitative studies. Am J Geriatr Psychiatry 14(1):6–20

Di Angelantonio E, Chowdhury R, Sarwar N, Ray KK, Gobin R, Saleheen D, Thompson A, Gudnason V, Sattar N, Danesh J (2009) B-type natriuretic peptides and cardiovascular risk: systematic review and meta-analysis of 40 prospective studies. Circulation 120(22):2177–2187

Diener ED, Emmons RA, Larsen RJ, Griffin S (1985) The satisfaction with life scale. J Pers Assess 49(1):71–75

Einarson TR, Machado M, Henk Hemels ME (2011) Blood glucose and subsequent cardiovascular disease: update of a meta-analysis. Curr Med Res Opin 27(11):2155–2163

Emerging Risk Factors Collaboration (2010) C-reactive protein concentration and risk of coronary heart disease, stroke, and mortality: an individual participant meta-analysis. Lancet 375 (9709):132–140

Franco OH, Kirkwood TB, Powell JR, Catt M, Goodwin J, Ordovas JM, Van Der Ouderaa F (2007) Ten commandments for the future of ageing research in the UK: a vision for action. BMC Geriatr 7(1):10

Franco OH, Karnik K, Osborne G, Ordovas JM, Catt M, van der Ouderaa F (2009) Changing course in ageing research: the healthy ageing phenotype. Maturitas 63(1):13–19

Gallucci M, Mazzuco S, Ongaro F, Di Giorgi E, Mecocci P, Cesari M, Albani D, Forloni GL, Durante E, Gajo GB, Zanardo A (2013) Body mass index, lifestyles, physical performance and cognitive decline: the "Treviso Longeva (TRELONG)" study. J Nutr Health Aging 17 (4):378–384

Gershon RC, Lai JS, Bode R, Choi S, Moy C, Bleck T, Miller D, Peterman A, Cella D (2012) Neuro-QOL: quality of life item banks for adults with neurological disorders: item development and calibrations based upon clinical and general population testing. Qual Life Res 21 (3):475–486

Hodes RJ, Insel TR, Landis SC (2013) The NIH Toolbox Setting a standard for biomedical research. Neurology 80(11 Supplement 3):S1

Idler EL, Benyamini Y (1997) Self-rated health and mortality: a review of twenty-seven community studies. J Health Soc Behav 38:21–37

Kaplan GA, Salonen JT, Cohen RD, Brand RJ, Leonard Syme S, Puska P (1988) Social connections and mortality from all causes and from cardiovascular disease: prospective evidence from eastern Finland. Am J Epidemiol 128(2):370–380

Kawachi I, Berkman LF (2001) Social ties and mental health. J Urban Health 78(3):458–467

Keyes CL, Shmotkin D, Ryff CD (2002) Optimizing well-being: the empirical encounter of two traditions. J Pers Soc Psychol 82(6):1007

Khaw KT, Wareham N, Bingham S, Welch A, Luben R, Day N (2008) Combined impact of health behaviours and mortality in men and women: the EPIC-Norfolk prospective population study. PLoS Med 5(1):e12

Langlois F, Vu TT, Chassé K, Dupuis G, Kergoat MJ, Bherer L (2013) Benefits of physical exercise training on cognition and quality of life in frail older adults. J Gerontol B Psychol Sci Soc Sci 68 (3):400–404

Lezak MD, Howieson DB, Loring DW, Fischer JS (2004) Neuropsychological assessment. Oxford University Press, New York

Moen P, Dempster-McClain D, Williams RM Jr (1992) Successful aging: a life-course perspective on women's multiple roles and health. Am J Sociol 97(6):1612–1638

Murray CJ, Barber RM, Foreman KJ, Ozgoren AA, Abd-Allah F, Abera SF, Aboyans V, Abraham JP, Abubakar I, Abu-Raddad LJ, Abu-Rmeileh NM (2015) Global, regional, and national disability-adjusted life years (DALYs) for 306 diseases and injuries and healthy life expectancy (HALE) for 188 countries, 1990–2013: quantifying the epidemiological transition. Lancet 386 (10009):2145–2191

Oeppen J, Vaupel JW (2002) Demography. Broken limits to life expectancy. Science 296:1029–1031

Prospective Studies Collaboration (2002) Age-specific relevance of usual blood pressure to vascular mortality: a meta-analysis of individual data for one million adults in 61 prospective studies. Lancet 360(9349):1903–1913

Prospective Studies Collaboration (2009) Body-mass index and cause-specific mortality in 900 000 adults: collaborative analyses of 57 prospective studies. Lancet 373(9669):1083–1096

Rantanen T, Masaki K, He Q, Ross GW, Willcox BJ, White L (2012) Midlife muscle strength and human longevity up to age 100 years: a 44-year prospective study among a decedent cohort. Age 34(3):563–570

Reiner M, Niermann C, Jekauc D, Woll A (2013) Long-term health benefits of physical activity—a systematic review of longitudinal studies. BMC Public Health 13(1):813

Reuben DB, Magasi S, McCreath HE, Bohannon RW, Wang YC, Bubela DJ, Rymer WZ, Beaumont J, Rine RM, Lai JS, Gershon RC (2013) Motor assessment using the NIH Toolbox. Neurology 80(11 Supplement 3):S65–S75

Richer SP, Stiles W, Graham-Hoffman K, Levin M, Ruskin D, Wrobel J, Park DW, Thomas C (2011) Randomized, double-blind, placebo-controlled study of zeaxanthin and visual function

in patients with atrophic age-related macular degeneration: the Zeaxanthin and Visual Function Study (ZVF) FDA IND# 78, 973. Optometry 82(11):667–680
Rowe JW, Kahn RL (1987) Human aging: usual and successful. Science 237(4811):143–149
Rowe JW, Kahn RL (2015) Successful aging 2.0: conceptual expansions for the 21st century. J Gerontol B Psychol Sci Soc Sci 70(4):593–596
Ryff CD (1989) Happiness is everything, or is it? Explorations on the meaning of psychological well-being. J Pers Soc Psychol 57(6):1069
Ryff CD, Keyes CLM (1995) The structure of psychological well-being revisited. J Pers Soc Psychol 69(4):719–727
Salthouse TA (2010) Selective review of cognitive aging. J Int Neuropsychol Soc 16(5):754–760
Sarwar N, Aspelund T, Eiriksdottir G, Gobin R, Seshasai SR, Forouhi NG, Sigurdsson G, Danesh J, Gudnason V (2010) Markers of dysglycaemia and risk of coronary heart disease in people without diabetes: Reykjavik prospective study and systematic review. PLoS Med 7(5): e1000278
Shipley BA, Der G, Taylor MD, Deary IJ (2007) Association between mortality and cognitive change over 7 years in a large representative sample of UK residents. Psychosom Med 69 (7):640–650
Sin NL, Lyubomirsky S (2009) Enhancing well-being and alleviating depressive symptoms with positive psychology interventions: a practice-friendly meta-analysis. J Clin Psychol 65 (5):467–487
Sirven N, Debrand T (2008) Social participation and healthy ageing: an international comparison using SHARE data. Soc Sci Med 67(12):2017–2026
Snowden JS, Neary D, Mann DM (2002) Frontotemporal dementia. Br. J. Psychiatry 180 (2):140–143
Stewart R (2012) Subjective cognitive impairment. Curr Opin Psychiatry 25(6):445–450
Studenski S, Perera S, Patel K, Rosano C, Faulkner K, Inzitari M, Brach J, Chandler J, Cawthon P, Connor EB, Nevitt M (2011) Gait speed and survival in older adults. JAMA 305(1):50–58
Thomas PA (2011) Gender, social engagement, and limitations in late life. Soc Sci Med 73 (9):1428–1435
Tucker-Drob EM, Salthouse TA (2011) Cognitive aging. In: The Wiley-Blackwell handbook of individual differences. Wiley Blackwell, Chichester, p 242
Van Dam RM, Li T, Spiegelman D, Franco OH, Hu FB (2008) Combined impact of lifestyle factors on mortality: prospective cohort study in US women. BMJ 337:a1440
Vaupel JW (2010) Biodemography of human ageing. Nature 464(7288):536–542
Verhaeghen P (2011) Aging and executive control: reports of a demise greatly exaggerated. Curr Dir Psychol Sci 20(3):174–180
Weuve J, Glymour MM, Hu H, Sparrow D, Spiro A, Vokonas PS, Litonjua AA (2011) Forced expiratory volume in 1 second and cognitive aging in men. J Am Geriatr Soc 59(7):1283–1292
Windle G, Markland DA, Woods RT (2008) Examination of a theoretical model of psychological resilience in older age. Aging Ment Health 12(3):285–292
Windle G, Bennett KM, Noyes J (2011) A methodological review of resilience measurement scales. Health Qual Life Outcomes 9(1):8
Wood V, Wylie ML, Sheafor B (1969) An analysis of a short self-report measure of life satisfaction: correlation with rater judgments. J Gerontol A Biol Sci Med Sci 24(4):465–469
World Health Organization (2015) World report on ageing and health. World Health Organization, Geneva
Yates LB, Djoussé L, Kurth T, Buring JE, Gaziano JM (2009) Exceptional longevity in men: modifiable factors associated with survival and function to age 90 years. Obstet Gynecol Surv 64(1):28–29
Zittermann A, Iodice S, Pilz S, Grant WB, Bagnardi V, Gandini S (2011) Vitamin D deficiency and mortality risk in the general population: a meta-analysis of prospective cohort studies. Am J Clin Nutr 95(1):91–100

Age-Related Diseases

3

Touqeer Ahmed, Abida Zulfiqar, and Sara Ishaq

Abstract

Age-related diseases occur frequently in the older population which signifies age as a major risk factor for old age disorders. Aging itself is not a disease but there are certain manifestations associated with aging that contribute to age-related diseases. There are many underlying causes that exacerbate age-related disorders, they include, inflammation, exposure to environmental pollutants, radiation, lifestyle, and dietary conditions. These factors tend to affect different people at a different rate and the disease progression depends on cellular aging and molecular pathways. Some of the prominent diseases associated with aging are cardiovascular, neurodegenerative, and metabolic in nature. Currently, many research models are proposed to determine the effect of aging on human body and how the rate of aging can be reduced by changes in lifestyle, specially controlling the social and mental stressors that catalyse aging by causing early cognitive impairment.

Keywords

Aging theory · Neurodegenerative diseases · Cognitive function

3.1 Normal Aging and Its Manifestations

Aging is an inevitable process. During this time, all organs tend to lose work as they age. Therefore, aging is basically a decline in the regenerative and reparative potential of tissues and organs. Aging takes place on various levels and there are

T. Ahmed (✉) · A. Zulfiqar · S. Ishaq
Neurobiology Laboratory, Department of Healthcare Biotechnology, Atta-ur-Rahman School of Applied Biosciences, National University of Sciences and Technology, Islamabad, Pakistan
e-mail: touqeer.ahmed@asab.nust.edu.pk

S. M. Nabavi et al. (eds.), *Nutrients and Nutraceuticals for Active & Healthy Ageing*,
https://doi.org/10.1007/978-981-15-3552-9_3

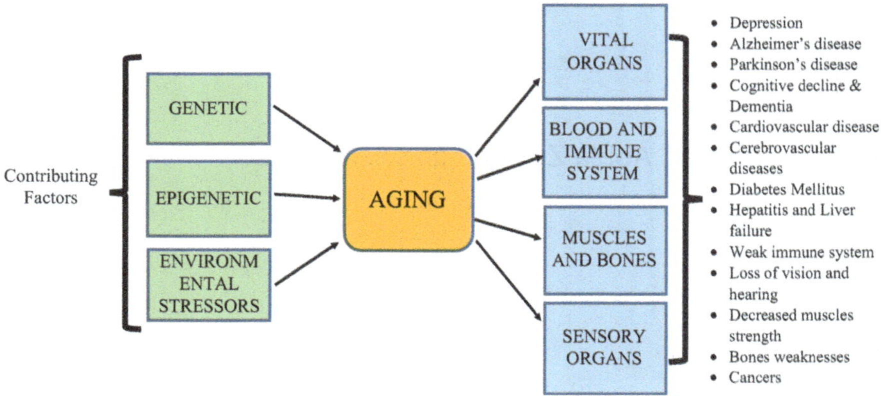

Fig. 3.1 Main contributors of aging associated diseases

factors like epigenetics and hereditary changes responsible for it (Roberts and Allen 2016). Along with this, many cells lose their function because there is an increase in the levels of pigments and fatty substances in the cells. As aging continues the waste materials build up inside the cells, airways and the organs become less efficient. This altogether leads to an altered cell membrane permeability (Gemma et al. 2007).

Aging is a slow and gradual process and continues throughout our lives. All our organs lose function with time. For example, the pumping ability of the heart of a 20 year old is about 10 times more efficient than that of a person aged 30–40 years. Organs like heart, kidneys, and lungs undergo substantial amount of changes, and such changes occur slowly over time (Gladyshev 2016). Whenever an organ undergoes a lot of stress or work, it results in abnormalities like sudden heart failure. Certain factors like medications, sudden increase in physical activity or change in environment tends to increase the work load on the organs. Aging also makes it difficult for the body to remove the drugs from the body as the kidneys and liver function at a slower rate (López-Otín et al. 2013) (Fig. 3.1).

3.1.1 The Aging Theory

Aging remains an enigma for the biologists (Gladyshev 2016). Various theories have been put forth to solve the mysteries associated with aging and its mechanisms. According to some theories, aging results due to the ultraviolet light exposure over time that causes injuries. Other studies have established that aging is controlled on the genetic level. There is no single process that can explain aging, as it is a complex phenomenon which varies and affects different people in entirely different patterns and diverse rates (Gemma et al. 2007).

3.1.2 Aging Cells and Their Replacement Mechanism

As cells age, they become less efficient in doing their normal duties. There has to be a mechanism where the old cells must die, as a normal feature of the cells. This process is called apoptosis where a sequence of reaction is initiated by the genes, that results in cell death. This is important because only in this way, the old cells will create room for the new ones (Saretzki and von Zglinicki 2002).

3.1.3 Telomeres and Aging

Telomeres are structures made from protein and DNA, present at the end of chromosomes which serve to protect the overall genome from degradation. Therefore, telomeres tend to preserve the genetic information. As part of normal cell division, a very small region of telomere is lost. Upon reaching a certain telomeric length, the cell senescence occurs and enters into the apoptotic phase. The telomere length therefore, determines the lifespan of a certain cell or organism. However, there are many external factors, i.e. lifestyle changes that may fasten this process and brings about early aging and health deterioration. The factors responsible for early aging and shortening of telomeres include obesity, exposure to pollution, unhealthy diet, and stress, etc. (Shammas 2011).

The evidence for the telomere aging hypothesis comes from the finding that the telomeres in the normal human cells (young individuals) shorten when they are grown in cell culture (Shay and Wright 2001). Some telomeric proteins like TRF1, TRF2, TIN2, TPP1, Rap1 are expressed almost all the time. Other proteins are expressed transiently, for example, the telomerase enzyme. Telomerase is a reverse transcriptase enzyme and it works by extending the 3′ end by adding TTAGGG repeats at chromosome ends. Therefore, it is important for a cell to have normal telomerase activity so that the integrity of chromosomes is maintained. In case of accelerated telomere shortening, manifestation of a number of age-associated diseases like diabetes, heart failure, osteoporosis, increased cancer risk takes place (Aubert and Lansdorp 2008).

According to studies, due to shorter telomere length, older people have three to eight times higher affinity to encounter infectious diseases or heart related disorders. In order to delay the signs of aging due to telomere shortening and to help protect telomeres, a number of antioxidants can be added to regular diet. These antioxidants are found in foods like salmon, tuna, sesame seeds, chia seeds, green tea, black raspberries, olive fruit, kiwi, red grapes, etc. (Shammas 2011).

3.1.4 The Role of Mitochondria in Aging

Mitochondria plays a significant role in the manifestation of age-related disorders, most importantly, the neurodegenerative diseases. Apart from energy production, mitochondria is involved in a number of functions, like calcium homeostasis,

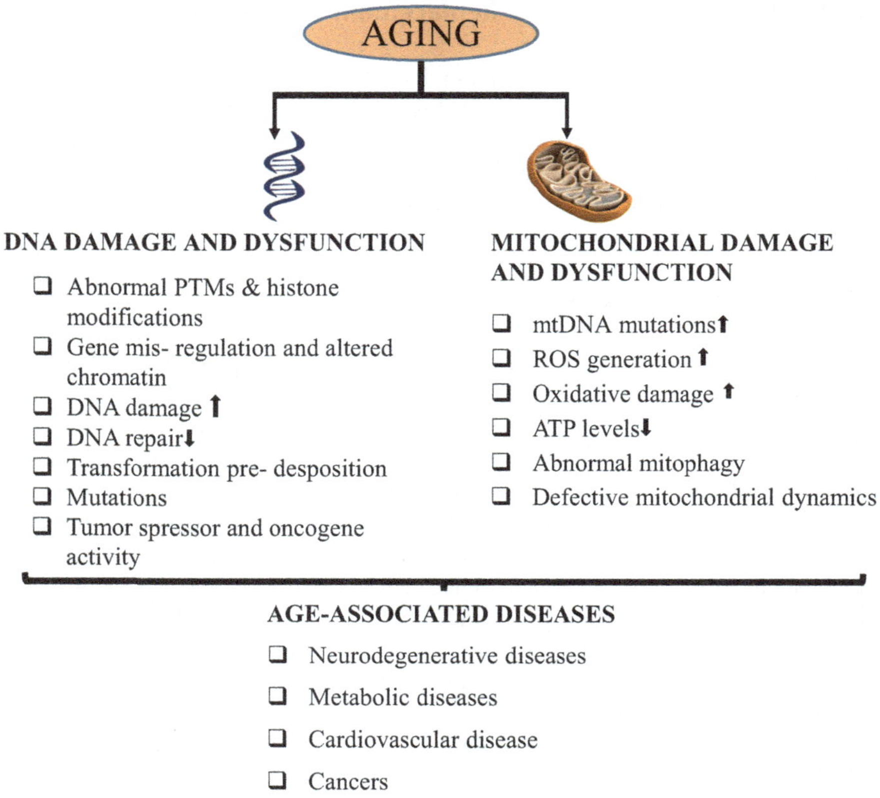

Fig. 3.2 Aging leads toward different diseases by cell structures damage

oxidative phosphorylation, and cell signaling to the nucleus in states of stress. It also participates in apoptosis and formation of cell's free radicals. The fact that mitochondria plays an important role in overall aging process is because it has a reduced DNA repair capacity and has an estimated tenfold greater rate of mutation than nuclear DNA (Haas 2019).

According to a study by Peter McGuire, the major player in aging of a cell are the mitochondrial damage-associated molecular patterns (mtDAMPs), which when released activates a number of proinflammatory cytokines like TNF-a, IL-1, IL-6 (McGuire 2019). This in turn results in an enhanced inflammation, declining function of T cells and susceptibility to viral infections in the older individuals (Jang et al. 2018) (Fig. 3.2).

3.1.5 Aging Organs

The optimal working of the organ depends on how well the resident cells function. The cells that are old have altered function compared to the new cells. Also in some organs, the number of total cells decrease as aging proceeds, for example, in the testes, kidneys, and liver. There are certain conditions in the elderly where the progressive loss of nerve cells leads to neurodegenerative disorders like the Parkinson's disease or the Alzheimer's disease (Wang and Bennett 2012). The loss of function of one organ may affect the efficiency of other organ and contribute altogether towards aging. For example, atherosclerosis results when the blood vessels are narrowed due to the accumulation of cholesterol, as a result the organ will not function well because the blood flow has been decreased (Gemma et al. 2007).

In the aged individuals, majority of the functions stay normal but the reduction in overall function of organs is due to the fact that they are less able to handle different stresses. These include extreme temperatures, strenuous physical activity, disorders, effects of drugs, etc. (López-Otín et al. 2013).

3.1.6 Signs of Aging

As aging proceeds, different people show different sings. But some of the common are as follows:

3.1.6.1 Sensory Changes Associated with Aging

As age advances, the risk of certain chronic hearing and vision related disabilities also increases which results in a decline in overall health and well-being. In the UK, about 27% (one in four) of aged adults 61–80 years suffer hearing disability (Crimmins 2018). Among adults aged 65 years, 13% have a vision impairment. Therefore, hearing and vision impairment is very common in aged population around the world (Cavazzana et al. 2018). One interesting finding suggests a link of hearing and vision impairment with disability, frailty, cardiovascular disease (CVD). However, these impairments are said to be preventable to some extent. The conditions like CVD, frailty, and disability can therefore be avoided if the hearing and vision impairments are addressed in time (Liljas 2017).

3.1.6.2 Hearing Loss

Hearing loss advances with age and is said to affect over half (55%) of the adults aged 75 years or more. Such losses can be prevented by the use of hearing aids. Hearing loss, other than aging, also occurs due to continuous exposure to loud noise. The loss which is age associated is called presbycusis (Huang and Tang 2010). During this phase it is also very difficult for the people to make sense out of the words because they cannot identify the consonants that are high pitched (k, t, s, p, and ch).

3.1.6.3 Vision Related Deficits

A change in a person's vision is usually the first sign of aging. As people age, the following vision related deficits takes place:

- The focal point of the eye becomes hard, which makes focusing on near objects harder.
- It becomes harder to see in the dim light as the focal point becomes denser.
- The eyes generally encounter dryness as the number of cells required for lubrication of the eye, decrease in number. Tear production is also reduced.
- The number of nerve cells also reduce in number, disabling the accuracy of perception.
- The focal point of the eye becomes yellow, it changes the way a color is perceived in normal conditions (Fine et al. 2000).

Loss of Close Vision The elderly population, in their late 1940s usually complain about a discomfort whenever they observe an object closer than 2 feet. This condition is called presbyopia, where the focal point of the eye hardens. Usually, in order to assist the eye to the center, the lens changes its shape. Because a stiffer focal point makes focusing on the near objects very difficult. At the later stages in life, almost everyone goes through presbyopia and needs amplifying glasses (Khan et al. 2017).

Color Perception Changes As the focal point tends to become yellow with time, colors are observed quite differently. It also becomes harder to differentiate between different colors. Specially the blue color may seem more gray and the blue background is indistinguishable (López-Otín et al. 2013).

Age Aged Individuals Need for Brighter Light to View Things Consequently, the pupil of eye reacts very slowly in response to light. A very slow reacting pupil in the elderly makes them unable to see when they first enter a very dark room or become blinded for few moments when they enter a very bright room. The pupil widens and also narrows bringing less amount of light inside. This increased sensitivity to the amount of light is because of cataracts or areas that have darkened in the lens. Older people often see many dark spots moving in their field of vision. These are called "floaters," which are small regions of normal fluids that have become solid over time. They do not usually interfere with the vision but can cause blurry images. Not only is the function of eye affected, its physical appearance also changes considerably (Stuen and Faye 2003).

- The sclera of the eye (white region) gradually turns brown or yellow. This happens due to continuous exposure to wind, dust, and UV light.
- People with a darker complexion may experience random blotches in the white region of the eye.
- Calcium and cholesterol accumulation may appear as a grayish white ring on eye surface called arcus senilis. However, vision is not affected by this.

- As the muscles around the eyes droop, and tendons stretch the lower eyelid tend to hang away. This results in a condition called ectropion, which tends to interfere with lubricating the eyeball thus making eyes dry (Liljas 2017).

3.1.6.4 Disturbances in Vestibular Functions During Aging

About 30–35% of older population suffers from the vestibular dysfunction. One common type of vestibular disorder in the elderly is benign paroxysmal positional vertigo (BPPV). Not only this but the elderly aged 65 face vestibular dysfunction due to injurious falls annually. This results due to balance impairments, and the older people experience more imbalances due to aging of the nervous system (Balatsouras et al. 2018). Such changes also impair walking capabilities, the elderly tend to walk with more care and slower pace to avoid any falls. The dysfunction of the vestibular system also leads to an altered spatial navigation. Aging altogether brings about rapid changes in the somatosensory function of an individual which manifests itself as visual impairments, decreased strength, and cognitive decline. The older adults are more prone to a decline in cognition and vestibular dysfunction mediates this decline (Shaffer and Harrison 2007).

3.1.6.5 Decline in Muscle Strength as a Consequence of Aging

As the aging progresses, usually by the age of 30, the overall muscle strength decreases. Major reduction in muscle strength takes place due to physical inactivity and reduced levels of growth hormones and testosterone. Moreover, muscles are unable to contract more frequently because the fast contracting muscle fibers are lost over a period of time. A case where extreme muscle loss takes place is called sarcopenia, which happens due to inactivity for a longer period of time, aging may or may not play a role in this situation. The regular decline in muscle strength accounts for about 10–15% in an adult's life and even this can be prevented with regular exercise and healthy lifestyle (Jaul and Barron 2017).

3.1.6.6 Immunosenescence (Age-Related Changes in the Immune System)

Immunity play a crucial role in age-related changes, which are mediated by a chronic inflammatory state that leads to aging of the immune cells. The immune cells like the B cell and T cells repertoire undergoes a decline in generation, activation and results in dysfunction. This results in an overall reduction in a body's strength to fight with different infections (Gruver et al. 2007). For instance, the infections of influenza and herpes zoster are common in the elderly and its vaccine is also not very effective. This implies the reason of slow healing of the wounds in older adults because of a delayed immune response. The total number of macrophages, neutrophils is also reduced resulting in an altered phagocytic activity (Jaul and Barron 2017).

3.2 Aging Associated Diseases

What is the reason for experiencing an overall decline in mental and physical health as aging progresses? A unit hypothesis clarifying the pathophysiology of various degenerative abnormalities in various organs, including Alzheimer's, Parkinson's, and other neurodegenerative issues, rheumatoid arthritis, atherosclerosis, and other cardiovascular sicknesses, macular degeneration, and diabetes, has been proposed to clarify how the physiological procedure of aging may prompt different diseases. This idea called the "free radical theory," at first proposed by Harman, gives the most conceivable underlying reason for aging (Harman 1992). The fundamental reason of this hypothesis is that aging and its related manifestations are the net result of free radical-instigated harm and the lack of ability to balance these progressions by anti-oxidative safeguards. The formation of reactive oxygen and nitrogen species (ROS furthermore, RNS) initiates the transcription elements that exacerbates inflammation. The constant expression of such elements over a prolonged period leads to targeting the cells and organs leading to aging related diseases. This oxidative pressure and consequent incessant aggravation is a major factor that leads to majority of the aging related illnesses (Sarkar and Fisher 2006). A summary of the aging associated diseases and their mechanisms is given in Table 3.1.

3.2.1 Contributing Factors for Aging Related Diseases

3.2.1.1 Social and Environmental Conditions

Two important predictors for a prolonged life are wealth and marriage. On the other hand, social seclusion envisages mortality and other hostile consequences in adults. Around 5% of older adults in their 80's are leading an isolated life devoid of basic health necessities, which makes them susceptible to different diseases (Waite 2018). Around 13% of women and 8% of men above 85 years are residing in nursing care or other healthcare locales. These rates are decreased in recent times probably as a result of a lesser amount of disability and enhanced care alternatives at home.

Around 17 million people worked as family caregivers and volunteers to older adults during 2011 in the USA. Caregiving is characteristically an extended obligation and the work hours may well vary in accordance with the requirements of the patient. Individuals suffering with dementia are in dire need of constant care and support. Therefore, the family caregivers must have proper training, counseling, and employment facilitation to help the elderly patients with daily chores (Jaul and Barron 2017).

3.2.1.2 Polypharmacy and Hospitalization

Polypharmacy is described as the simultaneous use of 5 or above prescriptions to a sole patient. Such high number of medications come with a high probability of hostile drug reaction or drug–drug interaction. It may escalate the danger of falls, disabilities, and other harmful consequences (Hajjar et al. 2007). The care providers should be careful while prescribing medicines and should make note of any side

Table 3.1 Different age-associated diseases and their mechanisms

Aging associated diseases	Aging induced disease mechanisms	References
Cardiovascular diseases	• Reduction of nitric oxide release • Reduced turnover of endothelial cells • Disrupted NADPH oxidases mediated ROS signaling pathways • Increased oxidative stress	Yang and Ming (2012), Sahoo et al. (2016), and Izzo et al. (2018)
Cerebrovascular diseases	• Increased oxidative stress • Amyloid beta accumulation • Neuronal damage due to increased ROS level • Reduced neurotransmitters levels • High iron accumulation in brain	Hirai et al. (1996), Patyar et al. (2011), and Izzo et al. (2018)
Hypertension	• Increased NO production due to reduced superoxide dismutase production • Increased expression of matrix metalloproteinases • Activation of transforming growth factor beta-1/SMAD, proinflammatory and profibrotic signaling pathways • Up-regulation of glacetin-3	Jun et al. (1996), Taddei et al. (2006), and Harvey et al. (2016)
Osteoarthritis	• Accumulation of advanced glycation end products (carbohydrate reacts with protein) in aged people lead to reduction in type II collagen of cartilage structure causing bone perforation • Reduced expression of Unc-51-like kinase, Beclin1, and microtubule-*associated protein 1 light chain 3 (all involved in regulation of autophagy)* leads to cartilage fragility and bone deformation • Reduced level of cartilage glycosaminoglycans leads to knee destabilization • Age-related increase in inflammatory cytokines and chemokines like IL-6 or TNF-α lead to joint tissue destruction and cell senescence	Hamerman (1993), DeGroot et al. (2004), and Greene and Loeser (2015)
Osteoporosis	• Increased ROS production causes reduced production and survival of osteoclasts, osteoblasts, and osteocytes by activating FoxO • Activation of peroxisome proliferator-activate receptor γ by ligands produced from lipid oxidation leads to age-associated reduced bone formation • Increased glucocorticoid production causes bone fragility	Riggs et al. (1998), Manolagas (2000), and Manolagas (2010)

(continued)

Table 3.1 (continued)

Aging associated diseases	Aging induced disease mechanisms	References
Diabetes mellitus	• Age-associated increased hypothalamic transforming growth factor beta induces hyperglycemia and glucose intolerance • Aging induced excessive production of inflammatory cytokines like IL-6 or TNF-α, C-reactive protein, and several cell adhesion molecules lead to increased recruitment of MHC and T cells and pancreatic beta cells leading to their destruction • Accumulation of ROS cause pancreatic cell damage • Accumulation of advanced glycation end products leading to kidney cells damage which ultimately cause diabetes	Davidson (1979), Vlassara (1996), Yin and Chen (2005), Morley (2008), Yan et al. (2014), Guarner and Rubio-Ruiz (2015), and Palmer and Kirkland (2016)
Cancers	• DNA methylation in CpG islands of genes like p16, p53, IB1, ER, IGF2, N33, and MyoD leads to tumor initiation and progression • Decline in the level of sirtuin protein family (involved in certain metabolic processes and also protects against certain stressors) leads to metabolic abnormalities and activation of oncogenes • Decreased levels of circulating B cells and T cells possibly due to age-related diminished follicular dendritic-cell function, leads to immunosenescence triggering tumor initiation • Reduced glutathione level leads to accumulation of ROS which can cause neoplasia	Ahuja and Issa (2000), Hakim et al. (2004), Hall et al. (2013), and Shay (2016)
Liver diseases like non-alcoholic fatty liver disease, hepatitis, etc.	• Decreased levels of serum albumin and bilirubin • Increased level of cholesterol, fats, gamma-glutamyltransferase, and alkaline phosphatase • Abnormal degradation of proteins damaged and denatured by oxidative stress cause accumulation of lipofuscin inside liver cells which cause increased levels of ROS, decreased mitochondrial number and function, and decline in ATP levels in liver cells • Defenestration of endothelial cells lead to deposition of lipoprotein-like chylomicron in liver triggering autoimmune diseases • Reduced number of Kupffer cells in old age leads to accumulation of antigen-antibody complexes and nanoparticles like senescent cell fragments in liver	Sheedfar et al. (2013), Kim et al. (2015), and Aravinthan and Alexander (2016)

Chronic obstructive pulmonary disease	• Neutrophils, macrophages and monocytes show enhanced ROS production in old age leading to increased oxidative stress • Oxidative stress causes shortening of telomeres which cause lungs inflammation by increasing the levels of proinflammatory cytokines • Accumulation of advanced glycation end products cause increase in levels of proinflammatory cytokines • B and T cell levels decrease with advanced age leading to immune deficiency in lungs • Changes in extracellular matrix	Faner et al. (2012) and Barnes (2014, 2016)
Age-related macular degradation	• Elevated accumulation of amyloid beta in retina with advanced age • Accumulation of ROS in retinal pigment epithelium leads to degradation of constantly shed apical photoreceptor outer segment leading to blindness • Accumulation of intracellular lipofuscin and extracellular drusens contained advanced glycation end product, high levels of oxidized low-density lipoproteins, and oxysterols leads to retinal inflammation • Abnormal activity of heat shock proteins and proteasomes due to oxidative stress lead to failure in removal of misfolded proteins causing retinal pigment endothelial degradation • Increased production of inflammatory cytokines cause autophagy in retinal pigment endothelium	Kinnunen et al. (2012), Wang et al. (2012), and Rickman et al. (2013)
Decreased blood cells production	• Hematocrit changes like salt depletion, capillary leakage, sudden release of vasoactive agents, and iatrogenic factors due to aging • Fibrinogen accumulation leading to RBC inflammation • RBC inflammation causes release of inflammatory cytokines and damage to other blood cells • Altered Na+, K+ ATPase activity of blood cell membranes • Altered viscosity of blood due to aging • Impaired autoimmune responses of suppressor T cells to altered self-antigens in old age	Naor et al. (1976), Ergen and Goodell (2010), and Simmonds et al. (2013)

(continued)

Table 3.1 (continued)

Aging associated diseases	Aging induced disease mechanisms	References
Immune disorders	• Increased production and accumulation of ROS in old age leads to excessive inflammatory substances release • Thymic atrophy in old age causes decreased production of B & T cells leading to immune compromised individuals • Reduced telomeres shortening in advanced age leads to DNA damage which ultimately causes elevation of proinflammatory cytokines which as a result affect nearby cells and continue an inflammatory process • During aging process, the process of autophagy cleansing declines leading to mitochondrial disordering and protein accumulation	Prelog (2006), Xia et al. (2016), and Weyand and Goronzy (2016)
Frailty syndrome	• Accumulation of ROS in advanced due to decreased levels of certain antioxidants like glutathione, etc. causes oxidative stress • Oxidative stress causes release of certain inflammatory cytokines like IL-6, TNF-alpha, etc. and their accumulation in cells • Increased inflammation causes more recruitment of immune cells which cause cell destruction, mitochondrial, and DNA damage • Increased serum levels of C-reactive protein, low serum levels of 25-hydroxyvitamin D and growth hormone IGF-1 cause decline in muscle mass and function leading to frailty	Ble et al. (2006), Topinková (2008), and Chen et al. (2014)
Cognitive decline and dementia	• Reduced function of mitochondrial uncoupling protein- 5 (UCP-5) responsible for neuronal metabolism and homeostasis leads to abnormal metabolism in neurons and accumulation of ROS • ROS accumulation causes increased oxidative stress which leads to mitochondrial autophagy • Increased oxidative stress also leads to excessive release of inflammatory cytokines which recruit more immune cells adding to neuronal autophagy • ROS accumulation also causes DNA and protein damage • Damaged and misfolded proteins start accumulating in neurons • Age-related reduction in the functions of insulin like growth factor-1 (IGF-1) & sirtuins, and mRNA abnormal translation further add to accumulation of misfolded proteins	Deary et al. (2009), Bishop et al. (2010), and Terrando et al. (2011)

Alzheimer's disease	• Age-related decline in anti-oxidant such as glutathione levels causes accumulation of ROS in brain leading to oxidative stress • Increased oxidative stress causes DNA damage and proteins misfolding • Amyloid beta and tau proteins start accumulating in the brain causing neurofibrillary tangles • Increased glucocorticoid accumulation in hippocampus leads to increased calcium influx through voltage-activated calcium channels which in turn causes excess neuronal excitability and further adds to misfolded proteins accumulation	Smith et al. (1992), Liu et al. (2004), and Samad et al. (2017)
Parkinson's disease	• Abnormal functioning of L-type calcium channels in substantia nigra pars compacta (SNc) dopaminergic neurons (DA) due to aging induced oxidative stress in nearby neurons causes increased influx of Ca2+ in these neurons • Increased calcium influx disturbs metabolic homeostasis of SNc DA leading to oxidative stress and DNA and mitochondrial damage • Increased oxidative stress causes increased production of proinflammatory cytokines which are released and recruit immune cells which destroy these DA recognizing them as abnormal • This leads to dopaminergic neuronal death in substanstia nigra	Tompkins et al. (1997), Calabrese et al. (2001), and Surmeier et al. (2010)
Clinical depression	• Aging causes low grade T-helper cells type-1 inflammation leading to release of inflammatory cytokines like interferon gamma, which transcriptionally induces the rate-limiting enzyme of tryptophan (Trp)-kynurenine (Kyn) pathway, indoleamine 2,3-deoxygenase (IDO) • Activation of IDO blocks Trp conversion to serotonin (substrate for antidepressant) and its derivatives: N-acetylserotonin (agonist to receptors of brain derived neurotropic factor), and melatonin (a pigment responsible for sleep regulation and circadian rhythms) • IDO activation leads to enhanced production of Kyn and its derivatives (anxiogenic, neurotoxic, and pro-oxidants), some of which upregulate nitric oxide synthase (NOS) • Increased activity of NOS leads to arginine metabolism to superoxide anions and other oxidative species leading to oxidative stress which exacerbate depression, anxiety and other cognitive impairments	Guidi et al. (1998), Oxenkrug (2011), and Costa et al. (2013)

effects, time of action, and proper dosage for the individuals over the age of 85. Hospitalization is considered as a normal practice amongst persons over the age of 85 mostly due to functional failure of the aged. As long as extensive acute care is taken at the home, older individuals can be saved from hospitalization and its difficulties like: further worsening of the condition due to iatrogenic infection. In order to avoid re-hospitalization, a routine medical checkup is important to minimize any errors that may lead to recurring disease manifestation (Singh 2016).

3.3 Types of Diseases Associated with Aging

3.3.1 Metabolic Diseases

3.3.1.1 Cardiovascular Diseases

Cardiovascular disease accounts for one of the most prevalent diseases in the older age. It affects the blood vessels and heart and is the leading cause of myocardial infarction (MI), coronary heart disease (CHD), and stroke. This disease is highly preventable if proper lifestyle measures are adopted, therefore reducing premature death ratio. In old age, a plethora of different conditions cause many diseases, similarly there has been a recent association of cardiovascular diseases with sensory impairments. A common mechanism of inflammation is implicated in both these conditions (Liljas 2017).

During the CVD, stiffness of the heart and blood vessels take place. This stiffness does not allow the arteries to carry more blood when pumped through the heart as a result causing increased blood pressure. This leads to an overall thickening of the heart muscle called hypertrophy (Jaul and Barron 2017). This makes it difficult for the heart to pump the blood, making physical activities more difficult as compared to that of a young heart. In CVD, the thickening of the capillary walls also takes place. This reduces the rate at which nutrients exchange takes place. The total volume of the blood also decreases, because there is a general reduction in total body water content as a person ages. In conditions of stress or illness, the pace at which red blood cell production takes place is lowered. Certain conditions that make the heart work harder include emotional stress, physical exertion, infections, and certain medicines (North and Sinclair 2012).

An overall change in the heart structure, with aging, leads to the formation of abnormal rhythms (arrhythmias) and are detected through ECG. Another common abnormality is the accumulation of lipofuscin—an aging pigment which is more common in the elderly. Because of stiffness of the heart valves, a certain disturbance in the heart sound is also observed called as the heart murmur. Apart from this, there are receptors in the blood vessels, called baroreceptors, these receptors help to normalize blood pressure. These receptors also undergo aging, which explains the reason of orthostatic hypotension in older people, a condition in which a person's blood pressure falls abruptly when a person goes from a lying position to standing or sitting (Strait and Lakatta 2012).

Common Problems Associated with Cardio Vascular Diseases (CVD)

- *Hardening of the arteries (Arteriosclerosis)*—fatty deposits in the blood vessels lead to this condition.
- *Angina* (a type of chest pain caused due to reduced flow of blood to the heart muscle).
- *Arrhythmias* (abnormal heart rhythm).
- *Anemia*—usually occurs due to nutritional deficits or chronic infections.
- High blood pressure/orthostatic hypotension.
- Deep vein thrombosis, blood clots, peripheral vascular disease, varicose veins are all manifestations of the CVD.
- In the older people, congestive heart failure is also very common.
- *Aortic stenosis* or the aortic valve is the most common valve disease in the elderly.
- *Aneurysms* may also develop in one of the major arteries from the heart or in the brain. They are caused by abnormal widening of the artery. If such an aneurysm bursts, it may lead to death.

3.3.1.2 Cerebrovascular Disease (Strokes)

Aging tend to bring about changes in the vascular endothelium and the arterial wall. According to recent studies, there is a reduction in vasodilatory capacity of the vascular endothelium. The arterial wall also undergoes changes due to calcium deposits and most importantly hypertrophy occurs that leads to an increase in vascular stiffness (Izzo et al. 2018).

Stroke is basically a vascular disease that directly affects brain and gives rise to a lot of manifestations. It is characterized by interrupted blood flow to the brain tissues due to breaking or closing of a cerebral artery. In the older adults, the prevalence is about 20–35%. In the western world, each year 10–12% deaths are caused due to stroke. Not only this, but it is also a leading cause of dementia and disability. Cerebral stroke tends to be a welfare issue, as the individuals are dependent upon others for their basic necessities (Gladyshev 2016). The protective function of the blood–brain barrier acts as a filter for the bloodstream in brain. As a person ages, the function of the BBB is distorted, the ability to selectively uptake nutrients is diminished. This in combination with inflammation like conditions leads to the formation of multi-infarct dementia (MID) and other neurodegenerative disorders (Farrall and Wardlaw 2009).

In the elderly population, the occasion of hemorrhagic events is more due a number of vascular structural changes like aneurisms, hypertension, increased cerebral trauma, amyloid deposits. It is quite evident from various studies that vascular dementia occurs in the older adults upon their exposure to a stroke or any hemorrhagic episode. In this type of dementia, the intellectual potential of a person is affected. The patients who are about 60, they are at a higher risk of experiencing a stroke associated vascular dementia, but here certain secondary conditions also exacerbate the disease, such as hypertension, diabetes, dyslipidaemias, etc. (Jaul and Barron 2017).

3.3.1.3 Hypertension

Another most common disease in the older adults is the hypertension which itself is the root cause of many other major diseases like atherosclerosis and CVD. However, its effects can be managed by bringing about positive changes in lifestyle and eating habits. Sedentary lifestyle and stress remains one of the prime causes for the diseases to persist and become severe (Logan 2011).

3.3.1.4 Bone and Joint Diseases

As aging proceeds bones become less dense, such moderate loss of bone density is called *osteopenia* and excessive loss of bone density is termed as *osteoporosis*. This excessive loss of bone density makes the bones become fragile and are susceptible to break even during a small accident. As estrogen (a female hormone) helps to main the bone density in women, upon menopause, less amount of estrogen results in further weakening of the bones (López-Otín et al. 2013).

Two other major factors that lead to the weakening of bones include reduction of calcium levels (as the body does not absorb maximum levels of calcium from dietary intake). Secondly, lower levels of Vitamin D further aggravate the disease. Some bones are more susceptible to weakening as compared to others. Most commonly affected bones include the spine bone, femur (thighbone), and radius and ulna (arm bones at wrist). As aging progresses, there are several changes taking place at the spine bone, which ends up in causing the head to slide forward which makes the throat compressed. This causes hard swallowing of food, and might as well result in occasional choking. The major reason that the older people become short is that their vertebra becomes dense and the tissues lose the lubricant fluid, making it thin and the spine becomes shorter (Jaul and Barron 2017).

Over the passage of time the joints become susceptible to injury because of many years of wear and tear. Such damage to the cartilage and use of joints leads to a condition called *osteoarthritis* which is the swelling and pain of joints. The joints of knee, hips, spine, and hands are usually affected. The elasticity of the ligaments which helps to bind the joints together is also lost with time and the tendons which binds muscle to bone also become less elastic. This as a result causes stiffness of joints and tissues weaken in the process contributing towards decrease flexibility and movements are altered while carrying on everyday tasks (Crimmins 2018).

3.3.1.5 Diabetes Mellitus

As people are becoming more and more overweight with increasing age, the diabetes rates are also increasing. The occurrence of diabetes amongst American adult population is likely to escalate more than 4 times by 2050. Cardiovascular diseases at the age of 85 are still influenced highly by Diabetes. Diabetes is also related directly to peripheral arterial disease and peripheral neuropathy, major contributing factor for diabetic foot ulcers and amputations. Every year around 6% and 0.5% of diabetic patients suffer from diabetic foot ulcers and amputations, respectively. Personalized management approaches are highly demanded in case of diabetes (Meneilly and Tessier 2001). Sulfonylureas and insulin transmit a considerable danger of hypoglycemia, therefore, the amount taken ought to be considered cautiously in susceptible older adults. Dosing needs may vary for patients treated with hypoglycemic agents, especially during post-acute care or when patients are sent from hospital to home in uncertain periods. Frequent foot inspections are crucial for individuals with diabetes to avoid amputations. Blood circulation in the legs likely improves with regular walking (Jaul and Barron 2017).

3.3.1.6 Cancer

Cancer tends to be the second chief reason of death in elderly. Though, by the age of 85, the death ratio of cancer starts to decrease. Still, slow growing tumors appear to occur frequently for this age group. People respond to cancer treatment in accordance to their overall health status rather than their age. People in their 1980s or 1990s must not be deprived of intensive cancer treatment just because of their physical weaknesses (Robinson and Turner 2003).

Screening is not endorsed for breast cancer beyond 75 years, because of inadequate evidence for a possible outcome, although this might be useful for women with higher life expectancy. Likewise, for people above the age of 75 years, colon cancer screening is suggested merely in circumstances when there is an extensive life expectancy and a seemingly resilient ability to bear cancer diagnosis. Life expectancy is relatively flexible in elderlies at any age, established on the basis of comorbidities and other elements. Numerous false positives like identification of slow growing tumors forbid the screening for prostate cancer (López-Otín et al. 2013).

3.3.1.7 Chronic Obstructive Pulmonary Disease

Evidence suggests that aging and illnesses due to prolonged inflammation are closely related. Chronic Obstructive Pulmonary Disease (COPD) is a very slowly developing disease due to prolonged inflammation of the lungs, thus, most of the patients are aged. Aging characteristically is the consistent decay of homeostasis which happens once the reproductive phase of life is concluded, which leads to aggregated risk of disease. Non programmed aging, i.e. failure of organs to heal DNA destruction due to oxidative stress or programmed aging, i.e. shortening of telomere due to repeated cell division may cause the disease (López-Otín et al. 2013). Pulmonary function gradually declines with aging due to increased pulmonary inflammation, along with physical changes, inferred as senile emphysema. Ecological fumes, for instance, smoke of cigarette or various pollutants, may well hasten the aging of lungs by the formation of molecules such as histone deacetylases and sirtuins which subsequently prompts accelerated development of COPD (Gan et al. 2004).

3.3.1.8 Age-Related Macular Degradation

This disease affects the vision of the older adults. In this disease macula is affected, which is the part of retina where all nerve cells exist in close connection and therefore this is the region where the eye focusses the images that we observe. Therefore, this area is responsible for developing a clear and sharp focused image with fine details. The macula is affected as we age and it affects people over 65 years of age causing blindness or vision loss (Jaul and Barron 2017).

3.3.1.9 Blood Production

Aging declines the amount of active bone marrow cells, where new blood cells are formed. So, the blood cells production reduces. However, the bone marrow typically produces sufficient blood cells all the way till death. Issues may surface at a time where there is an increased requirement for blood cells, for example, during anemia, infection and/or severe bleeding happens. Under such conditions, when large amount of blood cells are needed at old age, bone marrow might show limited capacity to meet the higher blood cells demand (López-Otín et al. 2013).

3.3.1.10 The Immune System

The action of cells of the immune system retards with aging. Immune cells have a capacity to recognize and then remove foreign substances, for instance, bacteria, viruses, or cancer cells. Due to the reduction in the number of immune cells, certain manifestations occur in the aging adults as following (Gruver et al. 2007).

- Cancer is found commonly amongst older people.
- Older people are more frequently effected by some infections, for instance: influenza and pneumonia resulting more often in death. Moreover, vaccines seem to be less protective for older people though influenza, pneumonia, and shingles vaccines may provide some level of protection (Khan et al. 2017).
- Allergic symptoms are not very severe. Eventually, due to the poor performance of the immune system, autoimmune disorders are less likely to occur.

3.3.2 Psychological and Cognitive Diseases

3.3.2.1 Frailty Syndrome

Frailty is a very common clinical syndrome associated with increasing age and leads to health conditions like: falls, hospitalization, incident disability, and mortality. In theory, frailty is a clinically detectable state of amplified vulnerability caused by aging related decay in reserve and function across multiple physiologic systems so that the capacity to handle daily or acute stressors is compromised (Xue 2011). According to Fried et al. frailty is a condition where the body of an adult meet three out of five phenotypic conditions representing compromised energetics: (Fried et al. 2001)

- Low grip strength.
- Low energy.
- Slowed waking speed.
- Low physical activity.
- Unintentional weight loss.

Modern efforts on frailty have enhanced the level of understanding on the aging process and its possible biological connection. The operational definition of frailty, its domains either cognitive or physical and its correlation with aging, disabilities, and prolonged sicknesses indicate that additional effort is required for improved understanding of frailty. Nonetheless, researchers and clinicians agree upon the strong effect of frailty on elder individuals, their caretakers and overall society. Though precise cures for frailty are still to be explored, the pre-existing medical methods offer convenient ways to identify prone people, which may lead to better treatment by keeping in view specific vulnerabilities and susceptibility for adverse health consequences (Xue 2011).

Regular exercising in order to strengthen the muscles can somehow delay loss of muscle mass and strength which contributes to frailty. On the contrary, being physically inactive, especially extensive bed rest during a disease, can significantly contribute to the muscle loss (Gladyshev 2016).

3.3.2.2 Cognitive Aging

By the age of 85, things like: difficulty in finding words, short-term memory loss, and slower processing of information are noticed quite often. Such regression in cognitive abilities can upset driving capabilities and may as well cause financial exploitation. Such manifestations can also lessen ability to cognize complex medical information. There is no uniform rate at which brain ages and typical cognitive aging may not lead to dementia. There are few functions like altruism, empathy, knowledge, and wisdom which generally increase with normal aging, giving the elderly a chance at society (Langa 2018).

3.3.2.3 Dementia

Chances of dementia rise with increasing age. Over the years, death toll from Alzheimer's disease have escalated than that for cardiovascular disease. The occurrence of dementia is estimated to increase to 131 million by 2050 over the globe. Even though dementia screening seems to possess inadequate clinical benefit as treatments are just marginally effective, however, provides alternates for improvement of public health. Several adults continue to live with dementia and might possibly be living precariously (Jaul and Barron 2017). The Folstein Mini-Mental State Examination though is frequently used tool to analyze dementia but it comprises numerous limitations. Individuals with dementia require chances for thinking incentive, caretaker's sustenance and perhaps evolved equipment to keep them safe (Waite 2018).

3.3.2.4 Dementia Including Alzheimer's Disease

In the USA, Alzheimer's disease is presently ranked to be the sixth foremost reason of death, however, recent assessments show that the disease may hike up to the third, after heart disease and cancer.

Dementia is most commonly caused by Alzheimer's in the elderly. Dementia is the loss of behavioral capabilities and cognitive activities like thinking, remembering, and reasoning so much so that it obstructs an individual's day-to-day life's activities. Severity of dementia may increase when it is mildly affecting a person's functioning, to the most severe stage, where the person essentially depends on someone else even for the very basic chores of everyday life (Karantzoulis et al. 2011).

Numerous studies are being conducted to understand excessively about various biological traits of Alzheimer's. Progresses in brain imaging allows scientists to comprehend the growth and extent of atypical amyloid and tau proteins in the living brain, along with alterations in brain structure and functioning. Researchers are also determining the very primitive stages in the development of sickness by understanding deviations in the brain and body fluids which can be used to diagnose Alzheimer's symptoms before they actually surface (Langa 2018).

The question that why Alzheimer's occur in the elderly is yet to be answered. Researchers are trying to find relevant reasons as to how the changes occur with age inside the brain like: inflammatory actions, reduction of specific parts of the brain, mitochondrial malfunctioning, creation of unstable free radicals and vascular damage that may cause harm to the neurons and affect other brain cells leading to Alzheimer's disease (Karantzoulis et al. 2011).

3.3.2.5 Parkinson's Disease

Age is one of the most important players for the manifestation of Parkinson's disease (PD). The exact source of PD is yet unidentified, still the researchers rely on the fact that it results due to the amalgamation of some genetic and other external factors which are yet to be specified. PD disturbs numerous regions of brain and body. The death of neurons in brain region called the substantia nigra pars compacta and consequently reduced levels of dopamine are mostly accountable for motor symptoms like: shivering, stiffness, and loss of impulsive movement (Nussbaum and Ellis 2003). Investigations have revealed that the substantia nigra pars compacta displays high damage of neurons than other areas of the brain. Furthermore, with aging, organelles' become nonproductive leading to the accumulation of waste products that cannot be recycled and hence destroy the neurons (Reeve et al. 2014).

It is assessed that individuals with PD lose around 80% or even more of the dopaminergic neurons establish the motor symptoms. Consequently, over the time, the altered functioning of mitochondria and other organelles lead to the buildup of abnormal forms of alpha-synuclein which can possibly cause loss of neurons leading to motor symptoms (Jaul and Barron 2017).

3.3.2.6 Depression

Depression is not a standard outcome of aging. Sorrow may be some usual reaction to life activities that befall with aging like: grief, pensiveness, damage to cognition, social and/or physical attributes that lead to sickness. Major depression is common all the way through maturity but their occurrence rates decline between the age of 60 and 80, rising again after that. Occurrence of depression above the age of 85 is twice the rate between 70 and 74. Depression is more likely in the elderly who are confined to in old houses and the individuals with disabilities (Alexopoulos 2005).

Various approaches are implied for the diagnosis and treatment in order to reduce the amount of suffering, improve overall attitude of an individual, and lower the chances of suicide. Suicide rates amongst 85-year-old white Americans are extremely common.

3.4 Conclusion

The process of aging is universal but not the same in all the individuals. Various manifestation of aging, for example, decline in vision and auditory patterns, slow processing of information, lack of coordination can be addressed in a better way if proper awareness is created among the elderly. This would not only help the patients but also prepare the caretakers to manage all sorts of unforeseen problems.

Social and mental stressors are also a major cause of early aging. Timely addressal of these issues may prevent isolation, cognitive decline, and depression leading to suicide. Regular exercising should be a part of routine to strengthen the muscles and prevent disabilities during later life. Maintaining a balanced body weight by taking care of the diet may increase the total well-being of an individual as the incidence of diseases like osteoarthritis, diabetes, and CVD become less likely. People should be facilitated in such a way that in coming times, individuals over the age of 85 profit from home-based facilities and technologies along with innovative conveyance and housing facilities prospects for social contribution, plus plans to support family caregivers.

References

Ahuja N, Issa JPJ (2000) Aging, methylation and cancer. Histol Histopathol 15(3):835–842

Alexopoulos GS (2005) Depression in the elderly. Lancet 365(9475):1961–1970

Aravinthan AD, Alexander GJ (2016) Senescence in chronic liver disease: is the future in aging? J Hepatol 65(4):825–834

Aubert G, Lansdorp PM (2008) Telomeres and aging. Physiol Rev 88(2):557–579

Balatsouras D, Koukoutsis G, Fassolis A, Moukos A, Apris A (2018) Benign paroxysmal positional vertigo in the elderly: current insights. Clin Interv Aging 13:2251

Barnes PJ (2014) Cellular and molecular mechanisms of chronic obstructive pulmonary disease. Clin Chest Med 35(1):71–86

Barnes PJ (2016) Inflammatory mechanisms in patients with chronic obstructive pulmonary disease. J Allergy Clin Immunol 138(1):16–27

Bishop NA, Lu T, Yankner BA (2010) Neural mechanisms of ageing and cognitive decline. Nature 464(7288):529

Ble A, Cherubini A, Volpato S, Bartali B, Walston JD, Windham BG et al (2006) Lower plasma vitamin E levels are associated with the frailty syndrome: the InCHIANTI study. J Gerontol Ser A Biol Med Sci 61(3):278–283

Calabrese V, Scapagnini G, Stella AG, Bates TE, Clark JB (2001) Mitochondrial involvement in brain function and dysfunction: relevance to aging, neurodegenerative disorders and longevity. Neurochem Res 26(6):739–764

Cavazzana A, Röhrborn A, Garthus-Niegel S, Larsson M, Hummel T, Croy I (2018) Sensory-specific impairment among older people. An investigation using both sensory thresholds and subjective measures across the five senses. PLoS One 13(8):e0202969

Chen X, Mao G, Leng SX (2014) Frailty syndrome: an overview. Clin Interv Aging 9:433

Costa IC, Carvalho HN, Fernandes L (2013) Aging, circadian rhythms and depressive disorders: a review. Am J Neurodegener Dis 2(4):228

Crimmins E (2018) Trends in mortality, disease, and physiological status in the older population. In: Future directions for the demography of aging: proceedings of a workshop. National Academies Press, Washington

Davidson MB (1979) The effect of aging on carbohydrate metabolism: a review of the English literature and a practical approach to the diagnosis of diabetes mellitus in the elderly. Metabolism 28(6):688–705

Deary IJ, Corley J, Gow AJ, Harris SE, Houlihan LM, Marioni RE et al (2009) Age-associated cognitive decline. Br Med Bull 92(1):135–152

DeGroot J, Verzijl N, Wenting-Van Wijk MJ, Jacobs KM, Van El B, Van Roermund PM et al (2004) Accumulation of advanced glycation end products as a molecular mechanism for aging as a risk factor in osteoarthritis. Arthritis Rheum 50(4):1207–1215

Ergen AV, Goodell MA (2010) Mechanisms of hematopoietic stem cell aging. Exp Gerontol 45 (4):286–290

Faner R, Rojas M, MacNee W, Agusti A (2012) Abnormal lung aging in chronic obstructive pulmonary disease and idiopathic pulmonary fibrosis. Am J Respir Crit Care Med 186 (4):306–313

Farrall AJ, Wardlaw JM (2009) Blood–brain barrier: ageing and microvascular disease–systematic review and meta-analysis. Neurobiol Aging 30(3):337–352

Fine SL, Berger JW, Maguire MG, Ho AC (2000) Age-related macular degeneration. N Engl J Med 342(7):483–492

Fried LP, Tangen CM, Walston J, Newman AB, Hirsch C, Gottdiener J, Seeman T, Tracy R, Kop WJ, Burke G (2001) Frailty in older adults: evidence for a phenotype. J Gerontol Ser A Biol Med Sci 56(3):M146–M157

Gan WQ, Man S, Senthilselvan A, Sin D (2004) Association between chronic obstructive pulmonary disease and systemic inflammation: a systematic review and a meta-analysis. Thorax 59 (7):574–580

Gemma C, Vila J, Bachstetter A, Bickford PC (2007) Chap. 15. Oxidative stress and the aging brain: from theory to prevention. In: Brain aging: models, methods, and mechanisms. CRC Press/Taylor & Francis, Boca Raton

Gladyshev VN (2016) Aging: progressive decline in fitness due to the rising deleteriome adjusted by genetic, environmental, and stochastic processes. Aging Cell 15(4):594–602

Greene MA, Loeser RF (2015) Aging-related inflammation in osteoarthritis. Osteoarthr Cartil 23 (11):1966–1971

Gruver A, Hudson L, Sempowski G (2007) Immunosenescence of ageing. J Pathol 211(2):144–156

Guarner V, Rubio-Ruiz ME (2015) Low-grade systemic inflammation connects aging, metabolic syndrome and cardiovascular disease. In: Aging and health—a systems biology perspective, vol 40. Karger Publishers, Basel, pp 99–106

Guidi L, Tricerri A, Frasca D, Vangeli M, Errani AR, Bartoloni C (1998) Psychoneuroimmunology and aging. Gerontology 44(5):247–261

Haas RH (2019) Mitochondrial dysfunction in aging and diseases of aging. Multidisciplinary Digital Publishing Institute, Basel

Hajjar ER, Cafiero AC, Hanlon JT (2007) Polypharmacy in elderly patients. Am J Geriatr Pharmacother 5(4):345–351

Hakim FT, Flomerfelt FA, Boyiadzis M, Gress RE (2004) Aging, immunity and cancer. Curr Opin Immunol 16(2):151–156

Hall JA, Dominy JE, Lee Y, Puigserver P (2013) The sirtuin family's role in aging and age-associated pathologies. J Clin Invest 123(3):973–979

Hamerman D (1993) Aging and osteoarthritis: basic mechanisms. J Am Geriatr Soc 41(7):760–770
Harman D (1992) Free radical theory of aging: history. In: Free radicals and aging. Birkhäuser, Basel, pp 1–10
Harvey A, Montezano AC, Lopes RA, Rios F, Touyz RM (2016) Vascular fibrosis in aging and hypertension: molecular mechanisms and clinical implications. Can J Cardiol 32(5):659–668
Hirai T, Korogi Y, Sakamoto Y, Hamatake S, Ikushima I, Takahashi M (1996) T2 shortening in the motor cortex: effect of aging and cerebrovascular diseases. Radiology 199(3):799–803
Huang Q, Tang J (2010) Age-related hearing loss or presbycusis. Eur Arch Otorhinolaryngol 267 (8):1179–1191
Izzo C, Carrizzo A, Alfano A, Virtuoso N, Capunzo M, Calabrese M, De Simone E, Sciarretta S, Frati G, Oliveti M (2018) The impact of aging on cardio and cerebrovascular diseases. Int J Mol Sci 19(2):481
Jang JY, Blum A, Liu J, Finkel T (2018) The role of mitochondria in aging. J Clin Invest 128 (9):3662
Jaul E, Barron J (2017) Age-related diseases and clinical and public health implications for the 85 years old and over population. Front Public Health 5:335
Jun T, ke-yan F, Catalano M (1996) Increased superoxide anion production in humans: a possible mechanism for the pathogenesis of hypertension. J Hum Hypertens 10(5):305–309
Karantzoulis S, Galvin JE, Braak, Salmon, Mckhann, McKhann, Jack, Sperling, Hodges, Butters (2011) Distinguishing Alzheimer's disease from other major forms of dementia. Expert Rev Neurother 11(11):1579–1591
Khan SS, Singer BD, Vaughan DE (2017) Molecular and physiological manifestations and measurement of aging in humans. Aging Cell 16(4):624–633
Kim H, Kisseleva T, Brenner DA (2015) Aging and liver disease. Curr Opin Gastroenterol 31 (3):184
Kinnunen K, Petrovski G, Moe MC, Berta A, Kaarniranta K (2012) Molecular mechanisms of retinal pigment epithelium damage and development of age-related macular degeneration. Acta Ophthalmol 90(4):299–309
Langa KM (2018) Cognitive aging, dementia, and the future of an aging population. In: Future directions for the demography of aging: proceedings of a workshop. National Academies Press, Washington
Liljas AEM (2017) The relationship of sensory impairments with cardiovascular disease and mortality, disability and frailty in older age: longitudinal cohort studies using the British Regional Heart Study and the English Longitudinal Study of Ageing, UCL (University College London)
Liu H, Wang H, Shenvi S, Hagen TM, Liu RM (2004) Glutathione metabolism during aging and in Alzheimer disease. Ann N Y Acad Sci 1019(1):346–349
Logan AG (2011) Hypertension in aging patients. Expert Rev Cardiovasc Ther 9(1):113–120
López-Otín C, Blasco MA, Partridge L, Serrano M, Kroemer G (2013) The hallmarks of aging. Cell 153(6):1194–1217
Manolagas SC (2000) Birth and death of bone cells: basic regulatory mechanisms and implications for the pathogenesis and treatment of osteoporosis. Endocr Rev 21(2):115–137
Manolagas SC (2010) From estrogen-centric to aging and oxidative stress: a revised perspective of the pathogenesis of osteoporosis. Endocr Rev 31(3):266–300
McGuire PJ (2019) Mitochondrial dysfunction and the aging immune system. Biology 8(2):26
Meneilly GS, Tessier D (2001) Diabetes in elderly adults. J Gerontol Ser A Biol Med Sci 56(1):M5–M13
Morley JE (2008) Diabetes and aging: epidemiologic overview. Clin Geriatr Med 24(3):395–405
Naor D, Bonavida B, Walford RL (1976) Autoimmunity and aging: the age-related response of mice of a long-lived strain to trinitrophenylated syngeneic mouse red blood cells. J Immunol 117(6):2204–2208
North BJ, Sinclair DA (2012) The intersection between aging and cardiovascular disease. Circ Res 110(8):1097–1108

Nussbaum RL, Ellis CE (2003) Alzheimer's disease and Parkinson's disease. N Engl J Med 348 (14):1356–1364
Oxenkrug G (2011) Interferon-gamma–inducible inflammation: contribution to aging and aging-associated psychiatric disorders. Aging Dis 2(6):474
Palmer AK, Kirkland JL (2016) Aging and adipose tissue: potential interventions for diabetes and regenerative medicine. Exp Gerontol 86:97–105
Patyar S, Prakash A, Modi M, Medhi B (2011) Role of vinpocetine in cerebrovascular diseases. Pharmacol Rep 63(3):618–628
Prelog M (2006) Aging of the immune system: a risk factor for autoimmunity? Autoimmun Rev 5 (2):136–139
Reeve A, Simcox E, Turnbull D (2014) Ageing and Parkinson's disease: why is advancing age the biggest risk factor? Ageing Res Rev 14:19–30
Rickman CB, Farsiu S, Toth CA, Klingeborn M (2013) Dry age-related macular degeneration: mechanisms, therapeutic targets, and imaging. Invest Ophthalmol Vis Sci 54(14):ORSF68–ORSF80
Riggs BL, Khosla S, Melton LJ III (1998) A unitary model for involutional osteoporosis: estrogen deficiency causes both type I and type II osteoporosis in postmenopausal women and contributes to bone loss in aging men. J Bone Miner Res 13(5):763–773
Roberts KL, Allen HA (2016) Perception and cognition in the ageing brain: a brief review of the short-and long-term links between perceptual and cognitive decline. Front Aging Neurosci 8:39
Robinson JD, Turner J (2003) Impersonal, interpersonal, and hyperpersonal social support: cancer and older adults. Health Commun 15(2):227–234
Sahoo S, Meijles DN, Pagano PJ (2016) NADPH oxidases: key modulators in aging and age-related cardiovascular diseases? Clin Sci 130(5):317–335
Samad N, Ishaq S, Bano S, Manzoor N (2017) Calcium regulation in Alzheimer's disease: mechanistic understanding. J Coll Physicians Surg Pak 27(9):566–571
Saretzki G, von Zglinicki T (2002) Replicative aging, telomeres, and oxidative stress. Ann N Y Acad Sci 959(1):24–29
Sarkar D, Fisher PB (2006) Molecular mechanisms of aging-associated inflammation. Cancer Lett 236(1):13–23
Shaffer SW, Harrison AL (2007) Aging of the somatosensory system: a translational perspective. Phys Ther 87(2):193–207
Shammas MA (2011) Telomeres, lifestyle, cancer, and aging. Curr Opin Clin Nutr Metab Care 14 (1):28
Shay JW (2016) Role of telomeres and telomerase in aging and cancer. Cancer Discov 6 (6):584–593
Shay JW, Wright WE (2001) Telomeres and telomerase: implications for cancer and aging. Radiat Res 155(1):188–193
Sheedfar F, Biase SD, Koonen D, Vinciguerra M (2013) Liver diseases and aging: friends or foes? Aging Cell 12(6):950–954
Simmonds MJ, Meiselman HJ, Baskurt OK (2013) Blood rheology and aging. J Geriatr Cardiol 10 (3):291
Singh I (2016) Assessment and management of older people in the general hospital setting. In: Challenges in elder care. InTech, London, p 37
Smith CD, Carney JM, Tatsumo T, Stadtman ER, Floyd RA, Markesbery WR (1992) Protein oxidation in aging brain a: protein oxidation in aging brain. Ann N Y Acad Sci 663(1):110–119
Strait JB, Lakatta EG (2012) Aging-associated cardiovascular changes and their relationship to heart failure. Heart Fail Clin 8(1):143–164
Stuen C, Faye E (2003) Vision loss: normal and not normal changes among older adults. Generations 27(1):8–14
Surmeier DJ, Guzman JN, Sanchez-Padilla J (2010) Calcium, cellular aging, and selective neuronal vulnerability in Parkinson's disease. Cell Calcium 47(2):175–182

Taddei S, Virdis A, Ghiadoni L, Versari D, Salvetti A (2006) Endothelium, aging, and hypertension. Curr Hypertens Rep 8(1):84–89

Terrando, N., Brzezinski, M., Degos, V., Eriksson, L. I., Kramer, J. H., Leung, J. M., et al. (2011). "Perioperative cognitive decline in the aging population." Mayo Clinic proceedings, Elsevier, Amsterdam 86(9): 885-893.

Tompkins MM, Basgall EJ, Zamrini E, Hill WD (1997) Apoptotic-like changes in Lewy-body-associated disorders and normal aging in substantia nigral neurons. Am J Pathol 150(1):119

Topinková E (2008) Aging, disability and frailty. Ann Nutr Metab 52(Suppl. 1):6–11

Vlassara H (1996) Protein glycation in the kidney: role in diabetes and aging. Kidney Int 49 (6):1795–1804

Waite LJ (2018) Social well-being and health in the older population: moving beyond social relationships. In: Future directions for the demography of aging: proceedings of a workshop. National Academies Press, Washington

Wang JC, Bennett M (2012) Aging and atherosclerosis: mechanisms, functional consequences, and potential therapeutics for cellular senescence. Circ Res 111(2):245–259

Wang J, Ohno-Matsui K, Morita I (2012) Elevated amyloid β production in senescent retinal pigment epithelium, a possible mechanism of subretinal deposition of amyloid β in age-related macular degeneration. Biochem Biophys Res Commun 423(1):73–78

Weyand CM, Goronzy JJ (2016) Aging of the immune system. Mechanisms and therapeutic targets. Ann Am Thorac Soc 13(5):S422–S428

Xia S, Zhang X, Zheng S, Khanabdali R, Kalionis B, Wu J et al (2016) An update on inflamm-aging: mechanisms, prevention, and treatment. J Immunol Res 2016:1

Xue Q-L (2011) The frailty syndrome: definition and natural history. Clin Geriatr Med 27(1):1–15

Yan J, Zhang H, Yin Y, Li J, Tang Y, Purkayastha S et al (2014) Obesity-and aging-induced excess of central transforming growth factor-β potentiates diabetic development via an RNA stress response. Nat Med 20(9):1001

Yang Z, Ming XF (2012) mTOR signalling: the molecular interface connecting metabolic stress, aging and cardiovascular diseases. Obes Rev 13:58–68

Yin D, Chen K (2005) The essential mechanisms of aging: irreparable damage accumulation of biochemical side-reactions. Exp Gerontol 40(6):455–465

Assessment of Nutritional Status in Older People

4

Giuseppe Annunziata, Angela Arnone, and Luigi Barrea

Abstract

Assessment of the nutritional status is a crucial item in the management of elderly people and a fundamental tool to improve life span. Elderly, indeed, are more prone to develop malnutrition, mainly due to a progressive cognitive decline, impaired metabolism pathways, reduced food intake, and physical activity. This causes severely affected body composition and clinical outcomes, resulting in vicious cycle that worsens the quality of life. An accurate nutritional assessment should be based on a step-by-step evaluation of all clinical outcomes, including evaluation of clinical signs, blood parameters, nutritional status, anthropometric parameters, and body composition. In this chapter, we eviscerated all these items, providing a useful and practical support for nutritional assessment in old subjects.

Keywords

Nutrition · Nutritional assessment · Elderly · Body composition · Malnutrition

G. Annunziata (✉)
Department of Pharmacy, University of Naples Federico II, Naples, Italy
e-mail: giuseppe.annunziata@unina.it

A. Arnone
Dipartimento di Medicina Clinica e Chirurgia, Unit of Endocrinology, Federico II University Medical School of Naples, Naples, Italy

L. Barrea
Dipartimento di Medicina Clinica e Chirurgia, Sezione di Endocrinologia, Università Federico II di Napoli, Naples, Italy

S. M. Nabavi et al. (eds.), *Nutrients and Nutraceuticals for Active & Healthy Ageing*,
https://doi.org/10.1007/978-981-15-3552-9_4

Malnutrition is a common pathological condition in elderly. It is characterized by a general imbalance of energy or single macro- or micronutrients, resulting in affecting body composition, functional and clinical outcomes. One of the main direct consequences of malnutrition is a decrease in both muscle and bone mass, accompanied with further complications, including immune dysfunction, cognitive decline, anemia, and higher hospitalization and mortality. An accurate evaluation of the nutritional status is necessary in order to increase life span. Nutritional status assessment in older people should include several outcomes, including clinical assessment, blood parameters, nutritional assessment, anthropometric parameters, and body composition, as discussed below.

4.1 Clinical Assessment

The general clinical assessment is mainly aimed to individuate eventually the presence of malnutrition, which is indicated by several clinical signs. In particular, subjects are generally thin with dry scaly skin and poor wound healing. In addition, patients present thin hair, spooned and depigmented nails, edema, and generalized bone and joint pain. In absence of these specific clinical signs, physician may deduce that patient is properly nourished.

4.2 Blood Parameters

During a clinic visit, evaluation of blood parameters is fundamental in order to elucidate both the metabolic assessment and the nutritional status of elderly patients (Table 4.1). Liver-derived proteins are recognized as direct markers of nutritional assessment, including albumin, transferrin, retinol-binding proteins, and total lymphocyte count. Among these, albumin is commonly used in the clinical practice, since it provides information about protein malnutrition and predicts mortality in elderly. Normal albumin values ranged from 3.5 to 5 g/dL; values less the normal range may indicate several pathological conditions, such as malnutrition, inflammation, infections, liver or renal disease, enteropathies, edema, hypothyroidism, or zinc deficit. However, albumin has a long half-life (about 20 days), thus it is not useful and sensitive to monitor short-term changes in the nutritional status. In this line, as prealbumin (normal range 16–36 mg/dL) has a shorten half-life (about 2–3 days)

Table 4.1 Normal range of selected markers related to the nutritional status

Parameter	Normal range
Albumin (g/dL)	3.5–5.0
Prealbumin (mg/dL)	16–36
Transferrin (mg/dL)	212–360
Retinol-binding protein (mg/dL)	2.6–7.6
Total lymphocyte count (n°/m^3)	≥1800

than albumin, monitoring its levels in addition to the general clinical assessment may be useful for the evaluation of the nutritional status in elderly.

Compared to albumin, transferrin is more sensitive for evaluation of protein–energy malnutrition in elderly. Normal values ranged from 212 to 360 mg/dL; higher values may indicate an iron deficit, while lower values are indicative of malnutrition, anemia, liver and renal disease, and enteropathies. Transferrin half-life is about 8–10 days, thus it is more sensitive than albumin to monitor short-term changes in the nutritional status.

4.3 Nutritional Assessment

Information regarding the eating habits in elderly may be obtained using the same dietary assessment methods used for young people, although strength and limitations are present, as shown in Table 4.2. However, these interviews should be performed by qualified Nutritionists, in order to minimize record errors and to obtain the best information.

Although these are the main standard methods used, some authors suggested and validated several adaptions in order to improve the quality of data obtained, including the Mediterranean Diet Adherence Screener (MEDAS), 14-item (in Spain) or 21-item (in Iceland) FFQ. These adaptions showed to be highly sensitive and quicker

Table 4.2 Strength and limitation of the main dietary assessment methods for evaluation of the nutritional status in elderly

Dietary assessment method	Description	Strength	Limitations
24-h recall (24 hrs)	During the interview subjects recall all foods eaten in the previous 24 h	Practice, rapid, simple	Only one day food intake is recorded and it does not reflect the patient's typical diet; in addition, cognitive impairment may affect the records
7- or 3-day food record	Subjects are asked to record for 7 or 3 days all foods and beverages consumed during the day. Foods and beverages should not to be measured, but house measures should be reported (i.e., spoon, glass, etc.)	More detailed, reflects the patient's dietary habits	Subjects might report foods and beverages at the end of the day, and this affects the record. Subjects should be instructed to complete the record as soon as they finish eating the meal
Food frequency questionnaires (FFQs)	This is a multiquestion questionnaire aimed to evaluate food intake during an established time period	Suitable for evaluation of groups or epidemiological studies	Not suitable for evaluation of individuals; a large number of questions

than the standard methods. Additionally, further questionnaires such as the "Appetite, Hunger and Sensory Perception Questionnaires (AHSPQ)" and "'Simplified Nutritional Appetite Questionnaire (SNAQ)" have been proposed in order to evaluate the dietary intake taking into account the possible age-related anorexia, which affect a large number of elderly subjects.

4.4 Evaluation of Anthropometric Parameters and Body Composition

After evaluation of clinical signs, blood parameters, and nutritional habits, the next step in the nutritional assessment is the valuation of anthropometric parameters, focusing the attention of both body weight and composition. In malnourished elderly subjects, indeed, a progressive body weight loss occurs, which may be due to:

- Wasting: involuntary weight loss due to reduced dietary intake.
- Cachexia: involuntary fat-free mass (FFM) or body cell mass (BCM) loss, mainly due to enhancement of catabolic pathways. This results from increased release of cytokines (mainly inflammatory interleukins) that impairs both hormonal and metabolic status, increasing the resting energy expenditure. In particular, a negative nitrogen balance occurs due to increased amino acids from muscle to the liver, gluconeogenesis, and a shift of the production of albumin to acute phase proteins.
- Sarcopenia: is a decline in muscle mass (MM), mainly due to reduced physical activity, altered hormonal status, and increased cytokine release. In particular, decreased levels of glucocorticoids, sex hormones, and catecholamine are accompanied by increased pro-inflammatory cytokines release and acute phase proteins levels, resulting in breakdown of muscle.

The first approach is represented by the evaluation of the body mass index (BMI), which results from the ratio between weight (expressed in kg) and the square of the height (expressed in m^2). According to the World Health Organization, subjects with a BMI < 18.5 are underweight, 18.5–24.9 are normal, 25–29.9 are overweight, and > 30 are obese. A J-shaped association has been established between BMI and mortality, resulting in increased risk for lower or higher BMI. Generally, BMI tends to increase from 18 to 65 years old, with a subsequent inverted trend in elderly, although they are more prone to loss MM and accumulate fat mass (FM). Elderly, thus, may have a normal BMI characterized by excessive FM, in a kind of *paradox obesity*, which significantly increases the risk of metabolic (such as diabetes or insulin-resistance) or cardiovascular diseases.

Although BMI is a rapid, quick, and direct method for an initial nutritional assessment, it is not free of limitations; mainly BMI does not allow identifying an unintentional weight loss and does not provide information about the body composition. In this context, further approaches, including circumference and skinfold measurement are useful in order to estimate the amount of FM and FFM, and to monitor them. Among these, tricipital skinfold and arm circumference can be used in the same formula to calculate the arm muscular area, which reflects the lean mass,

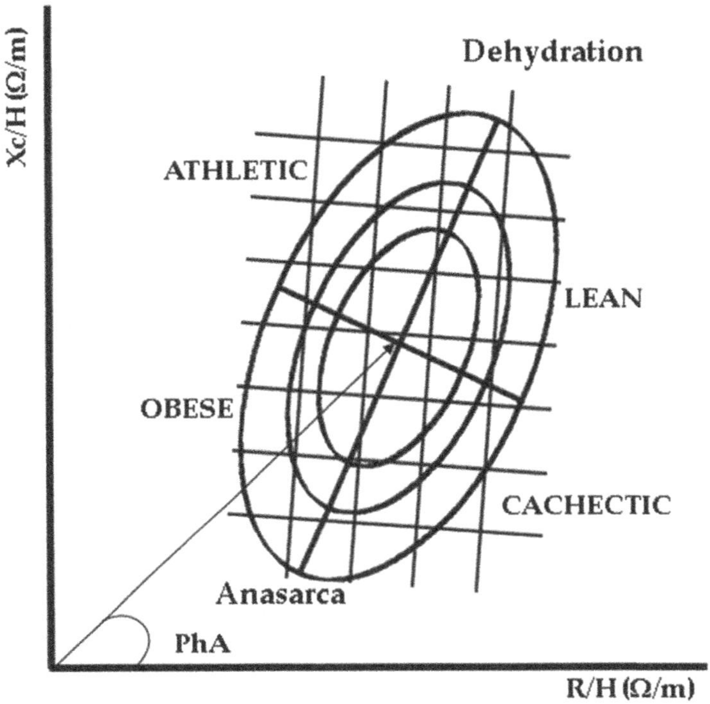

Fig. 4.1 BIVA graph

while mid-upper arm circumference is an indicator of malnutrition, using the Haboubi–Kennedy formula.

Bioelectrical impedance analysis (BIA) is recognized as the best method for evaluation of the body composition. As BIA is a useful, validated, rapid, inexpensive, and non-invasive method, its use is common in both clinical practice and large population studies, also for elderly. Specifically, BIA is based on the measurement of the resistance provided by the body against the passing of an electric current, providing a direct estimation of body water and its intra- and extracellular repartition; additionally, by indirect formulas, estimation of FFM, FM, and MM can be extrapolated. For nutritional assessment in older people bioelectrical impedance vector analysis (BIVA) has been licensed since it provides further information about hydration status and BCM. Differently from conventional BIA, BIVA provides a resistance (R)-reactance (Xc) graph, normalized by height, that categorizes subjects in four quadrants, as shown in Fig. 4.1.

In Fig. 4.2 are represented by way of example the BIVA of four subjects (two elderly and two middle-aged, M/F), BMI- and gender-matched. Importantly, among the parameters obtained by BIA and BIVA, phase angle (PhA), directly derived from R and Xc, represents a valid tool to the evaluation of nutritional assessment in elderly. PhA, indeed, is a valid indicator of inflammation and mortality in several chronic diseases; in addition small PhA values are indicative of malnutrition, cachexia, and/or anorexia.

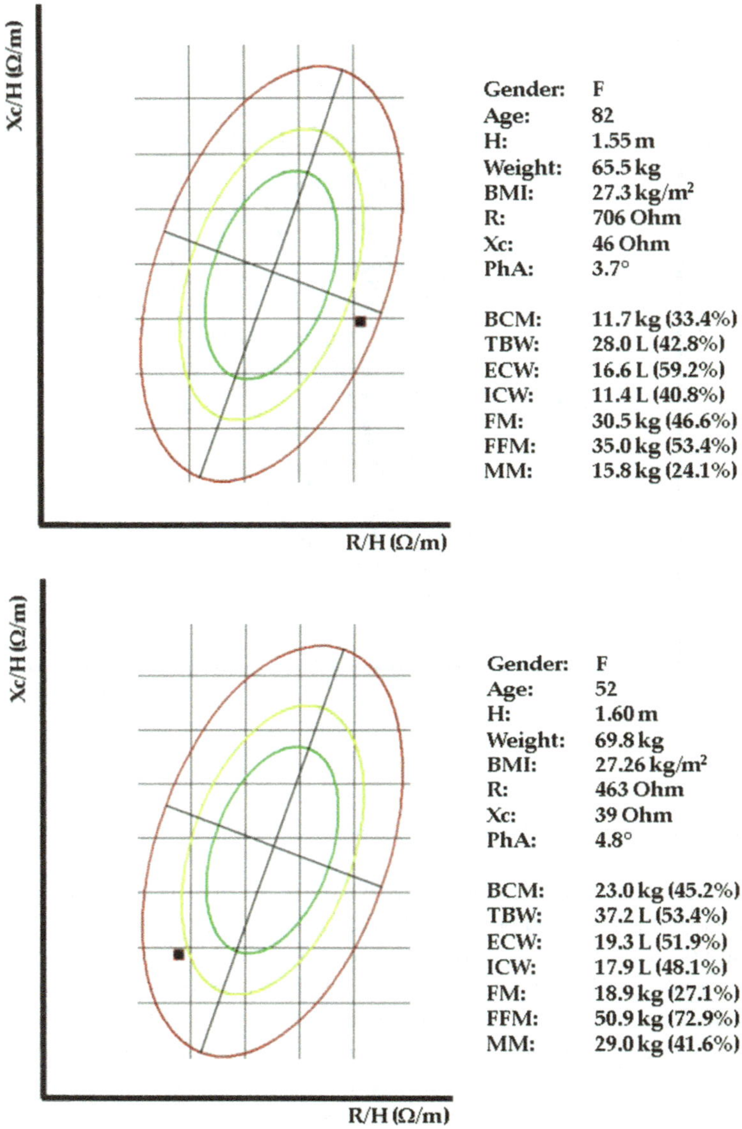

Fig. 4.2 Typical BIVA graph and BIA parameters of middle-aged and elderly of both gender

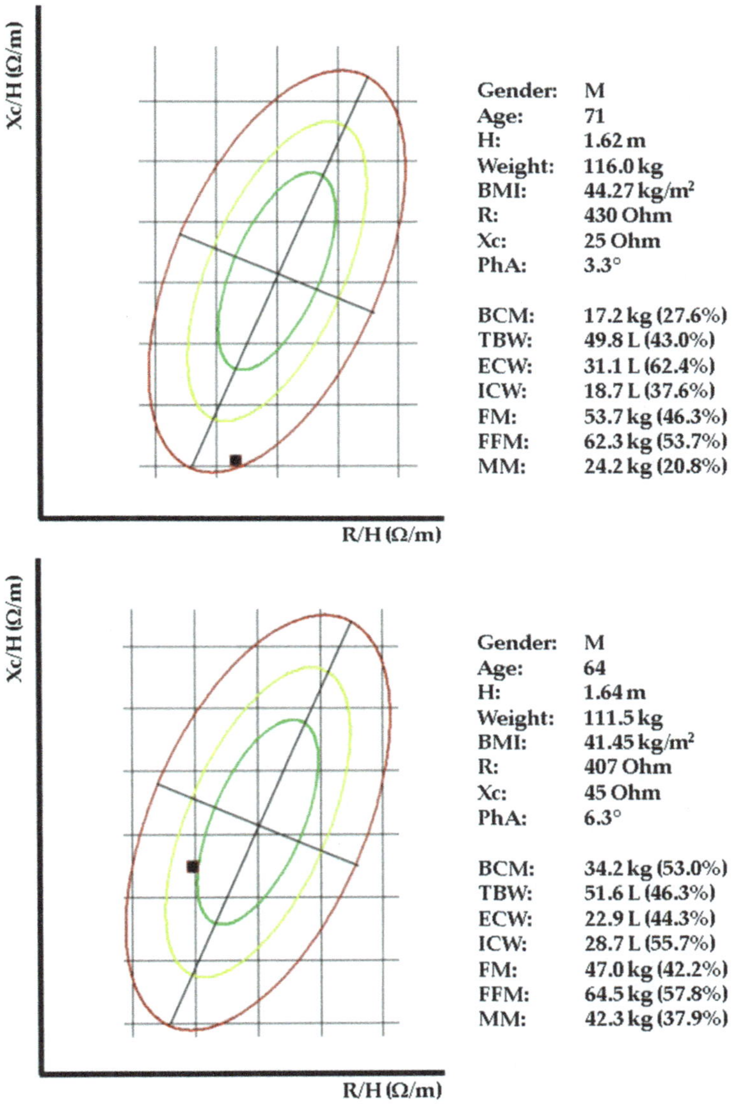

Fig. 4.2 (continued)

References

Adamson AJ, Collerton J, Davies K et al (2009) Nutrition in advanced age: dietary assessment in the Newcastle 85þ study. Eur J Clin Nutr 63(Suppl1):S6–S18

Ahmed T, Haboubi N (2010) Assessment and management of nutrition in older people and its importance to health. Clin Interv Aging 5:207–216

Alpers DH, Klein S (2003) Approach to the patient requiring nutritional supplementation. In: Yamada T (ed) Textbook of gastroenterology, 4th edn. Lippincott and Wilkins, Baltimore

Azad N, Murphy J, Amos SS, Toppan J (1999) Nutrition survey in an elderly population following admission to a tertiary care hospital. Can Med Assoc J 161:511–515

Barrea L, Macchia PE, Di Somma C, Napolitano M, Balato A, Falco A, Savanelli MC, Balato N, Colao A, Savastano S (2016) Bioelectrical phase angle and psoriasis: a novel association with psoriasis severity, quality of life and metabolic syndrome. J Transl Med 14(1):130. https://doi.org/10.1186/s12967-016-0889-6

Barrea L, Muscogiuri G, Macchia PE, Di Somma C, Falco A, Savanelli MC, Colao A, Savastano S (2017) Mediterranean diet and phase angle in a sample of adult population: results of a pilot study. Nutrients 9(2):E151. https://doi.org/10.3390/nu9020151

Bharadwaj S, Ginoya S, Tandon P, Gohel TD, Guirguis J, Vallabh H, Jevenn A, Hanouneh I (2016) Malnutrition: laboratory markers vs nutritional assessment. Gastroenterol Rep 4(4):272–280. Epub 2016 May 11

Calle EE, Thun MJ, Petrelli JM, Rodriguez C, Heath CW Jr (1999) Body-mass index and mortality in a prospective cohort of U.S. adults. N Engl J Med 341(15):1097–1105

Camina Martín MA, de Mateo SB, Redondo del Río MP (2014) Body composition analysis in older adults with dementia. Anthropometry and bioelectrical impedance analysis: a critical review. Eur J Clin Nutr 68(11):1228–1233. https://doi.org/10.1038/ejcn.2014.168. Epub 2014 Aug 13

Chapman IM (2006) Nutritional disorders in the elderly. Med Clin North Am 90(5):887–907

Cree MG, Newcomer BR, Katsanos CS, Sheffield-Moore M, Chinkes D, Aarsland A, Urban R, Wolfe RR (2004) Intramuscular and liver triglycerides are increased in the elderly. J Clin Endocrinol Metab 89(8):3864–3871

de Vries JH, de Groot LC, van Staveren WA (2009) Dietary assessment in elderly people: experiences gained from studies in the Netherlands. Eur J Clin Nutr 63(Suppl 1):S69–S74

Evans WJ (1995) Exercise, nutrition and aging. Clin Geriatr Med 11:725–734

Eysteinsdottir T, Thorsdottir I, Gunnarsdottir I, Steingrimsdottir L (2012) Assessing validity of a short food frequency questionnaire on present dietary intake of elderly Icelanders. Nutr J 11:12

Garibella SE, Sinclair AJ (1998) Nutrition, aging and ill health. Br J Nutr 80:7–23

Hanisah R, Suzana S, Lee FS (2012) Validation of screening tools to assess appetite among geriatric patients. J Nutr Health Aging 16:660–665

Houston DK, Nicklas BJ, Ding J et al (2008) Dietary protein intake is associated with lean mass change in older, community-dwelling adults: the Health, Aging, and Body Composition (Health ABC) Study. Am J Clin Nutr 87:150–155

Jeejeebhoy KN, Baker JP, Wolman S et al (1982) Critical evaluation of the role of clinical assessment and body composition studies in patients with malnutrition and after total parenteral nutrition. Am J Clin Nutr 35:1117–1127

Kesse-Guyot E, Andreeva VA, Lassale C et al (2013) Mediterranean diet and cognitive function: a French study. Am J Clin Nutr 97:369–376

Litchford MD (2006) Advanced practitioner's guide to nutrition and wounds. Case Software & Books, Greensboro

Mitchell DC, Tucker KL, Maras J et al (2012) Relative validity of the geisinger rural aging study food frequency questionnaire. J Nutr Health Aging 16:667–672

Norman K, Stobaus N, Pirlich M, Bosy-Westphal A (2012) Bioelectrical phase angle and impedance vector analysis: clinical relevance and applicability of impedance parameters. Clin Nutr 31:854–861

Pirlich M, Lochs H (2001) Nutrition in the elderly. Best Pract Res Clin Gastroenterol 15:869–884

Roubenoff R (1999) The path physiology of wasting in the elderly. J Nutr 121(Suppl 1):256–259S
Schroder H, Fito M, Estruch R et al (2011) A short screener is valid for assessing Mediterranean diet adherence among older Spanish men and women. J Nutr 141:1140–1145
Sergi G, Coin A, Enzi G, Volpato S, Inelmen EM, Buttarello M, Peloso M, Mulone S, Marin S, Bonometto P (2006) Role of visceral proteins in detecting malnutrition in the elderly. Eur J Clin Nutr 60(2):203–209
Stevens J, Cai J, Pumuk ER, Williamson DF, Thun MJ, Wood JL (1998) The effect of age on the association between body mass index and mortality. N Engl J Med 338:1–7
Talegawkar SA, Bandinelli S, Bandeen-Roche K et al (2012) A higher adherence to a Mediterranean-style diet is inversely associated with the development of frailty in community-dwelling elderly men and women. J Nutr 142:2161–2166
Tyrovolas S, Pounis G, Bountziouka V et al (2010) Repeatability and validation of a short, semi-quantitative food frequency questionnaire designed for older adults living in Mediterranean areas: the MEDIS-FFQ. J Nutr Elder 29:311–324
Vercambre MN, Boutron-Ruault MC, Ritchie K et al (2009) Long-term association of food and nutrient intakes with cognitive and functional decline: a 13-year follow-up study of elderly French women. Br J Nutr 102:419–427
Villareal DT, Apovian CM, Kushner RF, Klein S, American Society for Nutrition, NAASO, The Obesity Society (2005) Obesity in older adults: technical review and position statement of the American Society for Nutrition and NAASO, The Obesity Society. Am J Clin Nutr 82 (5):923–934
WHO (1989) Measuring obesity – classification and description of anthropometric date. Nutrition Unit. EUR/ICP/NUT 125. WHO Regional Office for Europe, Copenhagen
WHO (1995) Physical status: the use and interpretation of anthropometry. Report of a WHO Expert Committee. WHO Technical Report Series 854. World Health Organization, Geneva
Wilson MM, Thomas DR, Rubenstein LZ et al (2005) Appetite assessment: simple appetite questionnaire predicts weight loss in community-dwelling adults and nursing home residents. Am J Clin Nutr 82:1074–1081
Yeh SS, Schuster MW (1999) Geriatric cachexia: the role of cytokines. Am J Clin Nutr 70 (2):183–197

Nutrients and Nutraceuticals in Aging

5

Mahshid Hodjat, Madiha Khalid, Mona Asghari, Sepideh Atri,
Mahban Rahimifard, Solmaz Mohammadi Nejad,
and Maryam Baeeri

Abstract

Aging is accompanied with the incidence of wide range of disease. The best strategy to combat aging and related deteriorative pathologic conditions is nutraceutical interventions. Several dietary compounds and nutraceuticals (dietary fibre, probiotics, fatty acids, antioxidant vitamins, polyphenols, spices, and essential trace elements, etc.) have proven to be beneficial in abating age-related diseases and decelerating aging process. The protective mechanism of these compounds are related to their activities in improving intestinal microbial balance, reducing blood cholesterol, increasing defence systems against oxidative damage in living cells, as well as their role in normal cellular structure and function regulated by enzymes, transporters, sensors and metal-binding molecules.

In this chapter, we will discuss the most abundant researches focused on the effect of nutrients and nutraceuticals in prevention or ameliorating the progression of selected age-related disease including cardiovascular disease, diabetes, neurodegenerative disease, ocular aging, cancer, skin aging, respiratory disorders, immune restoration disease and tooth aging. We attempted to elucidate mechanistic information on the molecular targets of each dietary compound in contracting aging disease.

Keywords

Aging · Age-related disease · Dietary compounds · Nutraceuticals

M. Hodjat (✉) · M. Asghari · S. Atri
Dental Research Center, Dentistry Research Institute, Tehran University of Medical Sciences (TUMS), Tehran, Iran
e-mail: mhodjat@TUMS.Ac.Ir

M. Khalid · M. Rahimifard · S. M. Nejad · M. Baeeri
Toxicology and Diseases Group, The Institute of Pharmaceutical Sciences (TIPS), Tehran University of Medical Sciences (TUMS), Tehran, Iran

S. M. Nabavi et al. (eds.), *Nutrients and Nutraceuticals for Active & Healthy Ageing*,
https://doi.org/10.1007/978-981-15-3552-9_5

5.1 Introduction

The biological process of aging is associated with numerous diseases such as diabetes, cancer, neurodegenerative disease, cardiovascular problem, macular degeneration, and osteoporosis (Fulop et al. 2010). The age-related diseases are, in fact, the clinical manifestation of abnormal molecular changes that disrupt cellular and organismal function in aging population.

Nowadays the traditional concept of aging is changing and aging is considered as a disease rather than simply being a natural biological phenomenon. Based on the emerging concept, aging is a deteriorative pathologic condition characterized by a group of associated symptoms. Disregard of being a pathologic or a natural process, the important issue in medical science is to prevent age-related disease by treating the early stages of the pathology. Besides standard medical treatments and preventative medications, many nutrients and nutraceuticals have shown to influence aging symptoms (Meydani 2001, 2002). Recent review published by Bruce Ames (2018) has proposed specific nutrients such as plant and the marine antioxidant carotenoids, the fungal antioxidant ergothioneine, bacterial metabolites pyrroloquinoline quinone and queuine as the keys to longevity and that reduce risk of chronic disease and premature aging (Ames 2018).

Here we will discuss recent findings on most widely studied nutrients and nutraceuticals that assist to prevent age-related diseases and decelerate aging process. We categorized nutrients based on their contributions to different age-related symptoms (Fig. 5.1).

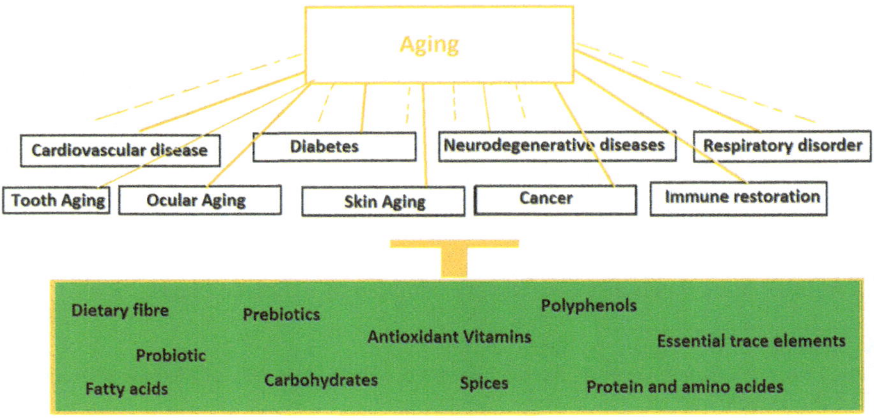

Fig. 5.1 Nutrients and nutraceuticals prevent age-related diseases and decelerate aging process

5.2 Nutrients and Nutraceuticals Interventions Influence Age-Related Cardiovascular Disease

Nutraceuticals or 'dietary supplement' influencing vascular aging and the cardiac vascular system can be categorized into the dietary fibre, probiotics, fatty acids, vitamins, polyphenols, spices and essential trace elements (Table 5.1).

5.2.1 Dietary Fibre

Source of dietary fibre are fruits, vegetables and whole grain capable of reducing the risk of coronary heart disease (CHD) (Greenwood et al. 2013), coronary artery disease (CAD) (Steffen et al. 2003), cardiovascular disease (CVD) (Lairon et al. 2005), hypertension (HTN) (Whelton et al. 2005) and reducing total serum low density lipoprotein (LDL) level (Das et al. 2012). Soluble dietary fibre (SDF) such as pectin, β-glucan, oat bran, guar gum and psyllium are known to reduce total and LDL cholesterol level (Brown et al. 1999). Oat function in controlling systolic blood pressure and diastolic blood pressure, blood lipid, fasting glucose and insulin level (Anderson et al. 2004), likewise, viscous fibres such as glucomannan is known to reduce indirectly the risk of heart diseases by reducing body weight (Birketvedt et al. 2005).

In general, dietary fibre executes protective effect against vascular ageing and cardiac problem by several mechanisms. SDF reduces blood cholesterol level by preventing bile salt reabsorption into enterohepatic circulation thereby promoting faecal excretion leading to hepatic bile acid depletion (Rideout et al. 2008). This stimulates hepatic cholesterol catabolism to replenish bile acid that ultimately induces cholesteryl esters metabolism and increases LDL surface membrane receptors leading to uptake blood LDL and thus decreases cholesterol level. Moreover, SDF reduces glucose and macronutrient absorption through negatively affecting their digestion, delaying gastric emptying and slowing transport of digestive enzyme. Moreover, SDF undergoes anaerobic bacterial fermentation to produce short-chain fatty acids (SCFA) and such fermentation products are also capable of reducing plasma cholesterol level by declining HMG-CoA reductase activity (Gunness and Gidley 2010). Dietary fibre also executes a protective effect on the cardiovascular system by reducing inflammation via reduced inflammatory cytokine production (North et al. 2009).

5.2.2 Probiotics

While imbalance of gastrointestinal microbiota has been associated with pathogenesis of CVD, a diet supplement of living microbes i.e. 'probiotic' could improve intestinal microbial balance (Thushara et al. 2016). Probiotics prevent vascular

Table 5.1 Mechanisms of nutraceuticals in reducing risk of CVD disease

Nutraceuticals	Example	Mechanisms
1. *Dietary fibre*		
1. Insoluble dietary fibre (IDF)	Celluloses, hemicelluloses, lignins	– Increase faecal bile acid excretion, reduce cholesterol, fatty acid absorption (Brown et al. 1999; Gunness and Gidley 2010) – Bacterial fermentation products inhibit HMG CoA reductase (Gunness and Gidley 2010) – Reduce glucose absorption, insulin secretion (Anderson et al. 2004; Keenan et al. 2002) – Reduce inflammation and BP (Gunness and Gidley 2010; North et al. 2009)
2. Soluble dietary fibre (SDF)	β-Glucans, pectins, gums, mucilages, hemicelluloses	
2. *Probiotics*		
1. Hypocholesterolemic probiotics	*Streptococcus thermophiles, Enterococcus faecium, L. acidophilus, Bifidobacterium longum, L. casei, Saccharomyces boulardii, L. curvatus, L. plantarum, L. reuteri*	– Reduce inflammation (Ejtahed et al. 2011) – Reduce BP (Gomez-Guzman et al. 2015) – Reduce blood glucose level (Saez-Lara et al. 2016) – Reduce BMI, fat mass (Saez-Lara et al. 2016) – Reduce LDL-cholesterol ratio (Kiessling et al. 2002) – Increase LDL/HDL ratio (Kiessling et al. 2002) – Inhibit ACE (Daliri et al. 2017; Chen et al. 2014)reduced NO and ROS production (Daliri et al. 2017; Chen et al. 2014)
2. Antidiabetic probiotics	*Lactobacillus reuteri, L. plantarum, L. casei, L. acidophilus, Bifidobacterium lactis, L. rhamnosus, B. bifidum*	
3. Antihypertensive probiotics	L. casei, Streptococcus thermophilus, L. plantarum, L. helveticus, Saccharomyces cerevisiae, L. acidophilus, L. fermentum, L. bulgaricus	
4. Antiobesity probiotics	Lactobacillus, Curvatus, L. plantarum, L. rhamnosus, Bifidobacterium animalis ssp. Lactis	
3. Prebiotics	Chicory roots, banana, tomato, alliums, beans, peas	– Promote Lactobacillus and Bifidobacterial growth (Tsai et al. 2014) – Improve lactose tolerance, blood lipids and cholesterol level (Das et al. 2012)
4. Polyunsaturated fatty acids	1. Omega 3 fatty acids (α-linolenic acid (ALA), eicosapentanoic acid (EPA), docosahexanoic acid (DHA)) 2. Omega 6 fatty acids	– Hypocoagulation (Leray et al. 2001) – Liver lipid peroxidation, inhibit lipogenesis (Leray et al. 2001) – Antiarrhythmic, reduce arterial fibrillation (Calo et al. 2005) – Anti-inflammatory (Endo and Arita 2016)

(continued)

Table 5.1 (continued)

Nutraceuticals	Example	Mechanisms
		– Promote fatty acid oxidation (Rodriguez-Cruz et al. 2005)
5. Antioxidant vitamins	Water soluble: Vitamin C Lipid soluble: Vitamin E and carotenoids (lycopene, β-carotene, lutein, zeaxanthin)	– Free radical scavengers (Lee et al. 2018) – Inhibit oxidative chain reaction initiation/propagation (Lee et al. 2018) – Vasodilation, improves blood flow (Axton et al. 2018)
6. Polyphenols	Flavonols, flavones, flavan-3-ols, flavanones, anthocyanins, flavonoids, phenolic acids, resveratrol, tea, legumes, soyabean, broccoli, red pepper	– Antioxidant, antiarrhythmic (Moskaug et al. 2005; Ganesan and Xu 2017) – Anti-inflammatory, antidiabetic and anti-hyperlipidaemic (Ganesan and Xu 2017)
7. Spices	Onion, garlic, fenugreek seeds, turmeric, capsaicin, curcumin, cumin, eugenol, ginger, gingerol, mustard, piperine, fennel, coriander, mint, asafoetida	– Reduce Ca^{2+} influx intracellularly (Esmaeili et al. 2017) – Anti-inflammatory, antioxidant (Oboh et al. 2019; Dhivya et al. 2017) – Antihypertensive, antithrombotic (Peixoto-Neves et al. 2015; Mnafgui et al. 2016) – Inhibit thromboxane B2 production (Wu et al. 2018) – NOS and cyclooxygenase suppression (Wu et al. 2018) – Inhibit ACE, α-amylase, α-glucosidase (Oboh et al. 2019) – PPAR-γ activation (Ma et al. 2017) – Inhibition AKT and GSK3β (Ma et al. 2017)
8. Essential trace elements	Ca, Mg, Na, K, P, Cl, S, Fe, Zn, Mn, Cu, I, Ni, F, V, Cr, Mo, Se, Su, Si	– Enzyme function, endothelial function, normal organ function (Baker 2017) – Regulates movement of Na^+, K^+ and Ca^{2+} through voltage-gated ion channels (Baker 2017)

ageing and risk of CVD by its hypocholesterolemic, antidiabetic, antihypertensive and antiobesity effects. Probiotic strains of lactic acid bacteria including *Pediococcus acidilactici, Bifidobacterium adolescentis, Lactobacillus rhamnosus* and *Lactobacillus acidophilus* reduce the risk of CVD by stimulating bile salt hydrolase leading thereby maintaining the normal concentration of cholesterol, LDL, and TG in plasma (Tsai et al. 2014). It was reported that daily consumption of 300 g daily yoghurt, supplemented with *L. acidophilus* and *Bifidobacterium longum* could increase serum HDL and LDL/HDL cholesterol ratio (Kiessling et al. 2002). Likewise, fermented milk products containing *L. acidophilus, S. thermophilus, L. rhamnosus, S. thermophilus, L. rhamnosus, S. thermophilus, Enterococcus faecium* could lower blood LDL level (Agerholm-Larsen et al. 2000). Moreover, hypocholesterolemic effect has been observed in healthy adult after weeks intervention with *L. rhamnosus* probiotic (Kekkonen et al. 2008).

Probiotic has shown to exert antihypertensive effect through inhibition of angiotensin-converting enzyme (ACE), reducing inflammation and optimizing serum lipid level. Consumption of dairy sour milk containing *L. helveticus and Saccharomyces cerevisiae* has shown to reduce systolic and diastolic blood pressure in elderly patients (Hata et al. 1996).

Obesity is another risk factor for vascular ageing and cardiac problem. Microbial imbalance in the gut leads to weight gain and aberrant gene expression related to lipid and carbohydrate metabolism resulted in metabolic syndrome (John and Mullin 2016). Probiotic has beneficial health effects on intestinal microbial ecology and protects against obesity mainly by affecting fat mass, improving insulin resistance syndrome, insulin sensitivity, carbohydrate metabolism, fasting blood glucose and metabolic disorder (Saez-Lara et al. 2016). Obese mice on a probiotic diet demonstrated a reduction in obesity by modulating genes associated with liver metabolism and tissue inflammation (Park et al. 2013a). Also in another study it is shown that probiotic supplemented with *L. rhamnosus* helps in obesity prevention by controlling appetite and related behaviors in obese individual (Sanchez et al. 2017). *Although probiotics are generally found to be safe for most of the populations, few side effects such as systemic infections, aberrant metabolic activity and immune overstimulation were reported in susceptible individuals (*Didari et al. 2014*).*

5.2.3 Prebiotic

Prebiotic dietary fibrs serves as a carbon sources for microbiota. The well-studied prebiotics are fructo-based oligosaccharides found in almost 15% of flowering plants. While they are not digested in gastrointestinal tract, prebiotic supplementation modulates the balance of gut microbiota generally by promoting Lactobacillus and Bifidobacterial growth, ultimately influencing human health (Das et al. 2012). Prebiotic supplementation reduces cardio-metabolic risk by improving dysbiosis of microbiota, and improving blood lipid profile and cholesterol level (Das et al. 2012).

5.2.4 Fatty Acids

Polyunsaturated fatty acids (PUFA) are essential fatty acids required for maintenance of good health that obtained externally from different food. While corn, safflower oil and soy are the rich source of Omega 6, fish, canola oil, flaxseeds are known for their high omega-3 content (Rodriguez-Cruz et al. 2005). Different studies have shown the cardioprotective effect of PUFAs as well as its preventive role in vascular ageing and overall risk of mortality (Das et al. 2012; Pigarevskii 1977). Omega 3 unsaturated fatty acid supplementation (2 g/day) before and after coronary artery bypass graft (CABG) has shown to reduce the incidence of postoperative atrial fibrillation (AF) and length of hospital stay in patients (Calo et al. 2005). Another study revealed that PUFAs reduced blood triglycerides (TG) level by promoting fatty acid oxidation and inhibiting lipogenesis (Cruz et al. 2005). PUFAs containing fish meals clearly showed a protective effect against coronary heart disease by decreasing blood viscosity, platelet aggregation, inflammation, TG and cholesterol blood level (Oelrich et al. 2013). There are also studies indicating the possible role of Omega 3 in preventing arrhythmias by modulating sodium and calcium channels in the cell membrane (Demaison and Moreau 2002). Eicosapentaenoic acid (EPA) and docosahexaenoic acid (DHA), the long-chain omega-3 fatty acids are known for their cardioprotective effects (Sherratt and Mason 2018).

5.2.5 Antioxidant Vitamins

The free radical theory of aging is considered as one of the most important and prominent theories of the aging process. According to this theory, aging is the result of prolonged life-long accumulation of cellular damage caused by reactive oxygen species, therefore, leading to increased risk of disease and eventually disability as the final unset of aging. There are two antioxidant defence systems against oxidative damage in living cells. The superoxide dismutase, catalase, and glutathione peroxidases are considered the first defence pathway function as antioxidant enzymes, and there are non-enzymatic antioxidants such as the thioredoxin, glutathione, vitamins A, C, E, lycopene, lutein, polyphenols, quercetin, that are taken from wide range of dietary nutrients and supplements (Prakash and Gupta 2014a; Obrenovich et al. 2011).

An imbalance of ROS/RNS generation resulting in cytokines release leads to inflammation, oxidative damage and thereby increased risk of vascular ageing and CVD (Salehi et al. 2018). Fruits and vegetables are rich in dietary antioxidants and vitamins. Water-soluble vitamin C and lipid-soluble vitamin E are free radical scavengers that were clearly associated with reduced risk of CVD development (Sesso et al. 2008). The natural pigments carotenoids including β-carotene, zeaxanthin, lutein, and lycopene are efficient singlet oxygen quencher (Nishino et al. 2016,

2017; Chang et al. 2013a; Kawata et al. 2018). Lutein prevents cardiac injury in diabetic rats by preventing oxidative stress markers (Sharavana et al. 2017). Dietary lycopene significantly reduces the risk of CHD and stroke (Song et al. 2017). Based on recent findings, ascorbic acid induces generation of NO, the inducer of vasodilation and blood flow improvement from glyceryl trinitrate (GTN), suggesting a role for ascorbic acid in nitrate therapy (Axton et al. 2018).

5.2.6 Polyphenols

The rich source of dietary polyphenols includes cereals, dill plants (barley, maize, nuts, oats, rice, cluster maize, wheat, seeds and beans), oilseeds (turnips, canola oil, Flax seeds and olive seeds), fruits and vegetables including grapes, strawberries, legumes, soybean, broccoli, red pepper and a variety of drinks (juices, tea, coffee, cocoa, beer and wine) (Katalinic et al. 2006; Prakash and Kumar 2011). Common polyphenols including flavonols, flavones, flavan-3-ols, flavanones, anthocyanin, flavonoids, phenolic acids and resveratrol are known for their beneficial health effects. Polyphenols with antioxidant properties are involved in prevention of certain degenerative diseases including vascular ageing and CVD (Scalbert et al. 2005a). It was shown that flavonoids, the plant polyphenol that modulate the expression of γ glutamylcysteine synthetase assist the synthesis of wide range of cellular antioxidant (Moskaug et al. 2005). Red grasp, eucalyptus, poplar, blueberries, white currant, peanuts and red wine are among rich herbal sources of resveratrol. Resveratrol could block the release of intracellular Ca^{2+} and maintain cardiomyocytes contractile function that accounts for its antiarrhythmic property (Stephan et al. 2017). Also, polyphenols in tea are considered beneficial to decrease risk of ischaemic heart disease (Li et al. 2017). Lentils and legumes are other sources of polyphenol compounds possessing antioxidant, anti-inflammatory, antidiabetic and anti-hyperlipidaemia properties (Ganesan and Xu 2017). Higher apple, pear, red wine and strawberries consumption are associated with reduced inflammation demonstrating the anti-inflammatory property of dietary flavonoids (Cassidy et al. 2015). Recent meta-analysis study revealed that food plants rich in anthocyanins could significantly reduce serum total cholesterol (TC), TG, LDL level in patients with dyslipidaemia patients (Liu et al. 2016).

5.2.7 Spices

Spices have been used as flavouring, colourant, appetite stimulant and medicine since long (Jessica Elizabeth et al. 2017). Asafoetida, capsaicin, cumin, curcumin, coriander, eugenol, fennel, fenugreek seeds, garlic, ginger, gingerol, mint, onion, piperine and turmeric have shown a protective effect against vascular ageing and CVD. Asafoetida acts as a potent vasodilator by reducing Ca^{2+} influx through plasma membrane Ca^{2+} channels (Esmaeili et al. 2017). Curcumin is a bioactive component of turmeric that demonstrated anti-inflammatory, antioxidant and healing properties as it helps in decreasing LDL and risk of atherosclerosis. It has a vasodilating effect via indirect impact on prostacyclins and directly on vascular endothelium (Zdrojewicz et al. 2017; Jiang et al. 2017). Several animal studies reported cardamom, coriander, turmeric and ginger for their important role in preventing and controlling CVD (Rastogi et al. 2017; Kulczynski and Gramza-Michalowska 2016). Clove, nutmeg, bay and basil leaf and cinnamon are sources of eugenol. Eugenol was recognized for antihypertensive properties as it reduces systemic blood pressure and causes vaso-relaxation by activating transient receptor potential cation channel (TRPV4), reduces inflammatory mediator proteins and increases heart activities of superoxide dismutase and GPx (Peixoto-Neves et al. 2015; Mnafgui et al. 2016).

Ginger has antithrombotic activity via inhibition of platelet aggregation and thromboxane B2 production. It increases vasoprotection via suppression of nitric oxide synthase and cyclooxygenase (Wu et al. 2018). White and purple onion similar to garlic exerts antidiabetic and antihypertensive properties and were considered as strong antioxidant agents (Oboh et al. 2019). Extract of *coriandrum sativum* was shown to be effective in prevention of myocardial infarction by inhibiting myofibrillar damage (Patel et al. 2012). Moreover, coriander the herb in the family Apiaceae was evaluated for its diuretic activity mediated through cholinergic and Ca (2+) antagonist mechanisms proposed it as a candidate for hypertension treatment (Jabeen et al. 2009). Piperine prevents cardiac fibrosis by peroxisome proliferator-activated receptor-γ (PPAR-γ) activation and inhibition of protein kinase B (AKT) and glycogen synthase kinase 2β (GSK3β) (Ma et al. 2017). It also possesses antioxidant and antidyslipidaemic action (Dhivya et al. 2017).

5.2.8 Essential Trace Elements

Essential trace elements are nutritionally vital minerals that have functional role in several biochemical and physiological processes. They are classified into macro and microelements based on their relative concentrations in plant tissue.

Nitrogen (N), phosphorus (P), potassium (K) and calcium (Ca), magnesium (Mg) and sulphur are among the group of microelements. Intake of high calcium, magnesium and potassium has been associated with prevention of hypertension, heart failure (HF) and diabetes (Houston and Harper 2008). Magnesium is the fourth abundant cation in the human body that regulates the movement of Na^+, K^+ and Ca^{2+} through voltage-gated ion channels. Magnesium supplementation was suggested for the treatment of arrhythmia (Baker 2017). It was found that sodium and calcium dysregulation leads to tachyarrhythmia and the onset of sudden cardiac death (Wagner et al. 2015). Consumption of white vegetables and potatoes, the rich source of dietary potassium was associated with reduced risk of stroke (Weaver 2013). Hypophosphatemia has been found to be associated with multiple organ dysfunction including acute HF (Keskek et al. 2015). Despite their vast majority important role in health, long-term imbalance of macro-element could initiate biochemical disturbances leading to cardiovascular abnormal pathophysiology.

Microelements include iron (Fe), zinc (Zn), manganese (Mn), copper (Cu), iodine (I), cobalt (Co), nickel (Ni), fluorine (F), vanadium (V), chromium (Cr), molybdenum (Mo), selenium (Se), tin (Su) and silicon (Si) (Prashanth et al. 2015). Despite the vital role of iron in blood haemostasis, many studies indicated the association between increased body iron level and incidence of a broad spectrum of chronic conditions including cardiovascular disease.

Adequate amount of zinc is required for normal cellular structure and function via cellular signalling regulated by Zn transporters, sensors and Zn binding molecules. Zn dyshomeostasis leads to myocardial ischaemia/reperfusion injury (Xu and Zhou 2013). Manganese, the cofactor for broad range of enzymes including manganese superoxide dismutase (MnSOD), arginine synthase, pyruvate carboxylase and other mitochondrial antioxidant enzymes is vital for normal cardiac contractility and function (Fang et al. 2016).

Copper is less abundant than iron though very important for many cellular processes e.g. mitochondrial respiration, antioxidant capacity and iron oxidation. Dietary Cu deficiency decreases cardiac cytochrome oxidase activity, O_2 consumption and mitochondrial membrane potential while increasing ROS formation causing cardiac hypertrophy (Rines and Ardehali 2013). Iodine is the essential element in the synthesis of thyroid hormones, therefore, its deficiency causes aberrant thyroid function affecting cardiovascular haemodynamics and cardiac function (Barbesino 2018).

Selenium regulates thyroid hormone metabolism and redox state of cells. It also acts as a chemo-preventive agent by reducing oxidative stress, DNA damage and by increasing the activity of natural killer (NK) cells (Das et al. 2012). Dietary Se deficiency was associated with Keshan disease, foetal cardiomyopathy (Chen 2012). While Se is of fundamental importance to human health, high blood Se level causes fatigue, hair loss, irritability and nerve damage (Das et al. 2012).

In short, aforementioned elements are essential in trace amount for normal and healthy cardiac health while excess or deficiency will lead to vascular aging and cardiovascular diseases.

5.3 Nutrients and Nutraceuticals Interventions in Preventing Respiratory Disorder and Immune Restoration

Unhealthy eating habits, low physical activity level, smoking and overall poor lifestyle increase the risk and progression of respiratory disorders. Nutrition is important for healthy lung development and prevents respiratory illness-related morbidity and mortality. Several nutritional interventions have proven to be beneficial in abating respiratory diseases via restoring immunity (Karim et al. 2017).

5.3.1 Dietary Fibre

Fermentation of soluble dietary fibre in large intestine helps in improving immunity (Galisteo et al. 2008). Complex saccharide-based molecules of dietary fibre generate various bioactive materials such as SCFAs that change the composition of gastrointestinal tract flora and biomass thus responsible for immunomodulation (Kaczmarczyk et al. 2012). SCFAs exhibit anti-inflammatory effect through the activation of free fatty acid receptors (Halnes et al. 2017). Consistently, taking soluble fibre meal in patients suffering from asthma showed decreased level of airway inflammation biomarkers and improved lung function. Adequate fibre diet demonstrated a balance of Th1/Th2 immunity that help attenuates allergic inflammatory response (Zhang et al. 2016a). Diet rich in fruits, vegetables and cereals is associated with healthy lung function and reduced risk of chronic obstructive pulmonary disease (COPD) and respiratory symptoms (Fonseca Wald et al. 2014; Hanson et al. 2016). Despite several benefits, there are limited reports on dietary fibre-induced food intolerances and respiratory allergies (Shewry 2009).

5.3.2 Probiotic

Intestinal microenvironment interacts far beyond local environment involved in physiology and pathology of many organs including immune and respiratory system (Marsland et al. 2015). Being efficient immunopotentiators, *Lactobacilli* and *Bifidobacteria* probiotics were shown to regulate immune response and reduce the allergic response (Kim et al. 2005).

The link between gut microbiome and microbial composition of lung has been confirmed in many studies. Indeed, probiotics help to prevent respiratory disease owing to their immunomodulatory capacity to exert beneficial effect on dendritic cell phenotype and other pathways involving T cells, natural killer cells and alveolar macrophages (Forsythe 2014). A recent study revealed the effect of probiotic consumption in preventing the risk of *Pseudomonas aeruginosa* induced ventilator-associated pneumonia (VAP) (Bo et al. 2014). Similarly, long-term consumption of fermented milk supplemented with *Lactobacillus casei* is reported to subside allergic rhinitis and asthmatic symptoms among children (Ivory et al. 2008). The possible mechanism for reducing allergic rhinitis attributes to reduced TNF-α, IFN-γ, IL-12 and IL-13 (Chen et al. 2010). Consumption of fermented milk fortified with *Lactobacillus paracasei* demonstrated improvement in the quality of life (QOL) of allergic rhinitis patients by reducing nasal swelling and symptoms (Ishida et al. 2005). In another study, oral administration of *Lactobacillus plantarum* demonstrated a reduction in airway hyperresponsiveness, asthma and systemic allergic response by lowering serum IgE level, regulating Th1 response and improving IL-2, IFN-γ and reducing IL-4, IL-5, IL-3 (Liu et al. 2014). Probiotic supplementation also improves vascular endothelial growth factor (VEGF) supporting lung microvascular development and angiogenesis (Ahmad et al. 2014).

5.3.3 Prebiotics

There are limited studies about the effect of prebiotics on alleviating respiratory diseases. The role of prebiotic supplementation in reducing allergy is yet uncertain (Cuello-Garcia et al. 2017). Based on a recent study, prebiotic bimuno-galactooligosaccharide (B-GOS) supplementation was effective in reducing airway inflammation markers among adults with asthma and hyperpnoea induced bronchoconstriction (HIB) (Williams et al. 2016). Prebiotics could reduce the level of proinflammatory cytokines while mostly act indirectly by changing the composition of gut microbiota. It should be noted that their overall effect is expected to have an individualized response (Dey 2017).

5.3.4 Fatty Acids

Several preclinical and clinical studies have demonstrated the preventive role of PUFA in airways inflammation, chronic respiratory diseases, exercise-induced bronchoconstriction (EIB) and HIB. Prenatal supplementation of omega 3 PUFA could protect against lung injuries by preventing hyperoxia and bronchopulmonary dysplasia in newborn rats (Sharma et al. 2015). Nutritional support with omega 3 rich diet was suggested for COPD as it was demonstrated to reduce LTB_4, *TNF-α,* IL-8 level (Matsuyama et al. 2005). Omega 3 PUFA precursors such as eicosapentaenoic acid monoacylglyceride, docosahexaenoic acid monoacylglyceride and docosapentaenoic acid monoacylglyceride revealed beneficial effect in chronic airway inflammatory diseases (Khaddaj-Mallat et al. 2016).

Resolvins, an active metabolite of omega 3, were shown to mediate resolution of acute lung injury indicating the therapeutic potential of these compounds in inflammatory disorders (Moro et al. 2016). The protective role of resolvin against allergic airway inflammation was related to its suppressing effect on IL-23 and IL-6 production, and its regulatory role on interferon γ level (Aoki et al. 2008, 2010).

EPA/DHA rich diet reduces neutrophil, chemokine and IL-6 levels therefore decreased the mortality rate by preventing pulmonary bacterial load and protecting lungs against pulmonary *Pseudomonas aeruginosa* infection in mice models (Tiesset et al. 2011). EPA in comparison to DHA is found to be a more potent inhibitor of the inflammatory response in human asthmatic alveolar macrophages (Mickleborough et al. 2009). Lipid extract of green lipped mussel is a rich source of n–3 PUFA that was shown to have a beneficial effect in patients with atopic asthma (Emelyanov et al. 2002). Fish oil supplementation showed a protective effect on EIB by reducing cytokine level (Mickleborough et al. 2006).

5.3.5 Antioxidant Vitamins

Vitamin C prevents and treats respiratory and systemic infection through different mechanisms. It acts as a cofactor for many enzymes involved in anti-microbial activity such as neutrophils chemotaxis, phagocytosis, ROS production. It facilitates infection clearance by macrophages, decreasing necrosis and preventing tissue damage. It is also responsible for the proliferation of B and T cells and contributes to innate and adaptive immunity (Carr and Maggini 2017). Several studies demonstrated that taking vitamin C reduces episodes of common cold among physically active people; it also exerts a protective role against pneumonia, upper respiratory tract infections (URTIs) and the development of bronchial asthma (Raposo et al. 2017; Hemila 2017; Johnston et al. 2014; Ginter and Simko 2016).

Vitamin E supplementation showed protective effect against common cold and pneumonia (Hemila 2016; Meydani et al. 2004). Low serum level of vitamin A, D and E is reported among children with recurrent respiratory tract infection (RRTI) (Zhang et al. 2016b). Maternal vitamin E intake during pregnancy is reported to protect children from developing asthma in the first 10 years (Allan et al. 2015). Vitamin E also restores T cell function, inhibits PGE2 production, and enhances immune function, therefore showed therapeutic effect in patients with influenza infection (Wu and Meydani 2014). Consumption of vitamin C and E demonstrated a beneficial effect on peak expiratory flow rate (PEFR) of asthmatic children (Su et al. 2013).

Lutein is a fat-soluble carotenoid found in green leafy vegetable and eggs. Lutein intake is associated with lower FEV1/FVC% in smokers probably due to its antioxidant property that showed to be helpful for respiratory health (Melo van Lent et al. 2016, 2017). Lycopene demonstrated the strong inhibitory effect on *C. trachomatis* and *C. pneumoniae* infections in alveolar macrophages (Zigangirova et al. 2017). Also, ranges of animal and human studies confirmed its protective effect against asthma and allergic inflammation in the lungs (Riccioni et al. 2007; Hazlewood et al. 2011). Lycopene-rich tomato extract was shown to be helpful in subsiding asthma symptoms (Wood et al. 2008). Some of the aforementioned examples support the therapeutic role of antioxidant vitamins for the healthy respiratory system.

5.3.6 Polyphenols

Flavonoids supplementation was shown to be effective in reducing the incidence of URTIs by 33% (Somerville et al. 2016). They exhibit an inhibitory effect on various enzymes involved in cell signalling via modulation of redox status and preventing a respiratory burst of phagocytes (Ciz et al. 2012). Flavones therapeutic effect on URTIs was also attributed to their antibacterial and anti-inflammatory potential. They modulate the respiratory innate immune system through increasing ciliary beating and mucociliary clearance in airways cells (Hariri et al. 2017). Studies have shown the bacteriostatic effect of resveratrol on *Haemophilus influenzae* (NTHi), the well-known gram-negative bacteria involved in the pathogenesis of COPD (Euba et al. 2017).

Polyphenol possesses a natural tendency to prevent and treat influenza infection through inhibition of viral adhesion, penetration, intracellular transduction signalling pathways, viral haemagglutination though suppressing viral replication cycle (Bahramsoltani et al. 2016). Polyphenols in tea and extract of *Chaenomeles sinensis* were also reported to have antiviral activity against influenza A and B virus (Yang et al. 2014; Sawai-Kuroda et al. 2013).

Gold kiwifruit, a rich source of a wide range of nutritional components such as vitamin C, E, folate, polyphenols and carotenoids was shown to be effective in reducing duration and severity of URTIs symptoms in older individuals (Hunter et al. 2012). Polyphenols of pomegranate have antioxidant and anti-inflammatory potential and were considered as a therapeutic option for respiratory diseases (Danesi and Ferguson 2017). Apple polyphenols inhibit cigarette smoke-induced acute lung injury and improve QOL among COPD patients (Bao et al. 2013). Consumption of nonalcoholic beer containing naturally occurring polyphenols demonstrated to reduce the incidence of URTIs before and after marathon (Scherr et al. 2012). Study on respiratory epithelial cells and nasal mucosa clearly showed the positive effect of polyphenols such as gingerol, quercetin and epigallocatechin gallate on normal nasal ciliary movement and inhibition of mucus secretion (Chang et al. 2010). Isorhapontigenin, the analogue of resveratrol found in Chinese herb Gnetum cleistostachyum, is another dietary polyphenol which is known as novel nutrition for COPD treatment based on its anti-inflammatory properties (Yeo et al. 2017).

5.3.7 Spices

Variety of spices have been reported for their therapeutic activity in modulating immune response and protecting the respiratory system (summarized in Table 5.2).

Capsaicin, the component in chili peppers possesses the antioxidant and anti-inflammatory effects that could attenuate lung ischaemia-reperfusion injury (Wang et al. 2012). Intranasal application of capsaicin was shown to be beneficial for patients suffering from non-allergic rhinitis (Stjärne et al. 1991). Also, oral capsaicin could be used to prevent cough sensitivity and symptoms (Ternesten-Hasseus et al. 2015).

Curcumin benefits the respiratory system through several mechanisms. It inhibits porcine reproductive and respiratory syndrome virus (PRRSV) and influenza A virus (Han et al. 2018; Du et al. 2017). Curcumin pretreatment also protects against *S. aureus* infection by activating plasminogen activator inhibitor-1 preventing neutrophil infiltration, lung oedema and vascular leakage (Xu et al. 2015). It possesses

Table 5.2 Mechanisms of nutraceuticals in preventing respiratory disorders

Nutraceuticals	Example	Mechanisms
1. Dietary fibre		
Insoluble dietary fibre (IDF)	Celluloses, hemicelluloses, lignins	– Balancing Th1/Th2 (Zhang et al. 2016a) – Activation of GPR41 & GPR43 (Halnes et al. 2017) – Inhibiting histone deacetylases (Maslowski and Mackay 2011) – Inhibiting activation of transcription factor NF-κB (Maslowski and Mackay 2011) – Inducing cytokines (e.g. IL-10) (Round and Mazmanian 2010) – Inducing Treg cells (Round and Mazmanian 2010)
Soluble dietary fibre (SDF)	β-Glucans, pectins, gums, mucilages, hemicelluloses	
2. Probiotics		
Hypocholesterolemic probiotics	*Streptococcus thermophiles, enterococcus faecium, L. acidophilus, Bifidobacterium longum, L. casei, Saccharomyces boulardii, L. curvatus, L. plantarum, L. reuteri*	– Induce epithelial cell mucin secretion (Mack et al. 2003) – *Reduce TNF-α, IFN-γ, IL-12, IL-13 IL-4, IL-5, IL-3* (Chen et al. 2010) – *Improve VEGF* (Ahmad et al. 2014) – *Facilitate microvascular development and angiogenesis* (Ahmad et al. 2014)
Antidiabetic probiotics	*Lactobacillus reuteri, L. plantarum, L. casei, L. acidophilus, Bifidobacterium lactis, L. rhamnosus, B. bifidum*	
Antihypertensive probiotics	L. casei, streptococcus thermophilus, L. plantarum, L. helveticus, Saccharomyces cerevisiae, L. acidophilus, L. fermentum, L. bulgaricus.	
Antiobesity probiotics	Lactobacillus, Curvatus, L. plantarum, L. rhamnosus, Bifidobacterium animalis ssp. Lactis	
3. Prebiotics	Chicory roots, banana tomato, alliums, beans peas	– *Increasing anti-inflammatory cytokines* – *Decreasing proinflammatory cytokines* – *Changing composition of gut microbiota* (Dey 2017; Shokryazdan et al. 2017)
4. Polyunsaturated fatty acids	1. Omega 3 fatty acids (α-linolenic acid (ALA), eicosapentanoic acid (EPA), docosahexanoic acid	– Reduce mucus secretion (Mickleborough et al. 2006) – Reduce LTC_4-LTE_4, PGD_2,

(continued)

Table 5.2 (continued)

Nutraceuticals	Example	Mechanisms
	(DHA)) 2. Omega 6 fatty acids	IL-1β, TNF-α, LTB_4 (Matsuyama et al. 2005) – Reduce pulmonary bacterial load (Tiesset et al. 2011) – Producing proinflammatory cytokines (Hsiao et al. 2013) – Upregulating anti-inflammatory cytokine (Khaddaj-Mallat et al. 2016)
5. Antioxidant vitamins	Water soluble: Vitamin C Lipid soluble: Vitamin E and carotenoids (lycopene, β-carotene, lutein, zeaxanthin)	– B and T cells proliferation (Carr and Maggini 2017) – Antioxidant (Melo van Lent et al. 2016, 2017) – Decrease cytokine, chemokine, ROS, serum IgE level (Walrand et al. 2005)
6. Polyphenols	Flavonols, flavones, flavan-3-ols, flavanones, anthocyanins, flavonoids, phenolic acids, resveratrol, tea, legumes, soyabean, broccoli, red pepper	– Maintain normal nasal ciliary movement (Hariri et al. 2017) – Inhibits mucus secretion (Chang et al. 2010) – Inhibit replication cycle of influenza virus (Bahramsoltani et al. 2016) – Bacteriostatic effect on NTHi (Euba et al. 2017)
7. Spices	Onion, garlic, fenugreek seeds, turmeric, capsaicin, curcumin, cumin, eugenol, ginger, gingerol, mustard, piperine, fennel, coriander, mint, asafoetida	– Inhibit Cox-2, proteasome, NF-κB activation (Obata et al. 2013) – Regulating Wnt/β-catenin signalling (Yang et al. 2017; Zhang et al. 2016c) – Inhibit caspase-3 signal activation (Zin et al. 2012) – Reduce proinflammatory chemokine and cytokines (Han et al. 2018) – SuppressTh2-mediated immune response (Podlogar and Verspohl 2012; Khan et al. 2015)
8. Essential trace elements	Ca, Mg, Na, K, P, Cl, S, Fe, Zn, Mn, Cu, I, Ni, F, V, Cr, Mo, Se, Su, Si	– Altered gene expression (Maciel-Dominguez et al. 2013) – Enzymatic detoxification of peroxides (Voicekovska et al. 2007) – Increase phagocytosis by neutrophils and macrophages (Maggini et al. 2007) – Increase NK cell activity (Maggini et al. 2007) – Increase Th1 response (Maggini et al. 2007)

therapeutic activity against lower respiratory tract infection caused by respiratory syncytial virus (RSV) (Obata et al. 2013). Animal studies also reported curcumin benefits in reducing asthma symptoms and COPD associated alveolar and epithelial injury by regulating Wnt/β-catenin signalling in lung tissues and dendritic cells (Yang et al. 2017; Zhang et al. 2016c). It also protects lung epithelial cells from oxidative stress and cellular damage associated with environmental carcinogen such as benzopyrene (Zhu et al. 2014).

Eugenol is a methoxyphenol component of clove oil that could inhibit NF-κB activation, Cox-2 and inflammatory cytokine expression in lipopolysaccharide-stimulated macrophages and is therefore shown to be effective in lipopolysaccharide-induced lung injury (Magalhaes et al. 2010). This compound also protects from pulmonary tissue damage associated with environmental diesel exhaust particles (Zin et al. 2012). Ginger extract possesses anti-hyperactivity, anti-inflammatory and bronchodilator activity on airways (Mangprayool et al. 2013). It reduces inflammation during infection by reducing proinflammatory chemokine secretion and suppressing the Th2-mediated immune response (Podlogar and Verspohl 2012; Khan et al. 2015). Fresh ginger has also demonstrated dose-dependent inhibition of human respiratory syncytial virus (HRSV) (Chang et al. 2013b). Despite several benefits, occupational exposure to spice could act as dust allergen responsible for chronic respiratory symptoms, such as a cough, chest tightness, irritated and dry throat (Zuskin et al. 1992, 2000).

5.3.8 Essential Trace Elements

Several clinical and preclinical studies have reported the importance of essential trace elements in modulating the immune system and protecting the respiratory system against environmental toxins and pathogens. An imbalance of trace elements e.g. Se, Mn and Zn leads to a longer duration of mechanical ventilation in COPD patient (El-Attar et al. 2009). Se positively supports the immune system of the body via various mechanisms. Indeed, Se is a component of selenoproteins and antioxidant enzymes such as glutathione peroxidases and thioredoxin reductase (Maciel-Dominguez et al. 2013). Its deficiency decreases immunoglobulin titers affecting cell-mediated immunity (Maggini et al. 2007).

Cu, Zn cytoplasmic antioxidant enzymes play a vital role in cellular cytosolic defence during oxidative stress (Maares and Haase 2016). Disturbed Zn homeostasis results in the impaired formation, activation, maturation of lymphocytes that could negatively affect overall innate hose defence mechanism (Bonaventura et al. 2015). Studies have shown the association between Cu deficiency as well as iron deficiency with unfavourable immune response indicating the significant role of these elements in normal immune function (Maggini et al. 2007).

5.4 Nutrients and Nutraceuticals Interventions in Preventing Aging of Brain and Neurodegenerative Disease

Aging is considered as the major risk factor for neurodegeneration and neuronal death disorders including Parkinson's disease, Alzheimer's disease (AD), Multiple sclerosis and Huntington's disease. The mechanisms involved in neurodegenerative diseases are multifactorial and complex. Previous studies have shown the importance of the circulatory system in accelerating brain aging and impaired cognitive function (Wyss-Coray 2016). Intracellular Ca^{2+} overload caused by endoplasmic reticulum, mitochondrial dysfunction, oxidative stress and excessive ROS production, inflammation, and mutation in nuclear and mitochondrial DNA are among the well-studied molecular mechanisms attributed to neurodegenerative disorders (Dadhania et al. 2016).

So far, the available therapeutic strategies for mitigating neurodegenerative processes including hormone replacement therapy, stem cell therapy and pharmacological therapy are often associated with long-term unknown adverse effects. Therefore, there is clearly an urgent need to find safer and risk-free medication option to conquer the age-related neurodegenerative disorders (Dadhania et al. 2016). Nutrients and nutraceuticals are bioactive ingredients with the potential to modulate different intracellular and extracellular signalling cascades that play role in prevention of age-related neurodegenerative disorders. The following are the most widely studied nutrient and nutraceutical mediators in this field.

5.4.1 Fatty Acids

Due to the diversity of lipids in structure and function, they could have different effects on physiologic and pathologic conditions (Creegan et al. 2015). It was shown previously that palmitic acid has negative health effect on the normal brain and could upregulate the production of ceramide, a pro-apoptotic lipid which is elevated in the AD brain tissue (Han et al. 2002). On the other hand, diets enriched with polyunsaturated fatty acids (PUFAs) such as docosahexaenoic acid (DHA) and eicosapentaenoic acid (EPA) in fish oil showed neuroprotective effects with anti-inflammatory and antithrombotic properties (Celik and Sanlier 2019). Indeed, studies have shown that the level of DHA in serum decreased through enhancing age.

5.4.2 Carbohydrates

There is a positive correlation between fructose content in sweetened food and enhancing the rate of obesity-metabolic disorders, which are risk factors for the progression of AD (Creegan et al. 2015). Based on studies by van der Borght et al. (2011) on animal models, the increased sucrose or fructose consumption in rats diet induces apoptosis in neural hippocampus that leads to the reduction of the number of mature neurons (Van der Borght et al. 2011).

5.4.3 Vitamins

Vitamin C, the essential nutrients with antioxidant properties plays a protective role against AD (Liu and Ames 2005). Numerous studies have shown decreased level of vitamin B_1, and thiamine in the brain of patients with neurodegenerative disease, particularly AD (Gibson and Blass 2007; Gibson et al. 1988). Thiamine deficiency has also been associated with increased lactic acid level in brain, resulted in decreased neurotransmitters activity (Amrein et al. 2011). Vitamin B_2 (riboflavin) has a key role in restoring glutathione antioxidant capacity and declining oxidative stress, therefore involved in inhibition of AD progression and development (Liu and Ames 2005). The other vitamin with known protective properties in the brain is B_3 (niacin) which has a vital role in dendritic growth and myelination as well as Ca2+ signalling (Morris et al. 2004).

Many studies have suggested the effectiveness of Vitamin B6 (pyridoxine), folate and Vitamin B12 consumption in reducing homocysteine levels in elderly people and decreasing risk of Alzheimer's disease (Luchsinger et al. 2008). Moreover, vitamin B12 (cobalamin) deficiency has shown to affect brain function via demyelination (Liu and Ames 2005).

Previous studies on the aged population with poor memory have shown a decrease in the serum level of vitamin E (Poulose et al. 2017). Vitamin E with antioxidant and anti-inflammatory features can regulate neurogenesis (Poulose et al. 2017). There are also studies on the implementation of vitamin A (retinoic acid) supplementation in the treatment of AD (Shudo et al. 2009). Vitamin A mediates regulation of microglial activation and production of α-secretase and acetylcholine transmission (Shudo et al. 2009).

Vitamin D protects the structure of neurons and upregulates synthesis of neurotrophin proteins, a family of proteins essential for neural survival (Buell and Dawson-Hughes 2008).

Vitamin K is effective in neuronal protection, myelinization and mitogenesis in the brain. It has been shown that in patients with AD, the level of vitamin K in serum decreased. Moreover, consuming a vitamin K-rich diet could be beneficial for improvement of memory in older adults (Celik and Sanlier 2019; Soutif-Veillon et al. 2016).

5.4.4 Essential Trace Elements

Selenoproteins, containing selenium in the form of selenocysteine, exert different positive effects on human health, such as detoxification of heavy metals, antioxidant properties and modulation of immune function. Deficiency of these proteins was associated with reduced cognitive function. Selenium has a significant role in the brain through modifying DNA methylation by inhibition of DNA methyltransferase (Athanasopoulos et al. 2016). The level of selenium in serum and hair of AD patients was lower compared to the normal cases (Koc et al. 2015).

In vivo study on zinc diet demonstrated that the zinc deficiency is a risk factor for neurodegenerative disease and AD (Loef et al. 2012). Magnesium can protect the brain against neurodegenerative diseases and is known as one of the best memory supplements for improving brain function. Also, magnesium has shown an inhibitory effect on amyloid-β formation and development of AD (Chui et al. 2011).

5.4.5 Key Nutraceuticals

A clinical link has been established between nutraceuticals and neurodegenerative disorders. Curcumin as a strong antioxidant compound is the major compound of turmeric, a plant native to southeast Asia has different therapeutic effects on neurodegenerative diseases along with diabetes, liver disease, arthritis and various cancers (Ghosh et al. 2015). Curcumin with the ability to cross the blood–brain barrier has shown to lessen the aggregation of amyloid-β, delay neuron degradation process, reduce microglia formation and finally improve memory in patients with AD (Mishra and Palanivelu 2008). Similar to curcumin, resveratrol can also attenuate the aggregation of amyloid-β (Sadhukhan et al. 2018). Most of the activity of this compound was detected in the medial cortex, hypothalamus and striatum. Resveratrol also acts as neuroprotective in Parkinson's disease through upregulating antioxidant status, reducing the level of thiobarbituric acid reactive substances (TBARS) and suppressing COX-2 expression (Ma et al. 2014). Mulberry extract is considered as a high source of anthocyanins and cyaniding, showed potential to slow down AD progression through the negative regulation of 1-phosphatidylinositol-4,5-bisphosphate phosphodiesterase, Beta-site APP cleaving enzyme 2 (BACE2) and apoptotic protease activating factor 1(Apaf-1) (Song et al. 2014). Also, there are many other nutraceuticals such as caffeic acid, berberine, silymarin, tannic acid, lipoic acid, etc. which has been indicated for their role in prevention and delay onset of neurodegenerative disorders. Table 5.3 focuses on groups of nutraceuticals and the related mechanisms of action involved in neurodegenerative diseases.

5.5 Nutrients and Nutraceuticals Interventions in Cancer Prevention

Cancer is one of the major public health problems worldwide with increasing incidence among aging population. Many genetic factors, as well as internal and external environmental modulators were shown to be involved in the emergence and development of several cancers. Diet and nutrition are receiving important attention as the crucial factors in preventing and treating different types of cancers. Many studies have supported the important role of dietary patterns in the prevention of cancer or inhibition of tumour growth (Banikazemi et al. 2018).

Table 5.3 List of nutraceutical and mechanisms involved in neurodegenerative diseases

Nutraceuticals	Example	Mechanisms
1. Carbohydrate	Fructose	Apoptosis in neural hippocampus (Van der Borght et al. 2011)
2. Fatty acids	α-Lipoic acid, docosahexaenoic acid (DHA) and eicosapentaenoic acid (EPA)	Anti-inflammatory effect (Celik and Sanlier 2019)
3. Antioxidant vitamins	Vitamin C, vitamin E, vitamin D Thiamine, vitamin B12, B6 Vitamin E, A and carotenoids, coenzyme Q10, Astaxanthin, B3 (niacin), vitamin E	– Modulate neurotransmitters activity (Amrein et al. 2011) – Targeting mitochondrial dysfunction (Liu and Ames 2005) – Nutraceuticals targeting inflammation, decrease cytokine, chemokine, ROS (Liu and Ames 2005; Morris et al. 2004; Shudo et al. 2009) – Regulation of microglial activation, producing α-secretase and acetylcholine (Morris et al. 2004) transmission (Poulose et al. 2017)
4. Polyphenols	Apigenin, isoflavones, soy isoflavones, rosmarinic acid, resveratrol, blueberry, carnosic acid, epigallocatechin-3-gallate, catechins, anthocyanins	– Nutraceuticals targeting calcium overload – Targeting oxidative stress (Ma et al. 2014) – Reducing the level of thiobarbituric acid reactive substances (TBARS) and suppressing COX-2 expression (Ma et al. 2014)
5. Spices	Curcumin, eugenol, ginsenosides in ginger	– Targeting oxidative stress – Nutraceuticals as neurotransmitter modulators (Briguglio et al. 2018)
6. Essential trace elements	Mg, S, Zn	– Targeting DNA methylation and histone modification (Athanasopoulos et al. 2016) – Act as cofactors of enzyme involved in neurotransmitter synthesis, many cellular process (Chen et al. 2016)

In this section, we represent the role of some nutrients and nutraceuticals that are effective in preventing cancer as an age-related disease.

5.5.1 Dietary Fibre

Fibres as probiotics stimulator have been implicated in health and prevention of few cancers particularly gastric and colorectal cancers (Aune et al. 2011). Grains are a major source of dietary fibre, variety of minerals, vitamins, and phytochemicals, which have demonstrated anticancer properties and could reduce the risk of colorectal cancer. Digested insoluble fibre increases the dry stool composition and assists in the prevention of constipation (Yang et al. 2012). Based on different meta-analysis of prospective studies, taking dietary fibre was inversely related to risk of oesophagus and breast cancer (Aune et al. 2012; Coleman et al. 2013).

5.5.2 Antioxidant Vitamins

A randomized controlled trial on 70,528 patients showed that consumption of Vitamin D and calcium could reduce the rate of mortality in the elderly and increase life expectancy (Rejnmark et al. 2012). Indeed, low vitamin D level and calcium deficiency are important risk factors for various types of cancer (Lappe et al. 2007). High dietary intake of β-carotene and vitamin A was associated with reduced risk of lung cancer (Yu et al. 2015). Studies have shown that taking Vitamin C and β-carotene, α-carotene, β-cryptoxanthin, lutein/zeaxanthin, lycopene and vitamin C appears to reduce the risk of lung cancer as well squamous cell carcinoma, and adenocarcinoma (Tanvetyanon and Bepler 2008; Shareck et al. 2017). Dietary carotenoids are derived from fruits and vegetables such as green leafy vegetables, spinach, carrots, peaches, plums and potatoes. Carotenoids as highly pigmented substances are responsible for red, yellow and orange colours in many fruits and vegetables. Based on epidemiological studies, a reverse correlation has been observed between intakes of high concentrations of dietary various carotenoids, such as β-carotene, β-cryptoxanthin, fucoxanthin, canthaxanthin and astaxanthin, and the risk of cancer in different tissue. Dietary intake of carotenoids has been suggested due to their cancer chemoprevention activity (Tanaka et al. 2012).

Among carotenoids, lycopene in tomato is the most powerful antioxidant (Di Mascio et al. 1989). Epidemiological, animal and in vitro studies have largely indicated the role of lycopene in cancer prevention (Story et al. 2010). Lycopene with antioxidant properties could protect cells against free radical-induced DNA damage and therefore inhibit neoplasm formation. In addition to its antioxidant properties, lycopene also interferes with cancer promote signalling pathways such as Ras and Akt pathways (Palozza et al. 2010).

Lutein and zeaxanthin, the members of the carotenoids family have been recognized for their protective role against the uterus, prostate, breast, large intestine gastrointestinal and lung cancers (Prakash and Gupta 2014b).

Soybean is a source of isoflavones that are structurally similar to oestrogen, and therefore, they are known as phytoestrogens. The soybean isoflavones (equol, genistein, daidzein, etc.) have potent antioxidant activity. Also by binding to oestrogen receptor, they could inhibit cell growth. Isoflavones anti-tumourigenic effects were attributed to their potential as endocrine disruptor and tyrosine kinase inhibitor that could inhibit cancer cell growth and induce cell apoptosis (Jian 2009). A recent study on Chinese women showed an inverse association between soybean food intake and breast cancer risk in adulthood (Baglia et al. 2016).

Coenzyme Q10 (Co Q10) is a vitamin-like compound that plays a crucial role in mitochondrial function and production of ATP, the major source of energy in the cells. In fact, Co Q10 is a powerful antioxidant against oxidative stress. Food supplements containing Co Q10 have anti-inflammatory effects associated with reduced IL-6 marker in coronary artery disease (Lee et al. 2012). Recent studies on Chinese women have shown that low levels of Co Q10 in plasma are associated with an increased risk of breast cancer (Cooney et al. 2011). Many studies indicated the efficacy of Co Q10 supplement as adjuvant therapy in the treatment of cancers and prevention of chemotherapy induced-cardiotoxicity (Narain et al. 2018).

5.5.3 Polyphenols

Polyphenols are the compounds rich in fruits, vegetables, dicotyledons, cereals and beverages. Dicotyledonous and chocolate plants play an important role in absorbing polyphenolic compounds in the body. As the secondary plant metabolites, they play an important role in defence systems against ultraviolet radiation or accumulation of cellular damage particularly DNA damage. Previous epidemiological studies have established a reverse relationship between the risk of the various cancer diseases and the use of polyphenol-rich diets (Kidd 2009).

Fruits such as apple, grapes, pears, cherries and coconut contain 200–300 mg of polyphenols in 100 mg of pure weight, and their dietary intake may reach around 1 mg per day, which is about 10 times more than vitamin C and 100 times more than vitamin E and carotenoids (Scalbert et al. 2005b). Citrus is also a rich source of flavones and hesperidin, which are abundant in natural fruit juices (120–250 mg/L) (Gupta et al. 2013).

Curcumin for many years was the centre of attention due to its wide biological activities, such as antioxidant activity, anti-inflammatory, anti-carcinogenicity and neuronal degeneration (Shanmugam et al. 2019). Indeed, curcumin has been considered as a promising natural compound applied potentially for the chemical prevention against multiple cancers (Park et al. 2013b). Results obtained from different clinical trial in patients with advanced pancreatic cancer as well as patients with pre-malignant lesions or breast cancer or clone cancer have clearly demonstrated the promising effect of curcumin in increasing health markers (Johnson and Mukhtar 2007; Bayet-Robert et al. 2010; Hsieh 2001).

Green tea is a common drink in Asia that is derived from Camellia sinensis. Pharmaceutical ingredients of green tea are the polyphenols antioxidants called catechins that contain four main compositions including Epicatechin, Epigallocatechin, Epicatechin trigallate, Epigallocatechin trigallate (EGCG) (Adhami and Mukhtar 2007). Laboratory and animal studies showed that EGCG, the main catechin in green tea mediated anticancer signalling process through 5α-reductase inhibition activity associated with human health and cancer prevention especially hormone stimulating cancers (Khan and Mukhtar 2016). Grape seed extract (GSE) is a rich source of polyphenols. Though clinical research on GSE for cancer and inflammation is not as advanced as that for research on turmeric, there are numerous studies indicating the protective role of GSE against cancer and inflammatory disease (Kidd 2009).

Resveratrol (3, 4′, 5-trihydroxy-trans-stilbene), a natural polyphenol, has antioxidants, anti-inflammatory and anti-aging properties. Resveratrol has an effect on all three different stages of carcinogenesis (onset, continuation and progression) through different signalling pathways that affect cell division and growth, apoptosis, inflammation, angiogenesis and metastasis (Kraft et al. 2009).

Blueberries are rich sources of antioxidants containing a high level of polyphenols that are extremely suggested for their anticancer activity (Neto 2007). Many studies at cellular and clinical levels have indicated polyphenols protective effect against a variety of cancer (Neto 2007). The pomegranate fruit is a rich source of antioxidants with anti-inflammatory and anticancer effects (Sharma et al. 2017). Also, it is a potent source of tannins and ellagitannins, catechins, gallocatechin and anthocyanin. Many studies have shown the role of pomegranate in the prevention and treatment of different cancers including breast, prostate, lung, skin and colon (Sharma et al. 2017).

5.5.4 Fatty Acids

Omega-3 fatty acids could reduce inflammation in middle-aged and elderly people (Prakash and Sharma 2014). Omega-3 consumption has been suggested in the prevention of cancer. Based on data from a large randomized controlled trial, it has been shown that vitamin D and omega-3 fatty acids supplements decreased the risk of cancer (Manson et al. 2012). A recent study has also shown that daily taking omega-3 is associated with a decreased risk of breast cancer in Mexican women (Chajès et al. 2012).

5.6 The Effects of Nutrients in the Management and Prevention of Diabetes

Diabetes is the progressive chronic disease, associated with reduced ability of the body to produce insulin or impairment of insulin function due to the tissue resistance to this hormone. Aging is known as a major risk factor for diabetes. Also, diabetes mellitus increases the incidence of accelerated aging, indicating that diabetes may contribute to the aging process (Yaffe et al. 2011). Although there is no definitive treatment available for diabetes, managing blood glucose is crucial in reducing the rate of progression and preventing its later complications. In addition to the commonly used medicinal treatments, lifestyle management and nutrition therapy are considered as the essential remedies for this disease (Galtier 2010).

5.6.1 Antioxidant Vitamins

It has been shown that Vitamin C reduces the accumulation of diabetes-induced sorbitol and lipid peroxidation in erythrocytes (Chen et al. 2006).

A reverse correlation between fasting insulin level and the main circulating vitamin D metabolite, 25-hydroxyvitamin D has been reported in many studies (Kayaniyil et al. 2010; Nikooyeh et al. 2011). Daily intake of this vitamin E reduces the fasting serum glucose and glycated haemoglobin (Nikooyeh et al. 2011). Vitamin E is a vital fat-soluble vitamin, acts primarily as an antioxidant. Low levels of vitamin E were shown to be associated with increased incidence of diabetes (Asmat et al. 2016).

The Co Q10 supplementation has been suggested for the treatment of type 2 diabetes (Shen and Pierce 2015). Studies on 120 diabetic patients have shown that the activity of the succinate dehydrogenase-coenzyme Q10 reductase was significantly lower compared to the normal group (Kishi et al. 1976).

Several studies have shown that biotin could alleviate hypertriglyceridemia and is supportive for people with diabetes (Xiang et al. 2015). Biotin directly activates the soluble guanylate cyclase, therefore increases the synthesis of cyclic GMP (cGMP). Biotin induced-cGMP exerts positive effects on beta cells through enhancing the release of glucose-stimulated insulin and improving the glucose tolerance mediated by modulating the gene expression of the key transcription factor involved in insulin secretion (De La Vega-Monroy et al. 2013). cGMP in the liver has the potential to suppress excessive production of hepatic glucose through increasing the expression of glucokinase and reducing the expression of phosphoenolpyruvate carboxykinase (Vilches-Flores et al. 2010). Based on clinical studies receiving high levels of biotin (9–16 mg daily) can lead to glycaemic control in diabetic patients (Fernandez-Mejia and Lazo-de-la-Vega-Monroy 2011). However, the optimal dosage of biotin-induced systemic activation of guanylate cyclase with fewer side effects is still unclear. L-Carnitine (β-Hydroxy-γ-Trimethyl ammonium Butyrate) is a natural vitamin-like compound, absorbed from food sources including meat and dairy products. Indeed, many in vivo and in vitro studies have shown the effect of L-Carnitine in improving insulin sensitivity and glucose disposal (Hajinezhad et al. 2016). The beneficiary effects of L-carnitine were reported in diabetic complications such as neuropathy, ventricular dispersion, diabetic eye and brain.

5.6.2 Carbohydrates

Carbohydrates are the largest group of energy substrate that has a direct impact on glycaemic level. They may induce high blood glucose and insulin concentrations, which leads to the glucose intolerance and increased risk of type 2 diabetes. Indeed, the low-carbohydrate diet was associated with glycaemic control, weight loss and lipid profile and compared with low-fat diets appeared to be superior on managing and prevention of diabetes (Ajala et al. 2013). However, still, the low-carbohydrate diet is not accepted as dietary treatment for diabetes. Some has proposed that good quality carbohydrate, defined as a lower ratio of carbohydrate to fibre, is associated with decreased risk of diabetes (AlEssa et al. 2015). Other study showed that starchy legumes, including beans, peas, and lentils have a positive effect on glycaemia (Bantle et al. 2008). The linear polysaccharide chitosan, a natural product derived from chitin, showed potential in reducing cholesterol absorption and improving insulin sensitivity (Hsieh et al. 2012). The guar gum is natural high-molecular weight polysaccharide implemented in lowering glycaemic index of the meal (Giri et al. 2017).

5.6.3 Fatty Acids

It is almost accepted that the fat-rich diet reduces glucose tolerance, modifies insulin resistance, induces obesity and dyslipidaemia (Risérus et al. 2009). Studies have shown the relationship between the portion of saturated fatty acids and increased insulin resistance and the incidence of diabetes (Vessby et al. 1994). On the other hand, canola, olive oil and peanut oil contents of monounsaturated fatty acid have beneficial effects in controlling triglyceride level and glycaemia in patients suffering from diabetes (Anderson et al. 1998). Omega-3 fatty acid in fish is well-known for its potential in reducing serum triglycerides (Kaushik et al. 2009). Studies have shown that diets rich in docosahexaenoic acid, eicosapentaenoic acid, arachidonic acid and ω-6 γ-linolenic acid inhibit the development of alloxan-induced diabetes mellitus (Mohan and Das 2001).

5.6.4 Dietary Fibres

Foods rich in fibre, particularly cereal is highly recommended to diabetic patients. Although the protective effect of fibres against some chronic diseases has been confirmed, the effectiveness of the fibres on the metabolism of glycaemia and lipids is still unknown (Higgins 2004). One of the most important soluble fibre is glucomannan, derived from konjac root that has been implicated in the management of diabetes (Onakpoya et al. 2014; Chearskul et al. 2007).

Cereals have high glycaemic index mostly related to their alpha-amylase inhibitor content (Stümpel et al. 2001).

Alpha-glucosidase inhibitors have the ability to prevent the absorption of carbohydrates from the intestine and were proposed for their beneficial effect in the treatment of patients with type 2 diabetes (van de Laar 2008). Among alpha-glucosidase inhibitors, acarbose and miglitol might be the first-line therapeutic approach in controlling glycated haemoglobin, fasting blood glucose and insulin level (Van de Laar et al. 2006).

5.6.5 Probiotic

Probiotic supplementation with L. casei Shirota prevents insulin resistance (Hulston et al. 2015). The role of probiotics in glycaemic control results by controlling metabolic changes, modulating oxidative stress and inflammation as well as improving gut bacterial flora (Shah and Swami 2017). Yoghurt probiotic containing L. acidophilus and Bifidobacterium lactis reduces total cholesterol, HDL/cholesterol and LDL/HDL ratio (Ejtahed et al. 2011). Many studies also concluded that consumption of probiotic yoghurt provides a potential intervention and prevention strategies against type2 diabetes (Asemi et al. 2013).

5.6.6 Trace Element

The relationship between magnesium intake and decreased risk of diabetes has been reported recently. Indeed, magnesium is necessary for insulin signalling and entering glucose to the cells (Takaya et al. 2004). Intake of high calcium, magnesium and potassium has been associated with the prevention of hypertension, heart failure (HF) and diabetes (Houston and Harper 2008).

There are limited studies that suggested the role of chromium in improving insulin sensitivity and glucose tolerance in patients with type 2 diabetes mellitus (Amoikon et al. 1995). Though based on meta-analysis study, there is no beneficial effect related to chromium supplementation on diabetes (Bailey 2014).

Vanadium, a plant-derived mineral acts similar to insulin in transferring glucose to the cells that was suggested in type 1 and type 2 diabetes treatment. Vanadium supplement also reduces fasting blood glucose, haemoglobin A1C and cholesterol levels (Halberstam et al. 1996).

5.6.7 Protein and Amino Acids

Evidence suggested that diabetic patients need the same amount of protein of about 0.86 g/kg BW per day as ordinary people (Cunningham 1998). Although dietary protein plays a role in stimulating the secretion of insulin, excessive use should be avoided as it may contribute to the pathogenesis of diabetic nephropathy. Evidence has shown that taking plant-based proteins can improve diabetes-induced hypercholesterolemia and nephropathy more effectively than animal proteins (Comerford and Pasin 2016). Phaseolamin protein extracted from white kidney beans is commercially available in concentrated form and has been widely indicated in the prevention of oxidative stress and improvement of diabetes complication (Oliveira et al. 2014).

5.6.8 Polyphenols

Various mechanisms have been proposed to explain the possible antidiabetic effects of polyphenols. Tea-specific phenols, especially green tea, including catechin and epigallocatechin gallate (EGCG), prevent oxidative stress and inflammation through activating the NRF2 pathway (Marinovic et al. 2015). Beyond their significant antioxidant capacity, polyphenols are present in cocoa and coffee and contain phenolic compounds such as caffeoyl derivatives, procardin and chlorogenic acid that have been proven to promote insulin sensitivity, vascular endothelial function, metabolism of lipids and carbohydrates, tory mediators. Antidiabetic effects of flavonoids including flavanones, flavonols, isoflavones and anthocyanidins have been confirmed in many studies (Lasa et al. 2014; Chen and Snyder 2013). Resveratrol also reported to prevent insulin resistance related metabolic disorders, dyslipidaemia, hyperglycaemia and hyperinsulinaemia via regulating gene expression and activity of rate-limiting enzymes (Abbasi Oshaghi et al. 2017).

Recent discoveries suggest consumption of coffee can lower the risk of diabetes (Jessica Elizabeth et al. 2017; Esmaeili et al. 2017). Such effects were attributed to chlorogenic acid or CGA, the important polyphenolic compounds of coffee beans. Evidence showed that CGA can reduce the absorption of carbohydrates by preventing glucose intestinal transfer via preventing the activity of glucose-6-phosphate translocase involved in glucose uptake (Rosengren et al. 2004; Hemmerle et al. 1997). The extract of green coffee beans is rich in GCA (55%) and is currently commercially available (Table 5.4).

5.7 Nutrients and Nutraceuticals Interventions Against Skin Aging

Skin aging is an unavoidable biological process in human life. Aging causes changes in all parts of the body system with subsequent signs on the skin. Aged skin is wrinkled with sensitive, dry pigmented and itchiness characteristics. Several studies reported that skin aging can be delayed by applying external and internal factors to improve skin health. In this regards, nutrition represents a promising approach in preventing, reducing and delaying skin aging (Tundis et al. 2015).

Exposure to some chemical agents, air pollutants, ultraviolet rays, smoking, heavy metals and organic solvents induce the formation of free radicals that could result in the activation of matrix metalloproteinases and inflammatory responses and thereby disruption of extracellular matrix involved in skin aging.

Daily intake of healthy food, fruit and vegetables containing antioxidant compounds play a critical supportive role against degeneration of skin cells (Petruk et al. 2018). Vitamin A (β-carotene, retinol) (Mukherjee et al. 2006), vitamin C (ascorbic acid) (Boyera et al. 1998), vitamin E (tocopherol) (Fryer 1993), vitamin D (Peric et al. 2009), polyphenols (green tea, resveratrol) (Nichols and Katiyar 2010; Menaa et al. 2014) and botanical antioxidants, ubiquinol (Co Q10) (Olivieri et al. 2013), prebiotics (Oligofructose and other oligosaccharides) and probiotics

Table 5.4 List of nutraceutical and mechanisms involved in the management of diabetes

Nutraceuticals	Example	Mechanisms
1. Dietary fibre/ carbohydrates	Glucomannan Guar gum Chitosan Beans, peas and lentils	– Decreased the rise of LDL (Chearskul et al. 2007) – Reducing cholesterol absorption and improving insulin sensitivity (Hsieh et al. 2012) – Lowering glycaemic index of the meal (Giri et al. 2017) – Alpha-glucosidase inhibitors (van de Laar 2008)
2. Fatty acids	Linoleic acid Cholesterol esters Omega-3 Eicosapentaenoic acid, Docosahexaenoic acid, Arachidonic acid, ω-6 linolenic acid	– Reducing serum triglycerides (Anderson et al. 1998) – Improves insulin sensitivity (ARIC Study Investigators Wang Lu Folsom Aaron R folsom@ epi. umn. edu Zheng Zhi-Jie Pankow James S Eckfeldt John H 2003) – Antioxidant status and suppressing production of cytokines (Mohan and Das 2001)
3. Antioxidant vitamins	Vitamin C Vitamin E Vitamin D Coenzyme Q10 Biotin L-Carnitine	– Protein glycation inhibitor, reduces the accumulation of lipid peroxidation in erythrocytes (Chen et al. 2006) – Antioxidant (Asmat et al. 2016) – Enhancing the release of glucose-stimulated insulin (De La Vega-Monroy et al. 2013), hypertriglyceridemia (Xiang et al. 2015; Fernandez-Mejia and Lazo-de-la-Vega-Monroy 2011)
4. Polyphenols	Tea (catechin and Epigallocatechin gallate) Resveratrol Flavonoids Chlorogenic acid	– Prevent oxidative stress and inflammation (Marinovic et al. 2015) – Alleviating insulin resistance by circulatory improving (Wong and Howe 2018) – Prevent metabolic disorders, dyslipidaemia, hyperglycaemia and hyperinsulinaemia (Abbasi Oshaghi et al. 2017)
5. Protein and amino acids	Phaseolamin	– Targeting oxidative stress (Oliveira et al. 2014)
6. Probiotics	*L. casei Shirota* *L. Acidophilus Bifidobacterium lactis* *Saccharomyces boulardii*	– *Controlling metabolic changes, inflammation and oxidative stress* (Shah and Swami 2017) – *Changing lipid parameters* (Asemi et al. 2013) – *Normalize insulin* Sensitivity (Shah and Swami 2017)
7. Essential trace elements	Mg, V, Ca, K	– Glucose utilization and insulin signalling (Takaya et al. 2004) – Reduces fasting blood glucose, haemoglobin A1C and cholesterol levels (Halberstam et al. 1996)

Table 5.5 List of nutraceutical and mechanisms in anti-skin aging medicine

Nutraceuticals	Example	Mechanisms	Malnutrition/ deficiency result
1. Antioxidant vitamins	Vitamin A Retinaldehyde	Protection in photo skin aging (Wong and Howe 2018) Essential for maintaining the health of epithelial cells (Wong and Howe 2018)	Disorders in skin reproduction/ skin aging
	Vitamin C	Cofactor for prolyl hydroxylase and lysyl hydroxylase in the synthesis of collagen (Tundis et al. 2015)	Scurvy
	Vitamin D	Expression induction of anti-microbial peptides in skin (Peric et al. 2009)	Skin aging
	Coenzyme Q10 Vitamin E	Antioxidant defence mechanism in epidermis (Peric et al. 2009) helps against collagen cross linking and lipid peroxidation (Petruk et al. 2018)	Skin aging
2. Polyphenols	Stilbenes and lignans Flavonoids	Changing the redox reaction Chelator (Boyera et al. 1998; Fryer 1993) Radical scavenger (Boyera et al. 1998; Fryer 1993)	UV protection Skin aging
3. Fatty acids	Eicosanoids Omega-6 Omega-3	Present in the epidermis and epidermal fatty acids Immunomodulatory role (Olivieri et al. 2013)	Skin atrophy and Skin dryness
4. Probiotics	Lactobacillus johnsonii	Modulate skin immune system (Nichols and Katiyar 2010; Menaa et al. 2014)	Increased transepidermal water loss

(enterococci, lactobacilli and bifidobacteria in the intestinal tract) (Cinque et al. 2017; Bouilly-Gauthier et al. 2010), essential fatty acids (long-chain and polysaturated fatty acids) (Miller et al. 1991) are among the nutrients and nutraceuticals that have been frequently reported for their role in skin protection and maintain its youth (Schagen et al. 2012) (Table 5.5).

5.8 Ocular Aging

The changes in the power of eye vision with aging defined as ocular aging. As the rates of metabolites are high in eye, there is an extra need for antioxidant protection in this organ. The retina is highly susceptible to oxidative degradation that was associated with the abundance of polyunsaturated fatty acids found primarily in the outer membrane of light receptors that are rapidly oxidized.

Many clinical studies have indicated the supportive effect of dietary nutrients in eye health and prevention of disease. High doses of vitamins C, E, zinc, beta-carotene supplements have shown to lower the risk of age-related macular degeneration (AMD) incidence in a randomized clinical trial (Group A-REDSR 2001). The

Table 5.6 List of nutraceutical and mechanisms in anti-ocular aging medicine

Nutraceuticals	Example	Mechanisms	Deficiency
1. Antioxidant vitamins	Vitamin A Vitamin E Retinaldehyde α-Tocopherol	– Free radical scavenger, protect against lipid oxidation (Miller et al. 1991) – Cofactor of enzymes – Effect on immune /inflammatory cells	Vision disorder AMD Progression (Miller et al. 1991)
	Vitamin C	Antioxidant	AMD progression (Miller et al. 1991)
	β-Carotene Lutein Zeaxanthin	– Conversion to retinol in lens, retina, macula – Absorbance of blue light – Antioxidant activity (Schagen et al. 2012)	Vision disorder AMD progression
2. Trace elements	Zinc (Zn) Copper	– Health of the retina – Optimal metabolic pathways of eye – Essential for enzymatic activity – Antioxidant – Involved in cellular membrane structure	– Disorder of neural impulse in eye – Vision disorders – Cataract – Increase risk of AMD
3. Fatty acids	Omega-3	– AMD prevention – Antioxidant activity (Group A-REDSR 2001) – Affecting inflammation (Group A-REDSR 2001)	Eye disorder

most important of nutrients for the maintenance of eye function are carotenoids the natural free radical scavenger. Lutein and zeaxanthin carotenoids represented a significant protective role against nuclear cataract and AMD in the patient over 65 years (Roberts et al. 2009). Moreover, it has been recently shown that retinal long-chain polyunsaturated fatty acids play important roles in normal retinal function and visual development (Liu et al. 2010). Based on data from a clinical study on 4513 participants, higher total fish consumption containing omega 3 polyunsaturated fatty acid was associated with decreased risk of AMD (SanGiovanni et al. 2003) (Table 5.6).

5.9 Tooth Aging

Gradual changes in the appearance and physiology of tooth occur during aging. The composition of dentin and the covering layer, enamel undergoes structural deteriorative changes partly attributed to the reduced pulp function and vitality. In fact, dental pulp is the main living part of the tooth inhabited by stem cells having odontoblastic differentiation capacity that mediate dentine regeneration in adult teeth. Decreased alveolar bone density was also associated with tooth loss as were reported in many controlled clinical trials on aging populations (May et al. 1995). Considering the vital role of tooth in food digestion system, oral health is of high importance to human health being. Moreover, periodontal disease has been linked to the pathology of age-related disease such as diabetes, cardiovascular disease (CVD) and chronic pulmonary disease (Borgnakke et al. 2013). Nutrients and nutraceuticals are bioactive ingredients with the potential to prevent oral disorders and preserve oral health.

Calcium is the main abundant trace element in tooth and bone. Studies have shown that increased intake of calcium and vitamin D could reduce the risk of oral bone loss (Uhrbom and Jacobson 1984). Dietary Vitamin C and Vitamin B-complex are also among supplements suggested for prevention and treatment of periodontal disease (Neiva et al. 2005). The oral disease is the early manifestation of trace elements deficiency such as selenium iodine, iron, zinc, copper (Bhattacharya et al. 2016). Although studies have shown no relation between fluoride supplement and prevention of dental caries (Ismail and Hasson 2008), topical application of fluoride has been highly recommended as a successful strategy for inhibiting tooth demineralization and control dental caries (Marinho et al. 2003).

The role of Bifidobacterium, and Lactobacilli probiotic have been suggested in oral health care and effective against periodontal disease, tooth caries and oral mucosal lesions. The underlying mechanism of oral protective role of probiotics were attributed to their competition with growth and adhesion of pathogens on oral tissue and their immunomodulatory role in reducing inflammatory response (Bhattacharya et al. 2016).

5.10 Conclusion

Nutrients and nutraceuticals could be the best therapeutic strategies for mitigating aging and age-related disorders. It is essential to establish healthy lifestyle with consuming healthful diet including wide range of healthy food, tea, fruits and vegetables containing antioxidant compounds, vitamins, minerals, polyphenols include flavonoids (anthocyanins, catechins, isoflavones), and non-flavonoids (resveratrol, etc.), ubiquinol (Co Q10), prebiotics, probiotics and essential fatty acids (long-chain and polysaturated fatty acids).

References

Abbasi Oshaghi E, Goodarzi MT, Higgins V, Adeli K (2017) Role of resveratrol in the management of insulin resistance and related conditions: mechanism of action. Crit Rev Clin Lab Sci 54 (4):267–293

Adhami VM, Mukhtar H (2007) Anti-oxidants from green tea and pomegranate for chemoprevention of prostate cancer. Mol Biotechnol 37(1):52–57

Agerholm-Larsen L, Raben A, Haulrik N, Hansen AS, Manders M, Astrup A (2000) Effect of 8 week intake of probiotic milk products on risk factors for cardiovascular diseases. Eur J Clin Nutr 54(4):288–297

Ahmad A, Cai CL, Kumar D, Cai F, D'Souza A, Fordjour L et al (2014) Benefits of pre-, pro- and Syn-biotics for lung angiogenesis in malnutritional rats exposed to intermittent hypoxia. Am J Transl Res 6(5):459–470

Ajala O, English P, Pinkney J (2013) Systematic review and meta-analysis of different dietary approaches to the management of type 2 diabetes. Am J Clin Nutr 97(3):505–516

AlEssa HB, Bhupathiraju SN, Malik VS, Wedick NM, Campos H, Rosner B et al (2015) Carbohydrate quality and quantity and risk of type 2 diabetes in US women. Am J Clin Nutr 102 (6):1543–1553

Allan KM, Prabhu N, Craig LC, McNeill G, Kirby B, McLay J et al (2015) Maternal vitamin D and E intakes during pregnancy are associated with asthma in children. Eur Respir J 45 (4):1027–1036

Ames BN (2018) Prolonging healthy aging: longevity vitamins and proteins. Proc Natl Acad Sci 115(43):10836–10844

Amoikon E, Fernandez J, Southern L, Thompson D Jr, Ward T, Olcott B (1995) Effect of chromium tripicolinate on growth, glucose tolerance, insulin sensitivity, plasma metabolites, and growth hormone in pigs. J Anim Sci 73(4):1123–1130

Amrein K, Ribitsch W, Otto R, Worm HC, Stauber RE (2011) Severe lactic acidosis reversed by thiamine within 24 hours. Crit Care 15(6):457

Anderson JW, Blake JE, Turner J, Smith BM (1998) Effects of soy protein on renal function and proteinuria in patients with type 2 diabetes. Am J Clin Nutr 68(6):1347S–1353S

Anderson JW, Randles KM, Kendall CW, Jenkins DJ (2004) Carbohydrate and fiber recommendations for individuals with diabetes: a quantitative assessment and meta-analysis of the evidence. J Am Coll Nutr 23(1):5–17

Aoki H, Hisada T, Ishizuka T, Utsugi M, Kawata T, Shimizu Y et al (2008) Resolvin E1 dampens airway inflammation and hyperresponsiveness in a murine model of asthma. Biochem Biophys Res Commun 367(2):509–515

Aoki H, Hisada T, Ishizuka T, Utsugi M, Ono A, Koga Y et al (2010) Protective effect of resolvin E1 on the development of asthmatic airway inflammation. Biochem Biophys Res Commun 400 (1):128–133

ARIC Study Investigators Wang Lu Folsom Aaron R folsom@ epi. umn. edu Zheng Zhi-Jie Pankow James S Eckfeldt John H (2003) Plasma fatty acid composition and incidence of diabetes in middle-aged adults: the Atherosclerosis Risk in Communities (ARIC) Study. Am J Clin Nutr 78(1):91–98

Asemi Z, Zare Z, Shakeri H, Sabihi S-S, Esmaillzadeh A (2013) Effect of multispecies probiotic supplements on metabolic profiles, hs-CRP, and oxidative stress in patients with type 2 diabetes. Ann Nutr Metab 63(1–2):1–9

Asmat U, Abad K, Ismail K (2016) Diabetes mellitus and oxidative stress—a concise review. Saudi Pharm J 24(5):547–553

Athanasopoulos D, Karagiannis G, Tsolaki M (2016) Recent findings in Alzheimer disease and nutrition focusing on epigenetics. Adv Nutr 7(5):917–927

Aune D, Chan DS, Lau R, Vieira R, Greenwood DC, Kampman E et al (2011) Dietary fibre, whole grains, and risk of colorectal cancer: systematic review and dose-response meta-analysis of prospective studies. BMJ 343:d6617

Aune D, Chan D, Greenwood D, Vieira A, Rosenblatt DN, Vieira R et al (2012) Dietary fiber and breast cancer risk: a systematic review and meta-analysis of prospective studies. Ann Oncol 23 (6):1394–1402

Axton ER, Cristobal E, Choi J, Miranda CL, Stevens JF (2018) Metabolomics-driven elucidation of cellular nitrate tolerance reveals ascorbic acid prevents nitroglycerin-induced inactivation of xanthine oxidase. Front Pharmacol 9:1085

Baglia ML, Zheng W, Li H, Yang G, Gao J, Gao YT et al (2016) The association of soy food consumption with the risk of subtype of breast cancers defined by hormone receptor and HER2 status. Int J Cancer 139(4):742–748

Bahramsoltani R, Sodagari HR, Farzaei MH, Abdolghaffari AH, Gooshe M, Rezaei N (2016) The preventive and therapeutic potential of natural polyphenols on influenza. Expert Rev Anti-Infect Ther 14(1):57–80

Bailey CH (2014) Improved meta-analytic methods show no effect of chromium supplements on fasting glucose. Biol Trace Elem Res 157(1):1–8

Baker WL (2017) Treating arrhythmias with adjunctive magnesium: identifying future research directions. Eur Heart J Cardiovasc Pharmacother 3(2):108–117

Banikazemi Z, Haji HA, Mohammadi M, Taheripak G, Iranifar E, Poursadeghiyan M et al (2018) Diet and cancer prevention: dietary compounds, dietary microRNAs, and dietary exosomes. J Cell Biochem 119(1):185–196

Bantle JP, Wylie-Rosett J, Albright AL, Apovian CM, Clark NG, Franz MJ et al (2008) Nutrition recommendations and interventions for diabetes: a position statement of the American Diabetes Association. Diabetes Care 31:S61–S78

Bao MJ, Shen J, Jia YL, Li FF, Ma WJ, Shen HJ et al (2013) Apple polyphenol protects against cigarette smoke-induced acute lung injury. Nutrition 29(1):235–243

Barbesino G (2018) Thyroid function changes in the elderly and their relationship to cardiovascular health: a mini-review. Gerontology 65:1–8

Bayet-Robert M, Kwiatowski F, Leheurteur M, Gachon F, Planchat E, Abrial C et al (2010) Phase I dose escalation trial of docetaxel plus curcumin in patients with advanced and metastatic breast cancer. Cancer Biol Ther 9(1):8–14

Bhattacharya PT, Misra SR, Hussain M (2016) Nutritional aspects of essential trace elements in oral health and disease: an extensive review. Scientifica 2016:1

Birketvedt GS, Shimshi M, Erling T, Florholmen J (2005) Experiences with three different fiber supplements in weight reduction. Med Sci Monit 11(1):PI5–PI8

Bo L, Li J, Tao T, Bai Y, Ye X, Hotchkiss RS et al (2014) Probiotics for preventing ventilator-associated pneumonia. Cochrane Database Syst Rev (10):CD009066

Bonaventura P, Benedetti G, Albarède F, Miossec P (2015) Zinc and its role in immunity and inflammation. Autoimmunity Reviews 14(4):277-285

Borgnakke WS, Ylöstalo PV, Taylor GW, Genco RJ (2013) Effect of periodontal disease on diabetes: systematic review of epidemiologic observational evidence. J Clin Periodontol 40: S135–SS52

Bouilly-Gauthier D, Jeannes C, Maubert Y, Duteil L, Queille-Roussel C, Piccardi N et al (2010) Clinical evidence of benefits of a dietary supplement containing probiotic and carotenoids on ultraviolet-induced skin damage. Br J Dermatol 163(3):536–543

Boyera N, Galey I, Bernard B (1998) Effect of vitamin C and its derivatives on collagen synthesis and cross-linking by normal human fibroblasts. Int J Cosmet Sci 20(3):151–158

Briguglio M, Dell'Osso B, Panzica G, Malgaroli A, Banfi G, Zanaboni Dina C et al (2018) Dietary neurotransmitters: a narrative review on current knowledge. Nutrients 10(5):591

Brown L, Rosner B, Willett WW, Sacks FM (1999) Cholesterol-lowering effects of dietary fiber: a meta-analysis. Am J Clin Nutr 69(1):30–42

Buell JS, Dawson-Hughes B (2008) Vitamin D and neurocognitive dysfunction: preventing "D" ecline? Mol Asp Med 29(6):415–422

Calo L, Bianconi L, Colivicchi F, Lamberti F, Loricchio ML, de Ruvo E et al (2005) N-3 fatty acids for the prevention of atrial fibrillation after coronary artery bypass surgery: a randomized, controlled trial. J Am Coll Cardiol 45(10):1723–1728

Carr AC, Maggini S (2017) Vitamin C and immune function. Nutrients 9(11):1211

Cassidy A, Rogers G, Peterson JJ, Dwyer JT, Lin H, Jacques PF (2015) Higher dietary anthocyanin and flavonol intakes are associated with anti-inflammatory effects in a population of US adults. Am J Clin Nutr 102(1):172–181

Celik E, Sanlier N (2019) Effects of nutrient and bioactive food components on Alzheimer's disease and epigenetic. Crit Rev Food Sci Nutr 59:102–113

Chajès V, Torres-Mejía G, Biessy C, Ortega-Olvera C, Angeles-Llerenas A, Ferrari P et al (2012) ω-3 and ω-6 polyunsaturated fatty acid intakes and the risk of breast cancer in Mexican women: impact of obesity status. Cancer Epidemiol Prev Biomarkers 21(2):319–326

Chang JH, Song KJ, Kim HJ, Kim JH, Kim NH, Kim KS (2010) Dietary polyphenols affect MUC5AC expression and ciliary movement in respiratory cells and nasal mucosa. Am J Rhinol Allergy 24(2):e59–e62

Chang CS, Chang CL, Lai GH (2013a) Reactive oxygen species scavenging activities in a chemiluminescence model and neuroprotection in rat pheochromocytoma cells by astaxanthin, beta-carotene, and canthaxanthin. Kaohsiung J Med Sci 29(8):412–421

Chang JS, Wang KC, Yeh CF, Shieh DE, Chiang LC (2013b) Fresh ginger (Zingiber officinale) has anti-viral activity against human respiratory syncytial virus in human respiratory tract cell lines. J Ethnopharmacol 145(1):146–151

Chearskul S, Sangurai S, Nitiyanant W, Kriengsinyos W, Kooptiwut S, Harindhanavudhi T (2007) Glycemic and lipid responses to glucomannan in Thais with type 2 diabetes mellitus. J Med Assoc Thail 90(10):2150–2157

Chen J (2012) An original discovery: selenium deficiency and Keshan disease (an endemic heart disease). Asia Pac J Clin Nutr 21(3):320–326

Chen R, Snyder M (2013) Promise of personalized omics to precision medicine. Wiley Interdiscip Rev Syst Biol Med 5(1):73–82

Chen H, Karne RJ, Hall G, Campia U, Panza JA, Cannon RO III et al (2006) High-dose oral vitamin C partially replenishes vitamin C levels in patients with Type 2 diabetes and low vitamin C levels but does not improve endothelial dysfunction or insulin resistance. Am J Phys Heart Circ Phys 290(1):H137–HH45

Chen YS, Jan RL, Lin YL, Chen HH, Wang JY (2010) Randomized placebo-controlled trial of lactobacillus on asthmatic children with allergic rhinitis. Pediatr Pulmonol 45(11):1111–1120

Chen Y, Liu W, Xue J, Yang J, Chen X, Shao Y et al (2014) Angiotensin-converting enzyme inhibitory activity of lactobacillus helveticus strains from traditional fermented dairy foods and antihypertensive effect of fermented milk of strain H9. J Dairy Sci 97(11):6680–6692

Chen P, Miah MR, Aschner M (2016) Metals and neurodegeneration. F1000Res 5:366

Chui D, Chen Z, Yu J, Zhang H, Wang W, Song Y et al (2011) Magnesium in Alzheimer's disease. In: Magnesium in the central nervous system [Internet]. University of Adelaide Press, Adelaide

Cinque B, Palumbo P, La Torre C, Melchiorre E, Corridoni D, Miconi G et al (2017) Probiotics in aging skin. In: Textbook of aging skin. Springer, Berlin, pp 1315–1327

Ciz M, Denev P, Kratchanova M, Vasicek O, Ambrozova G, Lojek A (2012) Flavonoids inhibit the respiratory burst of neutrophils in mammals. Oxidative Med Cell Longev 2012:181295

Coleman HG, Murray LJ, Hicks B, Bhat SK, Kubo A, Corley DA et al (2013) Dietary fiber and the risk of precancerous lesions and cancer of the esophagus: a systematic review and meta-analysis. Nutr Rev 71(7):474–482

Comerford K, Pasin G (2016) Emerging evidence for the importance of dietary protein source on glucoregulatory markers and type 2 diabetes: different effects of dairy, meat, fish, egg, and plant protein foods. Nutrients 8(8):446

Cooney RV, Dai Q, Gao Y-T, Chow W-H, Franke AA, Shu X-O et al (2011) Low plasma coenzyme q10 levels and breast cancer risk in chinese women. Cancer Epidemiol Prev Biomarkers 20(6):1124

Creegan R, Hunt W, McManus A, Rainey-Smith SR (2015) Diet, nutrients and metabolism: cogs in the wheel driving Alzheimer's disease pathology? Br J Nutr 113(10):1499–1517

Cruz MR, Tovar AR, del Prado M, Torres N (2005) Molecular mechanisms of action and health benefits of polyunsaturated fatty acids. Rev Investig Clin 57(3):457–472

Cuello-Garcia C, Fiocchi A, Pawankar R, Yepes-Nunez JJ, Morgano GP, Zhang Y et al (2017) Prebiotics for the prevention of allergies: a systematic review and meta-analysis of randomized controlled trials. Clin Exp Allergy 47(11):1468–1477

Cunningham JJ (1998) Micronutrients as nutriceutical interventions in diabetes mellitus. J Am Coll Nutr 17(1):7–10

Dadhania VP, Trivedi PP, Vikram A, Nand Tripathi D (2016) Nutraceuticals against neurodegeneration: a mechanistic insight. Curr Neuropharmacol 14(6):627–640

Daliri EB, Lee BH, Oh DH (2017) Current perspectives on antihypertensive probiotics. Probiotics Antimicrob Proteins 9(2):91–101

Danesi F, Ferguson LR (2017) Could pomegranate juice help in the control of inflammatory diseases? Nutrients 9(9):958

Das L, Bhaumik E, Raychaudhuri U, Chakraborty R (2012) Role of nutraceuticals in human health. J Food Sci Technol 49(2):173–183

De La Vega-Monroy ML, Larrieta E, German M, Baez-Saldana A, Fernandez-Mejia C (2013) Effects of biotin supplementation in the diet on insulin secretion, islet gene expression, glucose homeostasis and beta-cell proportion. J Nutr Biochem 24(1):169–177

Demaison L, Moreau D (2002) Dietary n-3 polyunsaturated fatty acids and coronary heart disease-related mortality: a possible mechanism of action. Cell Mol Life Sci 59(3):463–477

Dey M (2017) Toward a personalized approach in prebiotics research. Nutrients 9(2):92

Dhivya V, Priya LB, Chirayil HT, Sathiskumar S, Huang CY, Padma VV (2017) Piperine modulates isoproterenol induced myocardial ischemia through antioxidant and anti-dyslipidemic effect in male Wistar rats. Biomed Pharmacother 87:705–713

Di Mascio P, Kaiser S, Sies H (1989) Lycopene as the most efficient biological carotenoid singlet oxygen quencher. Arch Biochem Biophys 274(2):532–538

Didari T, Solki S, Mozaffari S, Nikfar S, Abdollahi M (2014) A systematic review of the safety of probiotics. Expert Opin Drug Saf 13(2):227–239

Du T, Shi Y, Xiao S, Li N, Zhao Q, Zhang A et al (2017) Curcumin is a promising inhibitor of genotype 2 porcine reproductive and respiratory syndrome virus infection. BMC Vet Res 13 (1):298

Ejtahed HS, Mohtadi-Nia J, Homayouni-Rad A, Niafar M, Asghari-Jafarabadi M, Mofid V et al (2011) Effect of probiotic yogurt containing lactobacillus acidophilus and bifidobacterium lactis on lipid profile in individuals with type 2 diabetes mellitus. J Dairy Sci 94(7):3288–3294

El-Attar M, Said M, El-Assal G, Sabry NA, Omar E, Ashour L (2009) Serum trace element levels in COPD patient: the relation between trace element supplementation and period of mechanical ventilation in a randomized controlled trial. Respirology 14(8):1180–1187

Emelyanov A, Fedoseev G, Krasnoschekova O, Abulimity A, Trendeleva T, Barnes PJ (2002) Treatment of asthma with lipid extract of New Zealand green-lipped mussel: a randomised clinical trial. Eur Respir J 20(3):596–600

Endo J, Arita M (2016) Cardioprotective mechanism of omega-3 polyunsaturated fatty acids. J Cardiol 67(1):22–27

Esmaeili H, Sharifi M, Esmailidehaj M, Rezvani ME, Hafizibarjin Z (2017) Vasodilatory effect of asafoetida essential oil on rat aorta rings: the role of nitric oxide, prostacyclin, and calcium channels. Phytomedicine 36:88–94

Euba B, Lopez-Lopez N, Rodriguez-Arce I, Fernandez-Calvet A, Barberan M, Caturla N et al (2017) Resveratrol therapeutics combines both antimicrobial and immunomodulatory properties against respiratory infection by nontypeable haemophilus influenzae. Sci Rep 7(1):12860

Fang X, Wang K, Han D, He X, Wei J, Zhao L et al (2016) Dietary magnesium intake and the risk of cardiovascular disease, type 2 diabetes, and all-cause mortality: a dose–response meta-analysis of prospective cohort studies. BMC Med 14(1):210

Fernandez-Mejia C, Lazo-de-la-Vega-Monroy M-L (2011) Biological effects of pharmacological concentrations of biotin. J Evid Based Complement Alternat Med 16(1):40–48

Fonseca Wald ELA, van den Borst B, Gosker HR, Schols A (2014) Dietary fibre and fatty acids in chronic obstructive pulmonary disease risk and progression: a systematic review. Respirology 19(2):176–184

Forsythe P (2014) Probiotics and lung immune responses. Ann Am Thorac Soc 11(Suppl 1):S33–S37

Fryer MJ (1993) Evidence for the photoprotective effects of vitamin E. Photochem Photobiol 58 (2):304–312

Fulop T, Larbi A, Witkowski JM, McElhaney J, Loeb M, Mitnitski A et al (2010) Aging, frailty and age-related diseases. Biogerontology 11(5):547–563

Galisteo M, Duarte J, Zarzuelo A (2008) Effects of dietary fibers on disturbances clustered in the metabolic syndrome. J Nutr Biochem 19(2):71–84

Galtier F (2010) Definition, epidemiology, risk factors. Diabetes Metab 36(6 Pt 2):628–651

Ganesan K, Xu B (2017) Polyphenol-rich lentils and their health promoting effects. Int J Mol Sci 18 (11):2390

Ghosh S, Banerjee S, Sil PC (2015) The beneficial role of curcumin on inflammation, diabetes and neurodegenerative disease: a recent update. Food Chem Toxicol 83:111–124

Gibson GE, Blass JP (2007) Thiamine-dependent processes and treatment strategies in neurodegeneration. Antioxid Redox Signal 9(10):1605–1620

Gibson GE, Sheu K-FR, Blass JP, Baker A, Carlson KC, Harding B et al (1988) Reduced activities of thiamine-dependent enzymes in the brains and peripheral tissues of patients with Alzheimer's disease. Arch Neurol 45(8):836–840

Ginter E, Simko V (2016) Deficiency of vitamin D and vitamin C in the pathogenesis of bronchial asthma. Bratisl Lek Listy 117(6):305–307

Giri S, Banerji A, Lele S, Ananthanarayan L (2017) Effect of addition of enzymatically modified guar gum on glycemic index of selected Indian traditional foods (idli, chapatti). Bioact Carbohydr Diet Fibre 11:1–8

Gomez-Guzman M, Toral M, Romero M, Jimenez R, Galindo P, Sanchez M et al (2015) Antihypertensive effects of probiotics lactobacillus strains in spontaneously hypertensive rats. Mol Nutr Food Res 59(11):2326–2336

Greenwood DC, Threapleton DE, Evans CE, Cleghorn CL, Nykjaer C, Woodhead C et al (2013) Glycemic index, glycemic load, carbohydrates, and type 2 diabetes: systematic review and dose–response meta-analysis of prospective studies. Diabetes Care 36(12):4166–4171

Group A-REDSR (2001) A randomized, placebo-controlled, clinical trial of high-dose supplementation with vitamins C and E, beta carotene, and zinc for age-related macular degeneration and vision loss: AREDS report no. 8. Arch Ophthalmol 119(10):1417

Gunness P, Gidley MJ (2010) Mechanisms underlying the cholesterol-lowering properties of soluble dietary fibre polysaccharides. Food Funct 1(2):149–155

Gupta C, Prakash D, Gupta S (2013) Relationships between bioactive food components and their health benefits. Introduction to functional food science textbook, 1st ed. Create Space Independent Publishing Platform, Scotts Valley, pp 66–85

Hajinezhad M, Hajian S, Saghayei S, Samzadeh-Kermani A, Nabavi R (2016) Comparison the protective effects of L-carnitine and acetyl L-carnitine on blood glucose and lipid peroxidation level in diabetic rats. Horiz Med Sci 22(3):229–235

Halberstam M, Cohen N, Shlimovich P, Rossetti L, Shamoon H (1996) Oral vanadyl sulfate improves insulin sensitivity in NIDDM but not in obese nondiabetic subjects. Diabetes 45 (5):659–666

Halnes I, Baines KJ, Berthon BS, MacDonald-Wicks LK, Gibson PG, Wood LG (2017) Soluble fibre meal challenge reduces airway inflammation and expression of GPR43 and GPR41 in asthma. Nutrients 9(1):57

Han X, Holtzman DM, McKeel DW Jr, Kelley J, Morris JC (2002) Substantial sulfatide deficiency and ceramide elevation in very early Alzheimers disease: potential role in disease pathogenesis. J Neurochem 82(4):809–818

Han S, Xu J, Guo X, Huang M (2018) Curcumin ameliorates severe influenza pneumonia via attenuating lung injury and regulating macrophage cytokines production. Clin Exp Pharmacol Physiol 45(1):84–93

Hanson C, Lyden E, Rennard S, Mannino DM, Rutten EP, Hopkins R et al (2016) The relationship between dietary fiber intake and lung function in the National Health and Nutrition Examination Surveys. Ann Am Thorac Soc 13(5):643–650

Hariri BM, McMahon DB, Chen B, Freund JR, Mansfield CJ, Doghramji LJ et al (2017) Flavones modulate respiratory epithelial innate immunity: anti-inflammatory effects and activation of the T2R14 receptor. J Biol Chem 292(20):8484–8497

Hata Y, Yamamoto M, Ohni M, Nakajima K, Nakamura Y, Takano T (1996) A placebo-controlled study of the effect of sour milk on blood pressure in hypertensive subjects. Am J Clin Nutr 64 (5):767–771

Hazlewood LC, Wood LG, Hansbro PM, Foster PS (2011) Dietary lycopene supplementation suppresses Th2 responses and lung eosinophilia in a mouse model of allergic asthma. J Nutr Biochem 22(1):95–100

Hemila H (2016) Vitamin E administration may decrease the incidence of pneumonia in elderly males. Clin Interv Aging 11:1379–1385

Hemila H (2017) Vitamin C and infections. Nutrients 9(4):339

Hemmerle H, Burger H-J, Below P, Schubert G, Rippel R, Schindler PW et al (1997) Chlorogenic acid and synthetic chlorogenic acid derivatives: novel inhibitors of hepatic glucose-6-phosphate translocase. J Med Chem 40(2):137–145

Higgins JA (2004) Resistant starch: metabolic effects and potential health benefits. J AOAC Int 87 (3):761–768

Houston MC, Harper KJ (2008) Potassium, magnesium, and calcium: their role in both the cause and treatment of hypertension. J Clin Hypertens 10(7):3–11

Hsiao HM, Sapinoro RE, Thatcher TH, Croasdell A, Levy EP, Fulton RA et al (2013) A novel anti-inflammatory and pro-resolving role for resolvin D1 in acute cigarette smoke-induced lung inflammation. PLoS One 8(3):e58258

Hsieh C (2001) Phase I clinical trial of curcumin, a chemopreventive agent, in patients with high-risk or pre-malignant lesions. Anticancer Res 21(2895):e2900

Hsieh Y-L, Yao H-T, Cheng R-S, Chiang M-T (2012) Chitosan reduces plasma adipocytokines and lipid accumulation in liver and adipose tissues and ameliorates insulin resistance in diabetic rats. J Med Food 15(5):453–460

Hulston CJ, Churnside AA, Venables MC (2015) Probiotic supplementation prevents high-fat, overfeeding-induced insulin resistance in human subjects. Br J Nutr 113(4):596–602

Hunter DC, Skinner MA, Wolber FM, Booth CL, Loh JM, Wohlers M et al (2012) Consumption of gold kiwifruit reduces severity and duration of selected upper respiratory tract infection symptoms and increases plasma vitamin C concentration in healthy older adults. Br J Nutr 108(7):1235–1245

Ishida Y, Nakamura F, Kanzato H, Sawada D, Hirata H, Nishimura A et al (2005) Clinical effects of Lactobacillus acidophilus strain L-92 on perennial allergic rhinitis: a double-blind, placebo-controlled study. J Dairy Sci 88(2):527–533

Ismail AI, Hasson H (2008) Fluoride supplements, dental caries and fluorosis: a systematic review. J Am Dent Assoc 139(11):1457–1468

Ivory K, Chambers SJ, Pin C, Prieto E, Arques JL, Nicoletti C (2008) Oral delivery of Lactobacillus casei Shirota modifies allergen-induced immune responses in allergic rhinitis. Clin Exp Allergy 38(8):1282–1289

Jabeen Q, Bashir S, Lyoussi B, Gilani AH (2009) Coriander fruit exhibits gut modulatory, blood pressure lowering and diuretic activities. J Ethnopharmacol 122(1):123–130

Jessica Elizabeth T, Gassara F, Kouassi AP, Brar SK, Belkacemi K (2017) Spice use in food: properties and benefits. Crit Rev Food Sci Nutr 57(6):1078–1088

Jian L (2009) Soy, isoflavones, and prostate cnacer. Mol Nutr Food Res 53(2):217–226

Jiang S, Han J, Li T, Xin Z, Ma Z, Di W et al (2017) Curcumin as a potential protective compound against cardiac diseases. Pharmacol Res 119:373–383

John GK, Mullin GE (2016) The gut microbiome and obesity. Curr Oncol Rep 18(7):45

Johnson JJ, Mukhtar H (2007) Curcumin for chemoprevention of colon cancer. Cancer Lett 255 (2):170–181

Johnston CS, Barkyoumb GM, Schumacher SS (2014) Vitamin C supplementation slightly improves physical activity levels and reduces cold incidence in men with marginal vitamin C status: a randomized controlled trial. Nutrients 6(7):2572–2583

Kaczmarczyk MM, Miller MJ, Freund GG (2012) The health benefits of dietary fiber: beyond the usual suspects of type 2 diabetes mellitus, cardiovascular disease and colon cancer. Metabolism 61(8):1058–1066

Karim T, Muhit M, Khandaker G (2017) Interventions to prevent respiratory diseases - nutrition and the developing world. Paediatr Respir Rev 22:31–37

Katalinic V, Milos M, Kulisic T, Jukic M (2006) Screening of 70 medicinal plant extracts for antioxidant capacity and total phenols. Food Chem 94(4):550–557

Kaushik M, Mozaffarian D, Spiegelman D, Manson JE, Willett WC, Hu FB (2009) Long-chain omega-3 fatty acids, fish intake, and the risk of type 2 diabetes mellitus. Am J Clin Nutr 90 (3):613–620

Kawata A, Murakami Y, Suzuki S, Fujisawa S (2018) Anti-inflammatory activity of beta-carotene, lycopene and tri-n-butylborane, a scavenger of reactive oxygen species. In Vivo 32(2):255–264

Kayaniyil S, Vieth R, Retnakaran R, Knight JA, Qi Y, Gerstein HC et al (2010) Association of vitamin D with insulin resistance and β-cell dysfunction in subjects at risk for type 2 diabetes. Diabetes Care 33(6):1379–1381

Keenan JM, Pins JJ, Frazel C, Moran A, Turnquist L (2002) Oat ingestion reduces systolic and diastolic blood pressure in patients with mild or borderline hypertension: a pilot trial. J Fam Pract 51(4):369

Kekkonen RA, Sysi-Aho M, Seppanen-Laakso T, Julkunen I, Vapaatalo H, Oresic M et al (2008) Effect of probiotic lactobacillus rhamnosus GG intervention on global serum lipidomic profiles in healthy adults. World J Gastroenterol 14(20):3188–3194

Keskek SO, Sagliker Y, Kirim S, Icen YK, Yildirim A (2015) Low serum phosphorus level in massry's phosphate depletion syndrome may be one of the causes of acute heart failure. J Nutr Sci Vitaminol (Tokyo) 61(6):460–464

Khaddaj-Mallat R, Morin C, Rousseau E (2016) Novel n-3 PUFA monoacylglycerides of pharmacological and medicinal interest: anti-inflammatory and anti-proliferative effects. Eur J Pharmacol 792:70–77

Khan N, Mukhtar H (2016) Green tea and cancer prevention. Inflammation, oxidative stress, and cancer. CRC Press, New York, pp 352–363

Khan AM, Shahzad M, Raza Asim MB, Imran M, Shabbir A (2015) Zingiber officinale ameliorates allergic asthma via suppression of Th2-mediated immune response. Pharm Biol 53(3):359–367

Kidd PM (2009) Bioavailability and activity of phytosome complexes from botanical polyphenols: the silymarin, curcumin, green tea, and grape seed extracts. Altern Med Rev 14(3):226–246

Kiessling G, Schneider J, Jahreis G (2002) Long-term consumption of fermented dairy products over 6 months increases HDL cholesterol. Eur J Clin Nutr 56(9):843–849

Kim H, Kwack K, Kim DY, Ji GE (2005) Oral probiotic bacterial administration suppressed allergic responses in an ovalbumin-induced allergy mouse model. FEMS Immunol Med Microbiol 45 (2):259–267

Kishi T, Kishi H, Watanabe T, Folkers K (1976) Bioenergetics in clinical medicine. XI. Studies on coenzyme Q and diabetes mellitus. J Med 7(3–4):307–321

Koc ER, Ilhan A, Aytürk Z, Acar B, Gürler M, Altuntaş A et al (2015) A comparison of hair and serum trace elements in patients with Alzheimer disease and healthy participants. Turk J Med Sci 45(5):1034–1039

Kraft TE, Parisotto D, Schempp C, Efferth T (2009) Fighting cancer with red wine? Molecular mechanisms of resveratrol. Crit Rev Food Sci Nutr 49(9):782–799

Kulczynski B, Gramza-Michalowska A (2016) The importance of selected spices in cardiovascular diseases. Postepy Hig Med Dosw (Online) 70:1131–1141

Lairon D, Arnault N, Bertrais S, Planells R, Clero E, Hercberg S et al (2005) Dietary fiber intake and risk factors for cardiovascular disease in French adults. Am J Clin Nutr 82(6):1185–1194

Lappe JM, Travers-Gustafson D, Davies KM, Recker RR, Heaney RP (2007) Vitamin D and calcium supplementation reduces cancer risk: results of a randomized trial. Am J Clin Nutr 85 (6):1586–1591

Lasa A, Miranda J, Bulló M, Casas R, Salas-Salvadó J, Larretxi I et al (2014) Comparative effect of two Mediterranean diets versus a low-fat diet on glycaemic control in individuals with type 2 diabetes. Eur J Clin Nutr 68(7):767

Lee B-J, Huang Y-C, Chen S-J, Lin P-T (2012) Coenzyme Q10 supplementation reduces oxidative stress and increases antioxidant enzyme activity in patients with coronary artery disease. Nutrition 28(3):250–255

Lee CH, Chan RSM, Wan HYL, Woo YC, Cheung CYY, Fong CHY et al (2018) Dietary intake of anti-oxidant vitamins A, C, and E is inversely associated with adverse cardiovascular outcomes in Chinese-a 22-years population-based prospective study. Nutrients 10(11):1664

Leray C, Wiesel ML, Freund M, Cazenave JP, Gachet C (2001) Long-chain n-3 fatty acids specifically affect rat coagulation factors dependent on vitamin K: relation to peroxidative stress. Arterioscler Thromb Vasc Biol 21(3):459–465

Li X, Yu C, Guo Y, Bian Z, Si J, Yang L et al (2017) Tea consumption and risk of ischaemic heart disease. Heart 103(10):783–789

Liu J, Ames BN (2005) Reducing mitochondrial decay with mitochondrial nutrients to delay and treat cognitive dysfunction, Alzheimer's disease, and Parkinson's disease. Nutr Neurosci 8 (2):67–89

Liu A, Chang J, Lin Y, Shen Z, Bernstein PS (2010) Long-chain and very long-chain polyunsaturated fatty acids in ocular aging and age-related macular degeneration. J Lipid Res 51 (11):3217–3229

Liu YW, Liao TW, Chen YH, Chiang YC, Tsai YC (2014) Oral administration of heat-inactivated Lactobacillus plantarum K37 modulated airway hyperresponsiveness in ovalbumin-sensitized BALB/c mice. PLoS One 9(6):e100105

Liu C, Sun J, Lu Y, Bo Y (2016) Effects of anthocyanin on serum lipids in dyslipidemia patients: a systematic review and meta-analysis. PLoS One 11(9):e0162089

Loef M, von Stillfried N, Walach H (2012) Zinc diet and Alzheimer's disease: a systematic review. Nutr Neurosci 15(5):2–12

Luchsinger JA, Tang M-X, Miller J, Green R, Mayeux R (2008) Higher folate intake is related to lower risk of Alzheimer's disease in the elderly. J Nutr Health Aging 12(9):648–650

Ma T, Tan M-S, Yu J-T, Tan L (2014) Resveratrol as a therapeutic agent for Alzheimer's disease. Biomed Res Int 2014:1

Ma ZG, Yuan YP, Zhang X, Xu SC, Wang SS, Tang QZ (2017) Piperine attenuates pathological cardiac fibrosis via PPAR-gamma/AKT pathways. EBioMedicine 18:179–187

Maares M, Haase H (2016) Zinc and immunity: an essential interrelation. Arch Biochem Biophys 611:58–65

Maciel-Dominguez A, Swan D, Ford D, Hesketh J (2013) Selenium alters miRNA profile in an intestinal cell line: evidence that miR-185 regulates expression of GPX2 and SEPSH2. Mol Nutr Food Res 57(12):2195–2205

Mack DR, Ahrne S, Hyde L, Wei S, Hollingsworth MA (2003) Extracellular MUC3 mucin secretion follows adherence of Lactobacillus strains to intestinal epithelial cells in vitro. Gut 52(6):827

Magalhaes CB, Riva DR, DePaula LJ, Brando-Lima A, Koatz VL, Leal-Cardoso JH et al (2010) In vivo anti-inflammatory action of eugenol on lipopolysaccharide-induced lung injury. J Appl Physiol (1985) 108(4):845–851

Maggini S, Wintergerst ES, Beveridge S, Hornig DH (2007) Selected vitamins and trace elements support immune function by strengthening epithelial barriers and cellular and humoral immune responses. Br J Nutr 98(S1):S29–S35

Mangprayool T, Kupittayanant S, Chudapongse N (2013) Participation of citral in the bronchodilatory effect of ginger oil and possible mechanism of action. Fitoterapia 89:68–73

Manson JE, Bassuk SS, Lee I-M, Cook NR, Albert MA, Gordon D et al (2012) The VITamin D and OmegA-3 TriaL (VITAL): rationale and design of a large randomized controlled trial of vitamin D and marine omega-3 fatty acid supplements for the primary prevention of cancer and cardiovascular disease. Contemp Clin Trials 33(1):159–171

Marinho VC, Higgins JP, Logan S, Sheiham A (2003) Topical fluoride (toothpastes, mouthrinses, gels or varnishes) for preventing dental caries in children and adolescents. Cochrane Database Syst Rev 2003(4):CD002782

Marinovic M, Morandi A, Otton R (2015) Green tea catechins alone or in combination alter functional parameters of human neutrophils via suppressing the activation of TLR-4/NFκB p65 signal pathway. Toxicol In Vitro 29(7):1766–1778

Marsland BJ, Trompette A, Gollwitzer ES (2015) The gut-lung axis in respiratory disease. Ann Am Thorac Soc 12(Suppl 2):S150–S156

Maslowski KM, Mackay CR (2011) Diet, gut microbiota and immune responses. Nat Immunol 12 (1):5–9

Matsuyama W, Mitsuyama H, Watanabe M, Oonakahara KI, Higashimoto I, Osame M et al (2005) Effects of omega-3 polyunsaturated fatty acids on inflammatory markers in COPD (Retraction in: Chest (2008) 134:4 (893)). Chest 128(6):3817–3827

May H, Reader R, Murphy S, Khaw K-T (1995) Self-reported tooth loss and bone mineral density in older men and women. Age Ageing 24(3):217–221

Melo van Lent D, Leermakers ETM, Darweesh SKL, Moreira EM, Tielemans MJ, Muka T et al (2016) The effects of lutein on respiratory health across the life course: a systematic review. Clin Nutr ESPEN 13:e1–e7

Melo van Lent D, Leermakers ET, Hofman A, Stricker BH, Brusselle GG, Franco OH et al (2017) Association between lutein intake and lung function in adults: the Rotterdam Study. Br J Nutr 117(5):720–730

Menaa F, Menaa A, Tréton J (2014) Polyphenols against skin aging. In: Polyphenols in human health and disease. Elsevier, Amsterdam, pp 819–830

Meydani M (2001) Nutrition interventions in aging and age-associated disease. Ann N Y Acad Sci 928(1):226–235

Meydani M (2002) The Boyd Orr lecture: nutrition interventions in aging and age-associated disease. Proc Nutr Soc 61(2):165–171

Meydani SN, Leka LS, Fine BC, Dallal GE, Keusch GT, Singh MF et al (2004) Vitamin E and respiratory tract infections in elderly nursing home residents: a randomized controlled trial. JAMA 292(7):828–836

Mickleborough TD, Lindley MR, Ionescu AA, Fly AD (2006) Protective effect of fish oil supplementation on exercise-induced bronchoconstriction in asthma. Chest 129(1):39–49

Mickleborough TD, Tecklenburg SL, Montgomery GS, Lindley MR (2009) Eicosapentaenoic acid is more effective than docosahexaenoic acid in inhibiting proinflammatory mediator production and transcription from LPS-induced human asthmatic alveolar macrophage cells. Clin Nutr 28 (1):71–77

Miller CC, Tang W, Ziboh VA, Fletcher MP (1991) Dietary supplementation with ethyl ester concentrates of fish oil (n-3) and borage oil (n-6) polyunsaturated fatty acids induces epidermal generation of local putative anti-inflammatory metabolites. J Investig Dermatol 96(1):98–103

Mishra S, Palanivelu K (2008) The effect of curcumin (turmeric) on Alzheimer's disease: an overview. Ann Indian Acad Neurol 11(1):13

Mnafgui K, Hajji R, Derbali F, Gammoudi A, Khabbabi G, Ellefi H et al (2016) Anti-inflammatory, antithrombotic and cardiac remodeling preventive effects of eugenol in isoproterenol-induced myocardial infarction in Wistar rat. Cardiovasc Toxicol 16(4):336–344

Mohan IK, Das U (2001) Prevention of chemically induced diabetes mellitus in experimental animals by polyunsaturated fatty acids. Nutrition 17(2):126–151

Moro K, Nagahashi M, Ramanathan R, Takabe K, Wakai T (2016) Resolvins and omega three polyunsaturated fatty acids: clinical implications in inflammatory diseases and cancer. World J Clin Cases 4(7):155–164

Morris MC, Evans DA, Bienias JL, Scherr PA, Tangney CC, Hebert LE et al (2004) Dietary niacin and the risk of incident Alzheimer's disease and of cognitive decline. J Neurol Neurosurg Psychiatry 75(8):1093–1099

Moskaug JO, Carlsen H, Myhrstad MC, Blomhoff R (2005) Polyphenols and glutathione synthesis regulation. Am J Clin Nutr 81(1 Suppl):277S–283S

Mukherjee S, Date A, Patravale V, Korting HC, Roeder A, Weindl G (2006) Retinoids in the treatment of skin aging: an overview of clinical efficacy and safety. Clin Interv Aging 1(4):327

Narain NR, Sarangarajan R, Gray TM, McCook JP, Jimenez JJ (2018) Methods of treatment of cancer by continuous infusion of coenzyme Q10. Google Patents

Neiva RF, Al-Shammari K, Nociti FH, Soehren S, Wang HL (2005) Effects of vitamin-B complex supplementation on periodontal wound healing. J Periodontol 76(7):1084–1091

Neto CC (2007) Cranberry and blueberry: evidence for protective effects against cancer and vascular diseases. Mol Nutr Food Res 51(6):652–664

Nichols JA, Katiyar SK (2010) Skin photoprotection by natural polyphenols: anti-inflammatory, antioxidant and DNA repair mechanisms. Arch Dermatol Res 302(2):71–83

Nikooyeh B, Neyestani TR, Farvid M, Alavi-Majd H, Houshiarrad A, Kalayi A et al (2011) Daily consumption of vitamin D–or vitamin D+ calcium–fortified yogurt drink improved glycemic control in patients with type 2 diabetes: a randomized clinical trial. Am J Clin Nutr 93 (4):764–771

Nishino A, Yasui H, Maoka T (2016) Reaction of paprika carotenoids, capsanthin and capsorubin, with reactive oxygen species. J Agric Food Chem 64(23):4786–4792

Nishino A, Yasui H, Maoka T (2017) Reaction and scavenging mechanism of beta-carotene and zeaxanthin with reactive oxygen species. J Oleo Sci 66(1):77–84

North CJ, Venter CS, Jerling JC (2009) The effects of dietary fibre on C-reactive protein, an inflammation marker predicting cardiovascular disease. Eur J Clin Nutr 63:921

Obata K, Kojima T, Masaki T, Okabayashi T, Yokota S, Hirakawa S et al (2013) Curcumin prevents replication of respiratory syncytial virus and the epithelial responses to it in human nasal epithelial cells. PLoS One 8(9):e70225

Oboh G, Ademiluyi AO, Agunloye OM, Ademosun AO, Ogunsakin BG (2019) Inhibitory effect of garlic, purple onion, and white onion on key enzymes linked with type 2 diabetes and hypertension. J Diet Suppl 16:105–118

Obrenovich ME, Li Y, Parvathaneni K, Yendluri BB, Palacios H, Leszek J et al (2011) Antioxidants in health, disease and aging. CNS Neurol Disord Drug Targets 10(2):192–207

Oelrich B, Dewell A, Gardner C (2013) Effect of fish oil supplementation on serum triglycerides, LDL cholesterol and LDL subfractions in hypertriglyceridemic adults. Nutr Metab Cardiovasc Dis 23(4):350–357

Oliveira RJ, de Oliveira VN, Deconte SR, Calábria LK, da Silva Moraes A, Espindola FS (2014) Phaseolamin treatment prevents oxidative stress and collagen deposition in the hearts of streptozotocin-induced diabetic rats. Diab Vasc Dis Res 11(2):110–117

Olivieri F, Lazzarini R, Babini L, Prattichizzo F, Rippo MR, Tiano L et al (2013) Anti-inflammatory effect of ubiquinol-10 on young and senescent endothelial cells via miR-146a modulation. Free Radic Biol Med 63:410–420

Onakpoya I, Posadzki P, Ernst E (2014) The efficacy of glucomannan supplementation in overweight and obesity: a systematic review and meta-analysis of randomized clinical trials. J Am Coll Nutr 33(1):70–78

Palozza P, Colangelo M, Simone R, Catalano A, Boninsegna A, Lanza P et al (2010) Lycopene induces cell growth inhibition by altering mevalonate pathway and Ras signaling in cancer cell lines. Carcinogenesis 31(10):1813–1821

Park DY, Ahn YT, Park SH, Huh CS, Yoo SR, Yu R et al (2013a) Supplementation of lactobacillus curvatus HY7601 and lactobacillus plantarum KY1032 in diet-induced obese mice is associated with gut microbial changes and reduction in obesity. PLoS One 8(3):e59470

Park W, Amin AR, Chen ZG, Shin DM (2013b) New perspectives of curcumin in cancer prevention. Cancer Prev Res 6(5):387–400

Patel DK, Desai SN, Gandhi HP, Devkar RV, Ramachandran AV (2012) Cardio protective effect of Coriandrum sativum L. on isoproterenol induced myocardial necrosis in rats. Food Chem Toxicol 50(9):3120–3125

Peixoto-Neves D, Wang Q, Leal-Cardoso JH, Rossoni LV, Jaggar JH (2015) Eugenol dilates mesenteric arteries and reduces systemic BP by activating endothelial cell TRPV4 channels. Br J Pharmacol 172(14):3484–3494
Peric M, Koglin S, Dombrowski Y, Groß K, Bradac E, Büchau A et al (2009) Vitamin D analogs differentially control antimicrobial peptide/"alarmin" expression in psoriasis. PLoS One 4(7): e6340
Petruk G, Del Giudice R, Rigano MM, Monti DM (2018) Antioxidants from plants protect against skin photoaging. Oxidative Med Cell Longev 2018:1
Pigarevskii VE (1977) Current trends in the study of phagocytosis and non-specific resistance. Arkh Patol 39(2):84–94
Podlogar JA, Verspohl EJ (2012) Antiinflammatory effects of ginger and some of its components in human bronchial epithelial (BEAS-2B) cells. Phytother Res 26(3):333–336
Poulose SM, Miller MG, Scott T, Shukitt-Hale B (2017) Nutritional factors affecting adult neurogenesis and cognitive function. Adv Nutr 8(6):804–811
Prakash D, Gupta C (2014a) Chap. 14. Role of antioxidant polyphenols in nutraceuticals and human health. In: Phytochemicals of nutraceutical importance. CABI, Wallingford, p 208
Prakash D, Gupta C (2014b) Chap. 12. Carotenoids: chemistry and health benefits. In: Phytochemicals of nutraceutical importance. Scitus Academics LLC, Valley Cottage, p 181
Prakash D, Kumar N (2011) Cost effective natural antioxidants. In: Nutrients, dietary supplements, and nutriceuticals. Springer, Totowa, pp 163–187
Prakash D, Sharma G (2014) Phytochemicals of nutraceutical importance. CABI, Wallingford
Prashanth L, Kattapagari K, Chitturi R, Baddam V, Prasad L (2015) A review on role of essential trace elements in health and disease. J Dr NTR Univ Health Sci 4(2):75–85
Raposo SE, Fondell E, Strom P, Balter O, Bonn SE, Nyren O et al (2017) Intake of vitamin C, vitamin E, selenium, zinc and polyunsaturated fatty acids and upper respiratory tract infection-a prospective cohort study. Eur J Clin Nutr 71(4):450–457
Rastogi S, Pandey MM, Rawat AKS (2017) Spices: therapeutic potential in cardiovascular health. Curr Pharm Des 23(7):989–998
Rejnmark L, Avenell A, Masud T, Anderson F, Meyer HE, Sanders KM et al (2012) Vitamin D with calcium reduces mortality: patient level pooled analysis of 70,528 patients from eight major vitamin D trails. J Clin Endocrinol Metab 97(8):2670–2681
Riccioni G, Bucciarelli T, Mancini B, Di Ilio C, Della Vecchia R, D'Orazio N (2007) Plasma lycopene and antioxidant vitamins in asthma: the PLAVA study. J Asthma 44(6):429–432
Rideout TC, Harding SV, Jones PJ, Fan MZ (2008) Guar gum and similar soluble fibers in the regulation of cholesterol metabolism: current understandings and future research priorities. Vasc Health Risk Manag 4(5):1023
Rines AK, Ardehali H (2013) Transition metals and mitochondrial metabolism in the heart. J Mol Cell Cardiol 55:50–57
Risérus U, Willett WC, Hu FB (2009) Dietary fats and prevention of type 2 diabetes. Prog Lipid Res 48(1):44–51
Roberts RL, Green J, Lewis B (2009) Lutein and zeaxanthin in eye and skin health. Clin Dermatol 27(2):195–201
Rodriguez-Cruz M, Tovar AR, del Prado M, Torres N (2005) Molecular mechanisms of action and health benefits of polyunsaturated fatty acids. Rev Investig Clin 57(3):457–472
Rosengren A, Dotevall A, Wilhelmsen L, Thelle D, Johansson S (2004) Coffee and incidence of diabetes in Swedish women: a prospective 18-year follow-up study. J Intern Med 255(1):89–95
Round JL, Mazmanian SK (2010) Inducible Foxp3+ regulatory T-cell development by a commensal bacterium of the intestinal microbiota. Proc Natl Acad Sci U S A 107(27):12204–12209
Sadhukhan P, Saha S, Dutta S, Mahalanobish S, Sil PC (2018) Nutraceuticals: an emerging therapeutic approach against the pathogenesis of Alzheimer's disease. Pharmacol Res 129:100–114
Saez-Lara MJ, Robles-Sanchez C, Ruiz-Ojeda FJ, Plaza-Diaz J, Gil A (2016) Effects of probiotics and synbiotics on obesity, insulin resistance syndrome, type 2 diabetes and non-alcoholic fatty liver disease: a review of human clinical trials. Int J Mol Sci 17(6):928

Salehi B, Martorell M, Arbiser JL, Sureda A, Martins N, Maurya PK et al (2018) Antioxidants: positive or negative actors? Biomol Ther 8(4):124
Sanchez M, Darimont C, Panahi S, Drapeau V, Marette A, Taylor VH et al (2017) Effects of a diet-based weight-reducing program with probiotic supplementation on satiety efficiency, eating behaviour traits, and psychosocial behaviours in obese individuals. Nutrients 9(3):284
SanGiovanni J, Chandra S, Chew E, Friberg T, Klein M, Kurinij N et al (2003) Dietary omega-3 long-chain polyunsaturated fatty acids and risk for age-related macular degeneration. Invest Ophthalmol Vis Sci 44(13):2112
Sawai-Kuroda R, Kikuchi S, Shimizu YK, Sasaki Y, Kuroda K, Tanaka T et al (2013) A polyphenol-rich extract from Chaenomeles sinensis (Chinese quince) inhibits influenza A virus infection by preventing primary transcription in vitro. J Ethnopharmacol 146(3):866–872
Scalbert A, Johnson IT, Saltmarsh M (2005a) Polyphenols: antioxidants and beyond. Am J Clin Nutr 81(1 Suppl):215S–217S
Scalbert A, Manach C, Morand C, Rémésy C, Jiménez L (2005b) Dietary polyphenols and the prevention of diseases. Crit Rev Food Sci Nutr 45(4):287–306
Schagen SK, Zampeli VA, Makrantonaki E, Zouboulis CC (2012) Discovering the link between nutrition and skin aging. Dermatoendocrinology 4(3):298–307
Scherr J, Nieman DC, Schuster T, Habermann J, Rank M, Braun S et al (2012) Nonalcoholic beer reduces inflammation and incidence of respiratory tract illness. Med Sci Sports Exerc 44 (1):18–26
Sesso HD, Buring JE, Christen WG, Kurth T, Belanger C, MacFadyen J et al (2008) Vitamins E and C in the prevention of cardiovascular disease in men: the Physicians' Health Study II randomized controlled trial. JAMA 300(18):2123–2133
Shah NJ, Swami OC (2017) Role of probiotics in diabetes: a review of their rationale and efficacy. Diabetes 5:104
Shanmugam MK, Arfuso F, Sng JC, Bishayee A, Kumar AP, Sethi G (2019) Epigenetic effects of curcumin in cancer prevention. In: Epigenetics of cancer prevention. Elsevier, London, pp 107–128
Sharavana G, Joseph GS, Baskaran V (2017) Lutein attenuates oxidative stress markers and ameliorates glucose homeostasis through polyol pathway in heart and kidney of STZ-induced hyperglycemic rat model. Eur J Nutr 56(8):2475–2485
Shareck M, Rousseau MC, Koushik A, Siemiatycki J, Parent ME (2017) Inverse association between dietary intake of selected carotenoids and vitamin C and risk of lung cancer. Front Oncol 7:23
Sharma D, Nkembi AS, Aubry E, Houeijeh A, Butruille L, Houfflin-Debarge V et al (2015) Maternal PUFA omega-3 supplementation prevents neonatal lung injuries induced by hyperoxia in newborn rats. Int J Mol Sci 16(9):22081–22093
Sharma P, McClees S, Afaq F (2017) Pomegranate for prevention and treatment of cancer: an update. Molecules 22(1):177
Shen Q, Pierce J (2015) Supplementation of coenzyme Q10 among patients with type 2 diabetes mellitus. Healthcare. Multidisciplinary Digital Publishing Institute
Sherratt SC, Mason RP (2018) Eicosapentaenoic acid and docosahexaenoic acid have distinct membrane locations and lipid interactions as determined by X-ray diffraction. Chem Phys Lipids 212:73–79
Shewry PR (2009) Wheat. J Exp Bot 60(6):1537–1553
Shokryazdan P, Faseleh Jahromi M, Navidshad B, Liang JB (2017) Effects of prebiotics on immune system and cytokine expression. Med Microbiol Immunol 206(1):1–9
Shudo K, Fukasawa H, Nakagomi M, Yamagata N (2009) Towards retinoid therapy for Alzheimer's disease. Curr Alzheimer Res 6(3):302–311
Somerville VS, Braakhuis AJ, Hopkins WG (2016) Effect of flavonoids on upper respiratory tract infections and immune function: a systematic review and meta-analysis. Adv Nutr 7(3):488–497
Song N, Yang H, Pang W, Qie Z, Lu H, Tan L et al (2014) Mulberry extracts alleviate Aβ25–35-induced injury and change the gene expression profile in PC12 cells. Evid Based Complement Alternat Med 2014:1

Song B, Liu K, Gao Y, Zhao L, Fang H, Li Y et al (2017) Lycopene and risk of cardiovascular diseases: a meta-analysis of observational studies. Mol Nutr Food Res 61(9):1601009

Soutif-Veillon A, Ferland G, Rolland Y, Presse N, Boucher K, Féart C et al (2016) Increased dietary vitamin K intake is associated with less severe subjective memory complaint among older adults. Maturitas 93:131–136

Steffen LM, Jacobs DR Jr, Stevens J, Shahar E, Carithers T, Folsom AR (2003) Associations of whole-grain, refined-grain, and fruit and vegetable consumption with risks of all-cause mortality and incident coronary artery disease and ischemic stroke: the Atherosclerosis Risk in Communities (ARIC) Study. Am J Clin Nutr 78(3):383–390

Stephan LS, Almeida ED, Markoski MM, Garavaglia J, Marcadenti A (2017) Red wine, resveratrol and atrial fibrillation. Nutrients 9(11):1190

Stjärne P, Lundblad L, Änggård A, Lundberg JM (1991) Local capsaicin treatment of the nasal mucosa reduces symptoms in patients with nonallergic nasal hyperreactivity. Am J Rhinol 5 (4):145–151

Story EN, Kopec RE, Schwartz SJ, Harris GK (2010) An update on the health effects of tomato lycopene. Annu Rev Food Sci Technol 1:189–210

Stümpel F, Burcelin R, Jungermann K, Thorens B (2001) Normal kinetics of intestinal glucose absorption in the absence of GLUT2: evidence for a transport pathway requiring glucose phosphorylation and transfer into the endoplasmic reticulum. Proc Natl Acad Sci 98 (20):11330–11335

Su HJ, Chang CH, Chen HL (2013) Effects of vitamin C and E intake on peak expiratory flow rate of asthmatic children exposed to atmospheric particulate matter. Arch Environ Occup Health 68 (2):80–86

Takaya J, Higashino H, Kobayashi Y (2004) Intracellular magnesium and insulin resistance. Magnes Res 17(2):126–136

Tanaka T, Shnimizu M, Moriwaki H (2012) Cancer chemoprevention by carotenoids. Molecules 17 (3):3202–3242

Tanvetyanon T, Bepler G (2008) Beta-carotene in multivitamins and the possible risk of lung cancer among smokers versus former smokers: a meta-analysis and evaluation of national brands. Cancer 113(1):150–157

Ternesten-Hasseus E, Johansson EL, Millqvist E (2015) Cough reduction using capsaicin. Respir Med 109(1):27–37

Thushara RM, Gangadaran S, Solati Z, Moghadasian MH (2016) Cardiovascular benefits of probiotics: a review of experimental and clinical studies. Food Funct 7(2):632–642

Tiesset H, Bernard H, Bartke N, Beermann C, Flachaire E, Desseyn JL et al (2011) (n-3) Long-chain PUFA differentially affect resistance to Pseudomonas aeruginosa infection of male and female cftr-/- mice. J Nutr 141(6):1101–1107

Tsai CC, Lin PP, Hsieh YM, Zhang ZY, Wu HC, Huang CC (2014) Cholesterol-lowering potentials of lactic acid bacteria based on bile-salt hydrolase activity and effect of potent strains on cholesterol metabolism in vitro and in vivo. ScientificWorldJournal 2014:690752

Tundis R, Loizzo M, Bonesi M, Menichini F (2015) Potential role of natural compounds against skin aging. Curr Med Chem 22(12):1515–1538

Uhrbom E, Jacobson L (1984) Calcium and periodontitis: clinical effect of calcium medication. J Clin Periodontol 11(4):230–241

van de Laar FA (2008) Alpha-glucosidase inhibitors in the early treatment of type 2 diabetes. Vasc Health Risk Manag 4(6):1189

Van de Laar FA, Lucassen PL, Akkermans RP, Van de Lisdonk EH, De Grauw WJ (2006) Alpha-glucosidase inhibitors for people with impaired glucose tolerance or impaired fasting blood glucose. Cochrane Database Syst Rev (4):CD005061

Van der Borght K, Köhnke R, Göransson N, Deierborg T, Brundin P, Erlanson-Albertsson C et al (2011) Reduced neurogenesis in the rat hippocampus following high fructose consumption. Regul Pept 167(1):26–30

Vessby B, Aro A, Skarfors E, Berglund L, Salminen I, Lithell H (1994) The risk to develop NIDDM is related to the fatty acid composition of the serum cholesterol esters. Diabetes 43 (11):1353–1357

Vilches-Flores A, Tovar AR, Marin-Hernandez A, Rojas-Ochoa A, Fernandez-Mejia C (2010) Biotin increases glucokinase expression via soluble guanylate cyclase/protein kinase G, adenosine triphosphate production and autocrine action of insulin in pancreatic rat islets. J Nutr Biochem 21(7):606–612

Voicekovska JG, Orlikov GA, Karpov Iu G, Teibe U, Ivanov AD, Baidekalne I et al (2007) External respiration function and quality of life in patients with bronchial asthma in correction of selenium deficiency. Ter Arkh 79(8):38–41

Wagner S, Maier LS, Bers DM (2015) Role of sodium and calcium dysregulation in tachyarrhythmias in sudden cardiac death. Circ Res 116(12):1956–1970

Walrand S, Farges MC, Dehaese O, Cardinault N, Minet-Quinard R, Grolier P et al (2005) In vivo and in vitro evidences that carotenoids could modulate the neutrophil respiratory burst during dietary manipulation. Eur J Nutr 44(2):114–120

Wang M, Ji P, Wang R, Zhao L, Xia Z (2012) TRPV1 agonist capsaicin attenuates lung ischemia-reperfusion injury in rabbits. J Surg Res 173(1):153–160

Weaver CM (2013) Potassium and health. Adv Nutr 4(3):368S–377S

Whelton SP, Hyre AD, Pedersen B, Yi Y, Whelton PK, He J (2005) Effect of dietary fiber intake on blood pressure: a meta-analysis of randomized, controlled clinical trials. J Hypertens 23 (3):475–481

Williams NC, Johnson MA, Shaw DE, Spendlove I, Vulevic J, Sharpe GR et al (2016) A prebiotic galactooligosaccharide mixture reduces severity of hyperpnoea-induced bronchoconstriction and markers of airway inflammation. Br J Nutr 116(5):798–804

Wong R, Howe P (2018) Resveratrol counteracts insulin resistance—potential role of the circulation. Nutrients 10(9):1160

Wood LG, Garg ML, Powell H, Gibson PG (2008) Lycopene-rich treatments modify noneosinophilic airway inflammation in asthma: proof of concept. Free Radic Res 42(1):94–102

Wu D, Meydani SN (2014) Age-associated changes in immune function: impact of vitamin E intervention and the underlying mechanisms. Endocr Metab Immune Disord Drug Targets 14 (4):283–289

Wu HC, Horng CT, Tsai SC, Lee YL, Hsu SC, Tsai YJ et al (2018) Relaxant and vasoprotective effects of ginger extracts on porcine coronary arteries. Int J Mol Med 41(4):2420–2428

Wyss-Coray T (2016) Ageing, neurodegeneration and brain rejuvenation. Nature 539(7628):180

Xiang X, Liu Y, Zhang X, Zhang W, Wang Z (2015) [Effects of biotin on blood glucose regulation in type 2 diabetes rat model]. Wei sheng yan jiu = Journal of hygiene research 44(2):185–189, 95

Xu Z, Zhou J (2013) Zinc and myocardial ischemia/reperfusion injury. Biometals 26(6):863–878

Xu F, Diao R, Liu J, Kang Y, Wang X, Shi L (2015) Curcumin attenuates staphylococcus aureus-induced acute lung injury. Clin Respir J 9(1):87–97

Yaffe K, Lindquist K, Schwartz A, Vitartas C, Vittinghoff E, Satterfield S et al (2011) Advanced glycation end product level, diabetes, and accelerated cognitive aging. Neurology 77:1351

Yang J, Wang H-P, Zhou L, C-F X (2012) Effect of dietary fiber on constipation: a meta analysis. World J Gastroenterol 18(48):7378

Yang ZF, Bai LP, Huang WB, Li XZ, Zhao SS, Zhong NS et al (2014) Comparison of in vitro antiviral activity of tea polyphenols against influenza A and B viruses and structure-activity relationship analysis. Fitoterapia 93:47–53

Yang X, Lv JN, Li H, Jiao B, Zhang QH, Zhang Y et al (2017) Curcumin reduces lung inflammation via Wnt/beta-catenin signaling in mouse model of asthma. J Asthma 54(4):335–340

Yeo SCM, Fenwick PS, Barnes PJ, Lin HS, Donnelly LE (2017) Isorhapontigenin, a bioavailable dietary polyphenol, suppresses airway epithelial cell inflammation through a corticosteroid-independent mechanism. Br J Pharmacol 174(13):2043–2059

Yu N, Su X, Wang Z, Dai B, Kang J (2015) Association of dietary vitamin A and beta-carotene intake with the risk of lung Cancer: a meta-analysis of 19 publications. Nutrients 7 (11):9309–9324

Zdrojewicz Z, Szyca M, Popowicz E, Michalik T, Smieszniak B (2017) Turmeric - not only spice. Pol Merkur Lekarski 42(252):227–230

Zhang Z, Shi L, Pang W, Liu W, Li J, Wang H et al (2016a) Dietary fiber intake regulates intestinal microflora and inhibits ovalbumin-induced allergic airway inflammation in a mouse model. PLoS One 11(2):e0147778

Zhang X, Ding F, Li H, Zhao W, Jing H, Yan Y et al (2016b) Low serum levels of vitamins A, D, and E are associated with recurrent respiratory tract infections in children living in northern China: a case control study. PLoS One 11(12):e0167689

Zhang M, Xie Y, Yan R, Shan H, Tang J, Cai Y et al (2016c) Curcumin ameliorates alveolar epithelial injury in a rat model of chronic obstructive pulmonary disease. Life Sci 164:1–8

Zhu W, Cromie MM, Cai Q, Lv T, Singh K, Gao W (2014) Curcumin and vitamin E protect against adverse effects of benzo[a]pyrene in lung epithelial cells. PLoS One 9(3):e92992

Zigangirova NA, Morgunova EY, Fedina ED, Shevyagina NV, Borovaya TG, Zhukhovitsky VG et al (2017) Lycopene inhibits propagation of chlamydia infection. Scientifica (Cairo) 2017:1478625

Zin WA, Silva AG, Magalhaes CB, Carvalho GM, Riva DR, Lima CC et al (2012) Eugenol attenuates pulmonary damage induced by diesel exhaust particles. J Appl Physiol (1985) 112 (5):911–917

Zuskin E, Kanceljak B, Mustajbegovic J, Godnic-Cvar J, Tonkovic-Lojovic M (1992) Comparative study of respiratory changes in female workers exposed to organic aerosols. Lijec Vjesn 114 (5–8):113–117

Zuskin E, Mustajbegovic J, Schachter EN, Kern J, Ivankovic D, Heimer S (2000) Respiratory function in female workers occupationally exposed to organic dusts in food processing industries. Acta Med Croatica 54(4–5):183–191

Dietary Fiber and Aging

6

Amir Hossein Abdolghaffari, Mohammad Hosein Farzaei, Naser-Aldin Lashgari, Nazanin Momeni Roudsari, Nazgol-Sadat Haddadi, Amit Kumar Singh, Harvesh Kumar Rana, Abhay K. Pandey, and Saeideh Momtaz

Abstract

Currently, healthy aging is a major public concern. In between, nutritional interventions and dietary patterns are supposed to play crucial roles in regulation/delaying the aging process. High intake of dietary fiber is inversely

A. H. Abdolghaffari
Department of Toxicology and Pharmacology, Faculty of Pharmacy, Tehran Medical Sciences, Islamic Azad University, Tehran, Iran

Medicinal Plants Research Center, Institute of Medicinal Plants, ACECR, Tehran, Iran

Toxicology and Diseases Group (TDG), Pharmaceutical Sciences Research Center (PSRC), The Institute of Pharmaceutical Sciences (TIPS), Tehran, Iran

Department of Toxicology and Pharmacology, School of Pharmacy, Tehran University of Medical Sciences, Tehran, Iran

Gastrointestinal Pharmacology Interest Group (GPIG), Universal Scientific Education and Research Network (USERN), Tehran, Iran

M. H. Farzaei
Pharmaceutical Sciences Research Center, Kermanshah University of Medical Sciences, Kermanshah, Iran

Medical Biology Research Center, Kermanshah University of Medical Sciences, Kermanshah, Iran

N.-A. Lashgari · N. M. Roudsari
Department of Toxicology and Pharmacology, Faculty of Pharmacy, Tehran Medical Sciences, Islamic Azad University, Tehran, Iran

N.-S. Haddadi
Brain and Spinal Cord Injury Research Center, Neuroscience Institute, Tehran University of Medical Sciences, Tehran, Iran

Experimental Medicine Research Center, Tehran University of Medical Sciences, Tehran, Iran

A. K. Singh · H. K. Rana · A. K. Pandey
Department of Biochemistry, University of Allahabad, Allahabad, India

S. M. Nabavi et al. (eds.), *Nutrients and Nutraceuticals for Active & Healthy Ageing*,
https://doi.org/10.1007/978-981-15-3552-9_6

associated with reduced risk of several age-related diseases. A bulk of evidence indicated that consumption of dietary fiber is implicated with lower concentrations of inflammatory and oxidative stress markers, resulting in improving health status during aging and extending the life span. Even, it was proven that inadequate amounts of dietary fiber are accelerating the aging process. Significant intake of dietary fiber is an essential parameter of a healthy aging. The current chapter reviews the specific roles of dietary fiber in conjunction with age-associated-gastrointestinal system, age-associated-metabolic status, age-associated cardiovascular diseases, age-associated neurodegenerative diseases, cancer, immune system, and allergic disorders.

Keywords

Dietary fiber · Age-associated gastrointestinal diseases · Age-associated metabolic status · Age-associated cardiovascular diseases · Age-associated neurodegenerative diseases · Cancer · Immune system · Allergic disorders

6.1 Introduction

The fact that dietary, genetic, and pharmacological interventions can postpone and/or ameliorate aging changed the world of preventive/therapeutic medicines for aging and aging-related diseases (Ofori-Asenso et al. 2016). Recent work confirmed the involvement of a number of molecular effectors able to regulate food-type-dependent longevity; the quality and quantity of diet, even the meal frequency and timing, are affirmative in life span and aging (Verma et al. 2018). In between, dietary fibers attribute to healthy diet; and are supposed to have different physiological effect (Jones et al. 2006). Definition of dietary fibers is complex and is evolving through the time (Redgwell and Fischer 2005). Accumulated metagenomic data pointed out that diet, particularly insoluble fibers, has determinative and profound effects on the structure, function, and secretion of factors (i.e., gut microbiota-derived factors) that control multiple inflammatory and metabolic events involved in the aging process; mainly those that are related to the increased risk of developing

S. Momtaz (✉)
Medicinal Plants Research Center, Institute of Medicinal Plants, ACECR, Tehran, Iran

Toxicology and Diseases Group (TDG), Pharmaceutical Sciences Research Center (PSRC), The Institute of Pharmaceutical Sciences (TIPS), Tehran, Iran

Department of Toxicology and Pharmacology, School of Pharmacy, Tehran University of Medical Sciences, Tehran, Iran

Gastrointestinal Pharmacology Interest Group (GPIG), Universal Scientific Education and Research Network (USERN), Tehran, Iran
e-mail: momtaz@imp.ac.ir

neurodegenerative and cardiovascular diseases (CVD), metabolic abnormalities, cancer, and allergic diseases (Clemente et al. 2012). This chapter reflects the role of dietary fibers in health and prevention of several aging-related diseases and disabilities.

6.2 Dietary Fiber

Dietary fiber constitutes our diet for centuries and was reported for their health beneficial effect. However, it is still difficult to define it. Hipsley in 1953 first referred the complex of cellulose, hemicellulose, and lignin as dietary fiber. Later Burkitt was the first who coined the term "dietary fiber" (Dhingra et al. 2012). The widely accepted definition is that "fiber is the analogous carbohydrates or the edible parts of plants that resists against digestion and absorption through small intestine, while finally undergoes complete or partial fermentation in the colon. Dietary fibers include lignin, oligosaccharides, polysaccharides, and associated plant substances" (Dhingra et al. 2012). They have water retention capacity, adsorption property, viscosity, gelling property as well as potential to alter the gut microbiota composition (Fåk et al. 2015). Chemical structures of dietary fibers are complex; comprises indigestible carbohydrates, lignin, and other associated non-digestible fibers of plant/animal or synthetic polymers. These are cellulose, hemicelluloses, β-glucan, gums, mucilage, inulin, pectin, and resistant starch (Mudgil and Barak 2013).

Dietary fibers are divided into insoluble and soluble fibers in the water. The former includes hemicellulose, cellulose, and lignin, while among soluble fibers are gums, mucilage, pectin, and storage polysaccharides. Interestingly, their viscosity and capacity to form gels, or their fermentation capability is beneficial for the gastrointestinal health. Intake of dietary viscous fibers is shown to reduce the blood cholesterol, normalize the blood glucose and insulin resistance. Fibers with slow or incomplete fermentation enhance laxation, thus is beneficial for intestinal disorders such as constipation, diverticulosis, and diverticulitis (Marlett et al. 2002). A large body of evidence support the preventive effects of fibers against certain types of cancers; e.g. colon, rectum, and breast cancers (Vieytes et al. 2019). An adequate fiber-rich diet contains other micronutrients such as vitamins, minerals, unsaturated fatty acids, phenols, and phytoestrogens that ensure the health.

Although fibers can be provided by natural sources or by supplements just in the case of certain therapeutic situations (Marlett et al. 2002). A balanced daily diet contains different mixtures of soluble and insoluble fibers. Cereals are the main source of insoluble fibers and a relatively low quantity of soluble fibers (about 25%). Exception is oat, which is a good source of soluble fibers (about 50%). Legumes contain both types of soluble and insoluble fibers, while fruits and vegetables contain soluble type. Moreover, fruits and vegetables contain less fibers than cereals and legumes (Fatani et al. 2007).

6.3 Nutrition and Aging

Aging is a biological process of gradual physiological and metabolic dysfunction leading to chronic age-related diseases including CVD, neurodegenerative disorders, neoplasia, etc., culminating to death. Along aging, the random damages to the DNA and the regulatory/structural proteins exceed the body's self-repair capabilities, resulting in impaired body function and physiology (Kirkwood 2017). The geriatric population (≥60 years) will rise to two billion by 2050, making up 21% of the world's population; more than the youth population for the first time (Shlisky et al. 2017). The growing old population would challenge the health care systems and impose a large burden on communities, especially those with low healthy life expectancy.

However, only 20–30% of the aging process attributed to heredity factors and 70–80% is secondary to external factors and lifestyle (e.g., environmental, behavioral, and societal factors), influencing the rate of aging through various mechanisms and determines the actual survival rate (Mathers 2015). Oxidative stress and inflammation, genomic instability, epigenetic changes, telomere attrition, mitochondrial dysfunction, cellular senility, stem cell and progenitor cell exhaustion, and altered intercellular communication are major contributors to the rate of aging (López-Otín et al. 2013).

The gap between the ideal and actual survival curves remains large while healthy aging pushes the actual survival curve to the right and decreases the gap (Hategan and Hirsch 2018). Healthy aging is having fine self-perceived health and the absence of premature chronic disease, physical or cognitive disability, or any mental disturbances. Having healthy dietary patterns providing adequate antioxidants, micronutrients, and fibers with controlling energy intake is one of the most significant effects on healthy aging. However, the growing global adaption of the Western lifestyle, lack of dietary fiber are primarily responsible for unhealthy aging (Cosco et al. 2014). Some factors may affect healthy diet in elderly people such as changes in sensation of taste and smell, chewing problems, loss of appetite, limitations in accessing high-quality foods, and inefficiencies in nutrient absorption (Dreher 2018a).

The recommended daily intake of fiber is about 14 g/1000 kcal (approximately 38 g for men and 25 g for women) (Karlsen et al. 2019). In addition, fiber-rich diets have almost lower energy density and higher nutrient quality. Taking enough fibers (particularly from vegetables) is associated with higher consumption of minerals, vitamins, phytoestrogens, and flavonoids, which are beneficial for various health outcomes (Slavin and Lloyd 2012). Low dietary fiber intake especially during midlife and older years increases the risk of obesity, unhealthy aging, and premature mortality (Assmann et al. 2015; Karlsen et al. 2019). Although, higher-fiber intake could facilitate to manage caloric intake, which translated into lower body mass index (BMI) (Gilhooly et al. 2008) and reduced the risk of developing several chronic diseases such as type 2 diabetes, metabolic syndrome, CVD, colorectal cancer, and premature death (Slavin 2005; Galisteo et al. 2008; EFSA Panel on Dietetic Products and Allergies 2010; Yao et al. 2014; Dahl and Stewart 2015; Aleixandre and Miguel 2016; Maćkowiak et al. 2016). Dietary fibers are known to be associated with lower serum levels of inflammatory biomarkers, and inflammation is known to be associated with unhealthy aging (King 2005). A higher consumption of fibers also is capable in altering intestinal microbiota composition and function, which is essential in regulating the immune system and might affect the aging process (Keenan et al. 2015; O'Toole and Jeffery 2015; Milani et al. 2016).

Other mechanism represented for the role of fibers in healthy aging is that the adequate fiber intake reduces epigenetic DNA methylation and slows the telomere shortening rate (Cassidy et al. 2010; O'Callaghan et al. 2012; de Mello et al. 2014; Triff et al. 2015). A cross-sectional analysis of the Nurses' Health Study involving 2284 women with the mean age of 59 years, have shown that dietary fiber is positively associated with leukocyte telomere length with the difference of 0.19 units between two extremes of fiber intake. Each 10 g increase in dietary fiber leads to 67 base pairs longer telomeres and a biologic aging difference of 4.3 years, after multivariate adjustment (Cassidy et al. 2010). Adequate fiber intake is also associated with successful aging and reduced risk of weakness, while low fiber consumption linked to increased risk of frailty in the elderly (Claesson et al. 2012; Wang et al. 2015; Gopinath et al. 2016; Jeffery et al. 2016). The Australian Blue Mountains Eye Study that included 1609 healthy individuals older than 49 years have declared that highest vs. lowest intake of dietary fibers has 79% increased multivariate-adjusted odds of aging (Gopinath et al. 2016).

Many studies suggested the adequate dietary fiber as an independent factor for reducing the overall mortality risks. A meta-analysis covering 25 cohort studies on 1,752,848 middle-aged subjects with an average of 12.4 years follow-up have stated that the mortality rate reduces, as the fiber consumption increases, 17% reduction for cancer, 23% for CVD, and 23% for all-cause mortality (Liu et al. 2015a). Several meta-analyses have estimated that increased consumption of fibers of 10 g/day significantly lowers all-cause mortality by 10–11%, coronary heart disease (CHD) by 11–20%, CVD by 9%, and cancer by 6–9% (Kim and Je 2014, 2016; Liu et al. 2015a; Yang et al. 2015b; Hajishafiee et al. 2016). Five prospective studies observed a significant direct association of fiber intake (25 and 35 g/day) with lower mortality risk (Streppel et al. 2008; Park et al. 2011; Chuang et al. 2012; Buil-Cosiales et al. 2014; Chan and Lee 2016). An European Prospective Investigation into Cancer and Nutrition (EPIC) prospective study on 452,717 participants with mean age of 51 years consistently found a 10% decrease in total mortality risk with each 10 g fiber/day (Chuang et al. 2012). A meta-analyses of 298 prospective observational studies have recently concluded that a higher consumption of dietary fiber is negatively associated with likelihood of early death and CVD (Veronese et al. 2018).

The Nurses' Health Study on 72,113 women, averagely 50 years old, have demonstrated that the adequate fiber intake significantly lowered all-cause mortality risk by 17% over 18 years of follow-up, while Western diet with very low fiber content significantly rises all-cause mortality risk by 21% (Heidemann et al. 2008). The large United State (US) National Institutes of Health (NIH)-AARP Diet and Health Study have reported that increased daily fiber consumption by 15 g significantly reduces 22% of all-cause mortality in both genders and CVD related mortality in men by 24% and in women by 34% (Park et al. 2011). Moreover, a cost-of-illness analysis observed that each 1 g increase of fiber consumption per day leads to an annual saving of $ 2.6–51.1 million and $ 4.6–92.1 million in type 2 diabetes and CVD care costs, respectively (Abdullah et al. 2015). The association between fiber consumption and all-cause mortality risk is reported to be attenuated with increasing age, such that, each additional 10 g dietary fiber consumption at age $\geq$ 80 years reduced the mortality risk by 1%, whereas at age 50 years the risk decrement was 29% (Streppel et al. 2008).

American Heart Association dietary guidelines recommended diets rich in fruits, vegetables, nuts, whole grains, and legumes for preventing the CVD, as the major cause of mortality in aged people. Several analyses have shown that each 7 g intake of fiber per day is associated with a significant 9% lower risk of CVD (Threapleton et al. 2013a, b). A meta-analysis of prospective studies reported that each 10 g/day increase in dietary fiber causes 8% reduction in risk of coronary events and the risk of CHD death by 24% (Wu et al. 2015). Meta-analyses of prospective studies have also demonstrated an inverse dose-response association between the fiber consumption and the stroke risk. A meta-analysis has suggested a 12% decline in stroke risk with each 10 g fiber consumption per day (Chen et al. 2013). Another meta-analysis has shown that the risk of stroke reduced by 17% for the highest compared with lowest fiber consumers (Zhang et al. 2013a). Interestingly, increased fiber intake reduces blood lipids (total cholesterol by 8.9 mg/dL, HDL cholesterol by 1.2 mg/dL, and LDL cholesterol by 5.4 mg/dL), diastolic blood pressure by 1.9 mmHg and HbA1c levels (Hartley et al. 2016). High fiber consumption is more effective in reducing blood pressure among older (>40 years) people than in younger ones (Streppel et al. 2005). The results of meta-analysis studies showed a nonlinear inverse association between the fiber intake and diabetes risk, and a linear inverse effect with consumption of $\geq$25 g fiber/day (Yao et al. 2014). Several large clinical trials also supported such negative association among high fiber consumption and diabetes risk. It has been shown that daily intake of 30 g/day of resistant starch fiber significantly improves insulin sensitivity by 16% in pre-diabetes women (Gower et al. 2016). A systematic review of 43 observational studies and randomized trials have also found a strong link between fiber-rich foods against weight gain and metabolic syndrome (Fogelholm et al. 2012). Increase of the fiber intake to $\geq$30 g/day is similar to the energy-restricted diet in losing weight, such regimen is more effective in weight loss compared with the lower-fiber diets of <20 g/day (Lindström et al. 2006; Parikh et al. 2012; Mollard et al. 2014; Ma et al. 2015).

Furthermore, the soluble fiber consumption is the most considerable contribution to reduced age-related cognitive impairment. A double-blind, placebo controlled, crossover randomized clinical trial (RCT) have reported that participants who used oligofructose-enriched inulin showed significant improvement in subjective well-being, mood, and cognitive scores compared with those consumed maltodextrin

(Smith et al. 2015). In addition, there is data on inverse association between the fiber intake and cancer death. Inadequate consumption of fibers could increase the incidence of various cancers such as colorectal, digestive, prostate, breast, and renal cancers. Previous meta-analysis studies have estimated that each 10 g/day increase of fiber consumption significantly reduces the risk of colorectal cancer by 13% (Murphy et al. 2012; Navarro et al. 2016), breast cancer by 5–7% (Dong et al. 2011; Aune et al. 2012), gastric cancer by 44% (Zhang et al. 2013a, b), and esophagus cancer by 31% (Sun et al. 2017). The adequate intake of fiber has also been associated with 11% decrease in risk of prostate cancer and 16% decline in renal cell carcinoma compared with a low-fiber diet (Huang et al. 2014; Sheng et al. 2015).

It was shown that fiber-rich diets prevented the periodontal diseases affecting up to 50% of the adult population, particularly elderlies with diabetes or systemic inflammation (Kaye et al. 2010; Schwartz et al. 2012; Kondo et al. 2014). The frequency of constipation also increases along aging, mainly due to the common aging-related conditions including diabetes, low physical activity, depression, and use of multiple medications (Gallegos-Orozco et al. 2012). Increasing the fiber intake to ≥25 g/day is recommended for promoting bowel regularity and function in elderly. In addition, fiber-related decrease in bowel diverticular diseases was attributed to improved digestive health such as better bowel movement, microbiota ecosystem, and the body weight regulation (Dreher 2018a, b).

Colonic microbiota does not obey the age-related physiological deterioration and it is regularly renewed with daily food consumption and the bowel movements (Brussow 2013; O'Toole and Jeffery 2015; Zapata and Quagliarello 2015). Fiber-rich healthy diet is the major contributor of preserving a healthy microbiota, which consequently promotes a healthier aging and lowers the risk of chronic diseases, frailty, and finally premature mortality (Mariat et al. 2009; Claesson et al. 2012; Janssen and Kersten 2015; Mazidi et al. 2016). Fiber is the major source of microbiota-accessible carbohydrates, which supports the healthier microbiota diversity and reduces the pathobionts (Milani et al. 2016). The healthy colon microbiota ecosystem promotes healthier function of the immune system, increases colonic pH, bowel transit time, and stool bulk to dilute harmful and carcinogenic compounds. The critical role of colon microbiota is to protect the colon from infections, inflammatory bowel disease (IBD), and colorectal cancer, as well as delaying the aging process (Nagpal et al. 2016).

Chronic disease, disability, and weakness are common in elderly population, thus contributors of healthy aging are the focus of recent investigations. Aging process is not entirely genetic-dependent, it is also modified by diet and lifestyle. High fiber containing foods enriched with minerals, vitamins, phytoestrogens, and flavonoids are the main determinant of healthy aging. Sufficient fiber intake (14 g/1000 kcal) is associated with decreased risk of obesity, type 2 diabetes and metabolic syndrome, healthier microbiota dysbiosis and colon movements, CVD, stroke, certain cancers, frailty, and all-cause mortality. Additionally, dietary fibers lower the concentrations of inflammatory and oxidative stress, thereby slowing the rate of aging, predominantly by reducing the rate of telomerase length (Zapata and Quagliarello 2015).

6.4 The Role of Dietary Fibers in Health and Prevention of Diseases

Nowadays, low dietary fibers are a major public health concern with high popularity of Western dietary pattern, in which fiber is a shortfall nutrient. It is evident that there is a strong link between the inadequate fiber intake and adverse health outcomes. While adequate fiber consumption provides healthier life and aging, and lowers the mortality risks by regulating metabolic status, body weight, colon microbiota, reducing CVD, colorectal cancer risk, and a number of other chronic diseases (Assmann et al. 2015).

6.4.1 Gastrointestinal System and Health

6.4.1.1 Colonic Microbiota

Today, there is no doubt about the important role of dietary fibers in good digestive health, which is mediating through accelerating the intestinal transit, and the preservation of a healthy colonic microbiota, as a result foods are fermented to bioactive short-chain fatty acids (SCFAs) such as butyrate, acetate, and propionate. SCFAs are the main energy source for the colon epithelial cells with anti-inflammatory properties that regulate the host metabolism and affect the cellular differentiation processes (Janssen and Kersten 2015; Deehan and Walter 2016; Milani et al. 2016).

The composition, diversity, and the activity of the gut microbiota and SCFAs productions are mainly influenced by heredity, lifestyle, diseases, antibiotic use, and most importantly by diet and its fiber content. High amount of fiber reduces the colonic pH. Fibers are known to be the major source of microbiota-accessible carbohydrates, which increase the healthy symbiotic bacteria diversity and decrease the pathogenic bacteria (Conlon and Bird 2014; Sonnenburg and Sonnenburg 2014; Milani et al. 2016). Fiber-rich intake vs. low fiber Western diet boosts healthy colonic microbiota. The colonic microbiota is the "symbiotic human organ" functioning in other processes such as absorption of nutrients, synthesis of vitamins, bile acid transformations, barrier effects against pathogens and lipopolysaccharide (LPS) absorption, immune system regulation, and immunoglobulin A and T-cell homeostasis, insulin sensitivity, release of satiety hormones, and metabolic risk factors, colorectal cancer initiation, progression and unhealthy aging (Parekh et al. 2015; Boulange et al. 2016; Milani et al. 2016).

6.4.1.2 Intestinal Transit Time and Constipation

Dietary fibers function as a potent laxative through their hydrophobicity and fermentation capacities, which alters the osmotic balance, regulates the gut microbiota, increases fecal bulk, and softens the stool. The increased amount of dietary fiber increases the number of *Bifidobacterium* and *Lactobacillus* species, which is associated with reduced constipation. In an RCT, it was shown that participants using low fiber diet including wheat bran and psyllium fibers, had shorter transit time and higher healthier stool weights. The constipation increased in the low-fiber diet group five times more than the high-fiber diet group, particularly bran and psyllium-rich diets (Pokusaeva et al. 2011; Gänzle and Follador 2012; Rivière et al. 2016). Coarse wheat bran increases the stool volume 21/2 times more than the same dose of fine bran, while fine bran is more fermentable, thus is more effective in increasing prebiotics and SCFAs. The US constipation medical costs stated that an increase in fiber intake by 9 g/day from bran (equivalent to one serving of high-fiber breakfast cereal/day) annually would save about a billion dollars in health system (Schmier et al. 2014). Water follows the unabsorbed and undigested carbohydrates of excessive fiber into the large intestine. Therefore, highly fermentable fiber or abruptly increased fiber intake may enhance the gastrointestinal distention and flatulence, cramping and bloating. Gradual enhancement in dietary intake of fibers could minimize or prevent such symptoms (Slavin 2008; Dahl and Stewart 2015).

6.4.1.3 Diverticulitis/Diverticular Disease

A limited number of prospective studies have suggested that there is a link between dietary fiber and red or processed meat with diverticular diseases. A large prospective study examined 46,295 men with mean age of 53 years for 26 years. It was reported that men with low diet healthy scores are more prone to diverticulitis by 55%, while those with higher healthy scores showed a lower risk of diverticulitis by 26–33% (Strate et al. 2017). In another study on 690,075 women with mean age of 60 years, fiber showed significantly decreased risk of diverticular disease, with cereal and fruits fibers having the most significant effects (Crowe et al. 2014). In addition, fruits and vegetable fibers have shown to be the most effective fiber sources, concerning lowering the risk of diverticular disease by 42%. It was also reported that cellulose consumption reduced the risk of diverticular disease by 48%. Interestingly, vegetarian and fiber-rich diets are both associated with a reduced risk of death due to diverticular disease (Crowe et al. 2011).

6.4.1.4 IBD

IBD is a chronic relapsing gastrointestinal tract inflammation that consists of two main phenotypes, Crohn's disease (CD) and ulcerative colitis (UC) (Loddo and Romano 2015). Diet-related microbial dysbiosis is an important risk factor for the development and progression of IBD. Colonic microbiota (fiber fermenters) with anti-inflammatory properties actively convert fibers to the SCFAs, promote gut microbiota diversity and activity that influence the immunological homeostasis, and lower the colonic pH as a defense against pathogens, which are all the mechanisms by which fibers could affect the development and progression of IBD (Chiba et al. 2010, 2015; Zimmer et al. 2012; Halmos and Gibson 2015).

Fiber-rich diet as adjunctive therapy together with the IBD medications seem to have a role in the long-term improvement of clinical symptoms and reducing the colonic inflammation (Copaci et al. 2000). In 2015, a meta-analysis study reported that fiber consumption lowers the mean risk for UC by 20% and for CD by 56% (Liu et al. 2015b). A 2017 meta-analysis showed that each 10 g/day addition in fiber consumption reduces CD risk by 15% (Zeng et al. 2017). Prospective studies and intervention trials have also proposed a link between dietary fiber intake and IBD.

The Nurses' Health Study has shown that a long-term consumption of fiber lowers the risk of CD flare-ups by 41%, while does not have any association with UC (Ananthakrishnan et al. 2013). However, diets with high-sugar and soft drinks containing low levels of vegetables reported to increase the risk of UC by 68% (Racine et al. 2016). A systemic review suggested fiber supplements such as psyllium as adjunctive therapies along with the IBD medications (Copaci et al. 2000). Previous interventional trials have shown that germinated barley-enriched foods, prebiotics, symbiotic, and psyllium are effective in maintenance of remission of UC (Ishikawa et al. 2011; Faghfoori et al. 2014). However, fiber should be restricted temporarily during the active period of IBD (Lane et al. 2017).

6.4.1.5 Irritable Bowel Disease

Irritable bowel syndrome (IBS) is the most common form of functional gastrointestinal disorder, affecting individuals younger than 45 years (Wilkins et al. 2012; Drossman 2016). Fibers could have a potential role on IBS symptoms due to their ability in regulating the gut microbiota, colonic permeability and inflammation, bowel movements, and lowering the colonic pH to defend against pathogens (Tetens 2011; Dahl and Stewart 2015). However, current evidence on the fiber's effectiveness in soothing the IBS symptoms is controversial (Ford et al. 2014). There is also limited data on the effects of different fiber sources on alleviating the IBS symptoms compared with placebo. Insoluble fibers like wheat bran cause minimal relief in the IBS symptoms, while soluble fibers such as psyllium may cause moderate relief (Ford et al. 2014). According to a meta-analysis, fiber consumption caused a 14% decrease in the mean risk of IBS (Moayyedi et al. 2014). A comprehensive systematic review and meta-analysis by analyzing the results of 22 RCTs and 1299 participants have stated that soluble fibers and total fibers could improve the global symptoms of IBS by 49% and 27%, respectively (Nagarajan et al. 2015). Various RCTs have shown that soluble fibers such as psyllium, partially hydrolyzed guar gum (Parisi et al. 2002; Russo et al. 2015; Niv et al. 2016), and pectin (Xu et al. 2015) are protective against IBS presentations. Kiwifruit could also be a safe and beneficial natural laxative for patients with constipation-dominant IBS (Chan et al. 2007, 2010).

6.4.2 Regulation of the Body Weight and Metabolic Status

6.4.2.1 Weight Regulation

Consuming >25 g fiber/day is effective on the body weight maintenance and stables the weight decrease in obese and overweight individuals (EFSA Panel on Dietetic Products and Allergies 2010). Several longitudinal studies also demonstrated that healthy diets and fiber supplementation are associated with lower body fat and BMI in children and adolescents (Davis et al. 2009; Altman et al. 2015). Fibers were found effective in suppressing appetite, improving glycemic control and colonic microbiota, reducing the amount of dietary energy and metabolizable energy, also restoring the IL-22-mediated enterocyte function (Zou et al. 2018). However, fiber supplementation does not displace high-energy foods, thus it would be less effective in promoting satiety and weight loss than healthy diets enriched in $\geq$30 g/fiber/day (Wanders et al. 2011). A systematic review of 32 blinded randomized-controlled trials reported that there was a significant increases in weight loss in groups treated by the fiber supplementation vs. the controls with the effect size of −0.42 (Chew and Brownlee 2018). The 2015 US Dietary Guidelines Advisory Committee reported that healthy fiber-rich diets could slow the weight regain (Dietary Guidelines Advisory Committee 2015a, c). A systematic review of 43 prospective cohort, case-control, randomized trials declared that increased fiber consumption might lead to less weight gain; however, higher-energy diets rich in refined grains are predictive of elevated weight gain (Fogelholm et al. 2012). Observational studies also support the positive effect of dietary fibers on weigh regulation, as fiber-rich diets help people to be slimmer than those consuming low-fiber diets (Slavin 2005; Davis et al. 2006; Lairon 2007; Fogelholm et al. 2012; Kieffer et al. 2016). The 2012 International Study of Macro−/Micronutrients and Blood Pressure (INTERMAP) cross-sectional study included 1794 adult subjects, demonstrated that normal-weight individuals have healthy diet lower in total energy, higher in nutrient-rich foods and fibers compared with overweight individuals (Shay et al. 2012). The Nurses' Health Study on 74,091 women with mean body mass index (BMI) of 25 showed that subjects who had the highest amount of fiber intake gained a mean of 1.5 kg less weight compared with those who have received the least. This effect was independent of age, the body weight at baseline, and the changes in covariate status. In addition, women consuming high amounts of fiber had 49% lesser risk of weight gain than those used the lower quantity (Liu et al. 2003). Consistently, an RCT have shown that fiber-rich diet (>30 g/day) is more effective in improving body composition and decreasing body weight compared with the lower-fiber diets (20 g/day) (Pal et al. 2011). Results from three long-term clinical trials have also shown that high-fiber diets (>30 g fiber/day) promote weight loss, similar to the reduced energy regimens (Esposito et al. 2004; Lindström et al. 2006; Ma et al. 2015). Additionally, 12-weeks RCT found that increased fiber (35 g/day) and lean protein intake (0.8 g/kg/day of individual's ideal body weight) lead to calorie and weight reduction (Zhang et al. 2018). A US longitudinal study on 170 overweight/obese children aged 7–11 years reported that a diet with higher-fiber quality led to a greater reduction of BMI (Altman et al. 2015). Overweight Latino youth who consumed low total fiber (mean of 3 g/1000 kcal) had a 21% gain in visceral adipose tissue volume (Davis et al. 2009).

6.4.2.2 Type 2 Diabetes

In addition to low-glycemic foods, adequate intake of fiber could lower the incidence of diabetes. Reduction of satiety, obesity, and visceral fat; promotion and maintenance of balanced gut microbiota, low systemic inflammation, control of postprandial glucose excursions and protection against insulin resistance, interfering with absorption of dietary protein have been proposed to contribute to the anti-diabetes effects of fiber-rich diets (Dreher 2018c).

The 2015 US Dietary Guidelines Advisory Committee Scientific Report analysis announced that healthy diets reduce the risk of diabetes by 21% in comparison with 44% increased risk by Western-type diets (Dietary Guidelines Advisory Committee 2015a, c). Accordingly, the American Diabetes Association recommends that fiber intake should be increased to 14 g fiber/1000 kcal daily or about >25 g/day in women and > 38 g/day in men with diabetes and the general population (Evert et al. 2014). A dose-response meta-analysis of 17 prospective studies on 488,293 individuals demonstrated that there is a significant nonlinear inverse association between diabetes risk and total fiber intake <25 g fiber/day and a linear decrease (> 25 g fiber/day), especially with cereal, fruits, and insoluble fibers consumption (Yao et al. 2014). The InterAct study meta-analysis of 19 cohort studies that included 617,968 participants found a 9% reduced risk of diabetes with each 10 g fiber/day increment in total fiber, while cereal fiber had the most protective effect, reduces the risk by 25%. Interestingly, there is no significant association between the fruits and vegetables fibers with diabetes risk (InterAct Consortium 2015). Another meta-analysis has emphasized on increased intake of total fiber with lower glycemic-index/load diets to protecting against type 2 diabetes (Bhupathiraju et al. 2014). Moreover, several RCTs consistently support the presence of a positive link between high-fiber diets and supplements in improving the metabolic status, thus reducing diabetes risk. They have showed that healthy lifestyle habits and fiber-rich diets could prevent the transition of a pre-diabetic or metabolic syndrome situation to diabetes, increasing weight loss, or improving glycemic control and lipoprotein profiles (Jenkins et al. 2008; Cugnet-Anceau et al. 2010; Wolfram and Ismail-Beigi 2010; Post et al. 2012; Silva et al. 2013; Gibb et al. 2015; Weinhold et al. 2015). Interestingly, 10 g β-glucan has been shown to be effective in delaying glucose response and lowering insulin levels in obese pre-diabetic women (Kim et al. 2009). Large prospective studies confirmed a positive correlation between high dietary fiber intake and enhanced insulin resistance, besides a 20–30% reduced risk of developing type 2 diabetes. The EPIC-InterAct Study on 26,088 participants showed that consuming 26.5 g fiber/day vs. 19 g fiber/day lowers the risk of diabetes by 18% (InterAct Consortium 2015).

Notably, the association between cereal fiber consumption and reduced diabetes risk remains significant, particularly after corrections for confounding factors, such as changes in the body, age, weight, exercise, smoking, alcohol intake, fat intake, or a family history of diabetes (Hu et al. 2001). No obvious mechanism for the beneficial effects of cereal fibers has been described though intake of purified insoluble fibers, which are the prevailing fraction of the cereal fibers, might improve the insulin sensitivity (Weickert et al. 2005). Intake of insoluble dietary fibers according to the recommended daily fiber consumption (Marlett et al. 2002) is found to improve the early insulin response, which is associated with an earlier elevation of the postprandial amounts of the incretins such as glucose-dependent insulinotropic polypeptide (GIP), whereas GLP-1 is spared. The effect of insoluble fibers on incretins level would not be the same as those produced by the soluble dietary fibers that are found to cause an increase in GIP and GLP-1 concentrations (Cani et al. 2005). These short-term effects of the cereal fibers are suggested to be independent of the colonic fermentation.

6.4.2.3 Fasting Blood Lipids

The effect of fibers on decreasing the high levels of lipids and lipoproteins depends on the fiber's properties with high viscosity being the major factor (Vuksan et al. 2011; McRorie 2015). The EPIC Norfolk cohort included 22,915 participants; with the mean age of 58 years; and mean BMI of 26 revealed an inverse association between the fiber consumption with *total cholesterol* (TC), low-density lipoprotein (LDL-C), and triglyceride (TG), and a positively association with low-density lipoprotein (HDL-C), regardless of the genetic profile (Wu et al. 2007). The Cochrane systematic review of 17 RCTs with 1067 participants found that there is a significant promising effect of increased fiber supplementation on reducing the TC by 8.9 mg/dL, LDL-C levels by 5.4 mg/dL, and HDL-C by 1.2 mg/dL, however, TG levels remained unchanged (Loveman et al. 2016). Diverse meta-analyses consistently highlighted the beneficial effects of the viscous, soluble fibers, psyllium, and β-glucans in supplements on lowering the TC and LDL-C, especially in adults with elevated blood lipids (Wei et al. 2009; Vuksan et al. 2011; Thies et al. 2014). Large RCTs demonstrated that high fiber diets (14 g fiber/1000 kcals) are associated with a significant reduction of lipid levels compared with lower-fiber diets. Mean intake of 5.5 β-glucan g/day results in 10 mg/dL and 8–11.5 mg/dL of decrease in the TC and LDL-C levels, respectively (Whitehead et al. 2014; Zhu et al. 2015). Oat bran β-glucan is also beneficial in lowering the TC and LDL-C levels in patients without hypercholesterolemia (Chen et al. 2006). A Canadian crossover RCT on 30 hyperlipidemic adults exhibited that consumption of cereals with 3 g high molecular weight β-glucan/day elevated *Bacteroides* and *Prevotella*, while reducing pathogenic *Dorea*, which are believed to reduce the TG levels, waist circumference, BMI, and blood pressure (Martinez et al. 2013). The mean intake of 10 g psyllium/day significantly lowers the TC by 14.5 mg/dL and LDL-C by 10.5 mg/dL (Wei et al. 2009). In an RCT, it was demonstrated that higher-viscosity soluble fibers are more effective to reduce LDL-C than psyllium at half the dose (Vuksan et al. 2011). Additionally, apple pectin and citrus at 6 g/day reduced the LDL-C level (approximately 7%) in mildly hypercholesterolemic adults (Brouns et al. 2012). While, rye whole grains or insoluble fiber-rich wheat does not significantly decrease the blood lipids more than diets rich in refined-grain (Ross et al. 2011; Ma et al. 2015). The mechanisms involved in the hypercholesterolemic effects of dietary fibers are unknown. Promotion of satiety and moderate or low bile acid binding capacities underlie the cholesterol-lowering properties of the insoluble fibers (van Bennekum et al. 2005). Soluble fibers increase the bile acid loss which results in reducing the intestinal absorption of lipids (Jenkins et al. 2000). Additionally, the physicochemical properties of soluble fibers in the intestine significantly affect the hepatic cholesterol metabolism and catabolism of lipoproteins. The main outcome of the fibers actions is lowering the hepatic cholesterol pool, accordingly, cholesterol diverted to the bile acid synthesis, resulting in lower cholesterol delivery to the liver by chylomicron remnants. Concerning the hypotriglyceridemia effects of the soluble fiber such as psyllium, although the actual mechanism is not fully explained, delayed intestinal absorption of triglycerides and sugars might be an explanation (Gao et al. 2015). Fibers with the glycemic index-lowering properties may also play a role in reduction of the plasma lipid levels (Liljeberg and Björck 2000). The hypotriglyceridemic effect of oligofructose is via inhibition of hepatic lipogenesis by modulating synthesis of fatty acid (Kulkarni and Shen 2019), and incretins contributes to regulating the insulin secretion and glucose homeostasis. The serum glucose also regulates the gene expression of fatty acid synthase in the liver and this effect is boosted by insulin (Mandaliya et al. 2018).

6.4.2.4 Metabolic Syndrome

Metabolic syndrome is a growing public health concern, affecting 20–40% of the adult population, characterized by increased abdominal fat, dyslipidemia (increased TG, apolipoprotein B, and reduced HDL cholesterol), high blood pressure, increased glucose levels, insulin resistance, pro-inflammatory and pro-thrombotic levels (Grundy 2008). Metabolic syndrome was implicated to 1.5 to 2-fold increase in risk of CVD mortality, stroke, and all-cause mortality in individuals over 5–10 years (Mottillo et al. 2010). High dietary fiber intake showed promising results on different components of metabolic syndrome. However, the pathogenesis of the disease is intricate, reduction of systemic inflammatory markers through maintaining of microbiota health, regulation of the body weight, and improvement of lipid and glucose homeostasis and insulin sensitivity are contributing to reduced risk of metabolic syndrome by fibers, especially viscous soluble fibers including psyllium (Wei et al. 2018).

Previous investigations showed prolonged consumption of fiber-rich diet leads to attenuation of metabolic syndrome risk, specifically through weight loss, while people suffering from metabolic syndrome often consume lower fiber (Ma et al. 2015). Former RCTs suggested that fiber-rich diets and balanced gut ecosystem reduce the risk of metabolic syndrome; however, observational studies reported varying results for the association between the dietary fibers and the risk of metabolic syndrome (Chen et al. 2017). Two recent meta-analysis (cross-sectional and cohort studies) exhibited an inverse association between dietary fiber consumption and the risk of metabolic syndrome, though their findings are limited by insufficient cohort data (Wei et al. 2018). A recent RCT on 72 overweight and obese non-diabetic participants with at least one further metabolic risk factor reported that high cereal fiber-diet improved the insulin resistance in overweight diabetic individuals after 6 weeks, but not in overweight participants (Weickert et al. 2018). Another RCT on 54 Italian adults with metabolic syndrome found that 29 g cereal fiber/day compared with the control group with 12 g cereal fiber/day significantly increased the fasting plasma propionate and lowered the postprandial insulin levels (Vetrani et al. 2016). Additionally, absorption of colonic propionate was shown to be effective on controlling the body weight of overweight individuals by potentiating the release of the anorectic gut hormones peptide YY and glucagon-like peptide 1 (Cani et al. 2004).

6.4.3 Cardiovascular Diseases

Cardiovascular disease (CVD) is one of the chief causes of mortality and morbidity throughout the world from infancy to old age. According to the data published by the Centre for Disease Prevention and Control that there is an increase in heart failure mediated deaths from 2012 to 2014 among the US citizens, i.e. 84 deaths/100,000. Approximately 17.5 million people died because of CVD in 2012 globally (Ofori-Asenso et al. 2016). Aging has a considerable harmful effect on the cardiac system of the body. It leads to the stage of CVD, which includes arterial blockade, hypertension, stroke, and myocardial infarction. The common symptoms of aging associated CVDs are impaired endothelial function, hypertrophy, increased arterial stiffness, and compromised diastolic and systolic function of left ventricle (Bai et al. 2018). The process of aging also affects the heart rate, which shows the decrease in both variability and heartbeat. The decrease in the number of cells of the sinoatrial node and structural alteration in hearts such as hypertrophy and fibrosis influences heartbeat as well as slows the propagation of electric impulse throughout the heart (Miyamoto 2019). Vascular tissue aging is associated with the dysfunction including ischemia, hypertension, and age-related macular degeneration. Endothelial tissue loss their proliferation and migration capability with the age (Trindade et al. 2017). Moreover, the endothelial barriers become porous which results in the migration of vascular cells to the sub-endothelial spaces and deposit the extracellular matrix proteins that cause thickening of tunica intima. It has been also reported that aging reduces the endothelial nitric oxide synthase (eNOS) activity of endothelial cells, results in a reduction in a number of nitric oxides (NO) which is a vasodilator. Reduction in the abundance of NO promotes senescence of endothelial cells (Bahadoran et al. 2019). As heart age, it undergoes the complex alteration, which affects the cellular composition as marked by the reduction in the number of cardiomyocytes because of increased apoptosis and necrosis. Oxidative stress is reported in old people. Cardiomyocytes are susceptible to the ROS; therefore, the overall increment in ROS production in old age people causes the increased cardiomyocyte death. Biomarkers associated with aging can be used in heart tissue to evaluate how variation of longevity genes affects the rate and level of cardiovascular aging at the cellular level (Singh et al. 2018). Regular intake of dietary fiber is shown to control the risk factors of CVD and hence provide protection against the CVD (Fuhrman et al. 2016).

Healthy diet significantly lowers the CVD risk by 22–59%. Adequate fiber intake is one of the most essential dietary factors in declining the risk of CVD. Wholegrains, fruits and vegetables, legumes and nuts are food sources of fibers contributing to reduced risk of CVD (Dietary Guidelines Advisory Committee 2015a, c). Several meta-analysis of prospective cohort studies have found that each 7–10 g fiber increment/day is associated with a 9–11% reduced CVD risk (Threapleton et al. 2013a, b; Kim and Je 2016). An umbrella review of 18 meta-analyses of 298 observational studies found that higher consumption of dietary fiber is significantly beneficial for CVD and CVD mortality (Veronese et al. 2018). Another umbrella review of 31 meta-analyses concluded that the highest versus the lowest dietary fiber intake is associated with a significant decrease in the relative risk (RR) of CVD and CVD mortality. Attenuating the elevated serum lipids, blood pressure, weight loss, and systemic inflammation are the key mechanism involving in the protective influences of fibers on CVD (McRae 2017).

Dietary fibers also affect the input/output and peripheral resistance of cardiac system either by controlling the sympathetic and parasympathetic nervous systems or by changing the level of systemic regulators (K^+, H^+, CO_2, prostaglandins, thromboxane, and NO) (Sánchez-Muniz 2012). It affects the activity of angiotensin converting enzyme which results in the formation of angiotensin II, a vasoconstrictor that encourages the production of aldosterone, noradrenaline, and vasopressin (Fernández-Vallinas et al. 2016). Dietary fibers make a complex structure with polyphenols, minerals. These minerals assist in modulating the systolic and diastolic blood pressure (Kaulmann et al. 2016). Marine dietary fibers show a similar promising result against the CVD (Wan-Loy and Siew-Moi 2016).

6.4.3.1 Coronary Heart Disease

The role of adequate fiber intake from 14 g/1000 kcals or 25 g/day for women and 38 g/day for men is established to reduce the CHD incidence and its risk factors such as fasting lipid concentration and inflammatory markers (Sánchez-Muniz 2012; Mahalle et al. 2016). A large randomized, double-blind, placebo trial on 22,000 men with the mean age of 57 years has reported that 35 g/day vs. 16 g/day consumption of the cereal and vegetable/fruits fibers reduced the incidence of coronary events by 16% and death by 32% over 6 years of follow-up. The cereal fibers are more effective than the vegetable and fruit sources of fibers (Fuhrman et al. 2016). Meta-analyses of prospective studies observed a reduced risk of all coronary events by 8–11% and CHD deaths by 24% by each 10 g/day increase of fibers (Wu et al. 2015; Kim and Je 2016). Umbrella reviews of meta-analyses have shown that dietary fibers lower the incidence of CHD (McRae 2017; Veronese et al. 2018). The 2015 US Dietary Guideline Advisory Committee Scientific analysis showed that healthy diets including Mediterranean Diet, US Dietary Guidelines pattern, and the Dietary Approaches to Stop Hypertension (DASH) diets could significantly reduce the CHD risk by 29–61%, 24–31%, and 14–27%, respectively (Dietary Guidelines Advisory Committee 2015b). Interestingly, two observational studies by multivariate-adjusted models have shown that those consuming $\geq$35 g fiber/day or high amount of soluble viscous fibers have a significant lower risk of atherosclerosis and carotid wall thickness compared with those consuming <25 g fibers per day (Dreher 2018d). Lowering CRP levels has been suggested as the possible mechanism underlying the observed fiber's effects (Veronese et al. 2018).

6.4.3.2 Hypertension

An adequate fiber intake modestly reduces the blood pressure (BP), especially in hypertensive individuals due to a significant reduction of the risk of central obesity and weight gain, increasing the insulin sensitivity, promoting the healthier LDL-C and HDL-C profiles, lowering systemic inflammation and LDL oxidation, and promoting the healthier microbiota composition, which all support the healthy vascular and endothelial functions (EFSA Panel on Dietetic Products and Allergies 2010; Lattimer and Haub 2010; Sánchez-Muniz 2012; Dahl and Stewart 2015).

Two 2005 meta-analyses of RCTs reported a significant decrease in the mean of diastolic blood pressure (DBP) by 1.3–1.7 mm Hg and non-significant reduction of systolic blood pressure (SBP) mean by 1.14 mm Hg by 11 g/day increase in fiber consumption in all individuals. Even though, such effects were more prominent in hypertensive patients whom showed a reduced SBP by 6 mm Hg and DBP by 4.2 mm Hg; and in trials with a duration of intervention $\geq$8 weeks (Streppel et al. 2005; Whelton et al. 2005). Later, a 2014 systematic review revealed that oat products and oat bran significantly lower the BP (Thies et al. 2014). A 2015 meta-analysis on 18 RCTs including five β-glucan RCTs found that each 6 g/day increase in total fibers is associated with a 0.7 mm Hg and 0.9 mm Hg lower DBP and SBP, respectively. While, β-glucans enriched foods significantly decreased the DBP by 1.5 mm Hg and SBP by 2.9 mm Hg compared with similar wheat-based foods (Evans et al. 2015). A 2016 Cochrane systematic review of 8 RCTs, including 661 adults reported that increased total fiber intake is capable of a significant reduction in mean SBP by 1.9 mm Hg and mean DBP by 1.8 mm Hg (Loveman et al. 2016). A greater effect of soluble—than insoluble—fibers on BP is of more interest. Accordingly, the recent meta-analysis of 22 (1430 participants) for SBP and 21 RCTs (1343 participants) for DBP found that each 8.7 g/day increment in viscous soluble fibers leads to a 1.59 mm Hg and 0.39 mm Hg, lowering of the SBP and DBP, respectively, with psyllium fiber being the most effective fiber source (Khan et al. 2018). Consistently, several observational studies have demonstrated that the elevated fiber intake (by 7–15 g/day above the usual amount) is linked with decreased BP and thus risk of hypertension (Aljuraiban et al. 2015). The US Dietary Guidelines Advisory Committee Scientific analysis also have recommended that healthy dietary patterns, especially the DASH diet, could have a significant BP lowering effect (reducing SBP by 6 mm Hg and DBP by 3 mm Hg) compared with unhealthy Western diet (Dietary Guidelines Advisory Committee 2015a, c).

6.4.3.3 Dietary Fiber and Prothrombotic State

Adults with metabolic syndrome have abnormal prothrombotic conditions, such as high plasma levels of fibrinogen, factor VII, and plasminogen activator inhibitor-1 (PAI-1). The latter is the main inhibitor of endogenous fibrinolysis. These hemostatic changes are associated with the elevated risk of coronary heart disease. In this regard, there are evidence that dietary treatments may be effectual (Giglia et al. 2019), albeit, nearly all the cross-sectional analysis stated that there is a negative correlation between the cereal fiber consumption and the fibrinogen plasma level (Fuhrman et al. 2016).

A randomized clinical trial of 201 subjects reported that 4 weeks consumption of a diet high in cereal fibers had no effect on fibrinogen compared with low diet cereal fibers. Another randomized analysis of the National Heart, Lung, and Blood Institute Family Heart Study, a multicenter population-based study including 883 men and 1116 women, did not show any significant association between the fiber intake and fibrinogen level (Boban et al. 2019). This study indicated that there is an opposite association between PAI-1, age- and energy adjusted indexes of the dietary fiber found for both men and women. Following adjustment for anthropometric, lifestyle, and metabolic factors, two extremes of dietary fiber decreased (ln) PAI-1 values in men. However, this reduction was less pronounced in women group. Similarly, another cross-sectional study on 260 subjects reported an inverse dose–response association between the fiber-rich diet and PAI-1 levels (Palaniswamy et al. 2019).

Results of another study on 11 healthy subjects whose diets were supplemented with 10 g/day oat husk for 2 weeks reported an inverse association between the dietary fiber intake and PAI-1 activity (Siyame 2018). There were no changes in other thrombotic parameters. Intake of fermentable soluble fiber increases the production of SCFAs that prevent the hepatic synthesis of coagulation factors by inhibiting the release of fatty acid (Cheah et al. 2019). The most likely mechanism behind the effect of dietary fibers on PAI-1 synthesis is reducing the fat absorption and the flow of triacylglycerol-rich lipoproteins to the liver. Moreover, SCFAs reduce serum fatty acids (Cheah et al. 2019), leading to increased insulin sensitivity with subsequent decrease of PAI-1 levels, (Cheah et al. 2019).

6.4.4 Brain Function and Neurodegenerative Diseases

Dietary fibers guarantee the brain function and development. The advocative explanation in this matter is mainly pointed to the supportive effect of fibers on maintaining the healthy gut microbial symbiosis. The neuroimmune modulation of microbiota and its influence on the gastrointestinal physiology, liver metabolism, and blood–brain barrier integrity contribute to the pathogenesis of neurobehavioral and neurodegenerative disorders, such as autism spectrum disorder, anxiety disorders, depression, Parkinson's disease, and Alzheimer's disease (Fung et al. 2017). Gut microbiota ecosystem also communicates with central nervous system (CNS) via vagus nerve, parasympathetic motor and sensory fibers that convey information from gastrointestinal tract to the CNS. In addition, fiber-rich foods are high in antioxidants (alpha-tocopherol, ascorbic acid, and βeta-carotene) and flavonoids (polyphenol and anthocyanin), which are all advantageous for the brain function (Farooqui and Farooqui 2018).

It has been revealed that children on a high fiber diet show better multitasking, working memory, and retaining focus vs. children on a lower fiber diet (Khan et al. 2015). Furthermore, probiotics have shown to lower the psychological stress and anxiety in human objects (Messaoudi et al. 2011). Interestingly, there is conflicting data on different microbiome of autistic patients compared with those without autism. It has been hypothesized that abnormal microbiota could be deleterious for autistic children (Finegold et al. 2010; Adams et al. 2011; Kang et al. 2013). Peripheral injections of propionic acid, butyric acid, and acetic acid during the development result in autistic like behavior in rats (El-Ansary et al. 2012).

A limited number of studies have proposed the fibers as protective factors for preventing from cognitive decline (Vercambre et al. 2009; Smith et al. 2015). A large-scale French longitudinal study that monitored cognitive alterations, lifestyle, and dietary habits of 4809 elderly women concluded that soluble fibers are the major contributors of the age-related cognitive impairment (Vercambre et al. 2009). The results of a double-blind, placebo controlled, crossover RCT showed that oligofructose-inulin improved patients well-being, mood, and cognitive scores; especially, recognition memory, and improved the immediate and delayed recall performance on episodic memory tasks (Smith et al. 2015). Various prospective and interventional studies including the Health and Retirement Study, a large US population-based cross-sectional study included 5907 participants, and 2015 US Memory and Aging Project showed a dose dependent association of fiber-rich diets, Mediterranean and MIND (hybrid of Mediterranean diet and DASH) diets, on better cognitive function and the risk of dementia such as Alzheimer's disease (Morris et al. 2015; McEvoy et al. 2017).

6.4.4.1 Brain Stroke

Stroke is the root cause of 11% deaths worldwide and is a major cause of disability (Threapleton et al. 2013a, b). Adequate fiber prevents the stroke by modulating obesity, hypertension, dyslipidemia, type 2 diabetes, and systemic inflammation; the main risk factors associated with stroke. Several meta-analyses of prospective studies confirmed the inverse dose–response association between the fiber intake and the stroke risk (Casiglia et al. 2013; Chen et al. 2013). A 17% reduction of the stroke risk for the highest- vs. the lowest-fiber consumers was shown in a meta-analysis of 21 cohorts including 1,023,131 participants (Zhang et al. 2013a, b). Another meta-analysis of six prospective studies including 314,864 subjects revealed that each 10 g/day fiber could decrease the stroke risk by 12%, of which the cereal fibers being the most effective fiber source (Chen et al. 2013). Another meta-analysis found that 7 g/day total fiber could significantly reduce the risk of stroke by 7%, whereas 4 g/day of soluble fibers are capable of decreasing the stroke risk by 6% (Threapleton et al. 2013a, b). Moreover, a 50% decrease in the stroke incidence by consuming >25 g soluble fibers or >47 g insoluble fibers are reported in population-based study that included 1347 adults (Casiglia et al. 2013). A Swedish study with involvement of 66,677 subjects has reported that multivariate risk of stroke decreases with a higher intake of the fruit fibers by 15%, the vegetable fibers by 10%, and the cereal fibers by 6% (Larsson and Wolk 2014).

6.4.4.2 Chronic Kidney Disease

Chronic kidney disease (CKD) is a considerable global public health issue that increases the rate of mortality by 59% (McCullough et al. 2012). The adequate consumption of fibers could improve CKD by reducing the micro-inflammatory levels and proteolytic fermentation metabolites, and also by lowering the weight gain risk (Evenepoel and Meijers 2012; Dreher 2018e). The increased amount of the fiber intake was shown to reduce the systemic inflammation in CKD patients and lowers the overall mortality (Krishnamurthy et al. 2012; Xu et al. 2014). A National Health and Nutrition Examination Survey (NHANES III) found that in patients with CKD, a 10 g/day increment of total fibers reduced the elevated serum C-reactive protein (CRP) levels by 38% and overall mortality by 17%, while it does not have any significant association with the mortality rate in individuals without CKD (Krishnamurthy et al. 2012). Lowering CRP linked to the reduced incidence and the complication of CKD, and improved the estimated glomerular filtration rate (eGFR) in hypertensive adults (Xu et al. 2014; Shu et al. 2015). Of note, each 10 g/day of fiber consumption is associated with eGFR improvement (adjusted difference, 2.6 mL/min/1.73 m^2). A meta-analysis of 14 RCTs involving 143 CKD patients has shown that the adequate fiber intake decreases the serum creatinine by 22.8 mmol/L and serum urea by 1.8 mmol/L (Jiao et al. 2015). Whole grains, fruits, and vegetables were found to amend the renal function and lessen the metabolic acidosis compared with diets without enough fruits and vegetables (Adeva and Souto 2011).

Patients with CKD have a dysbiotic colonic microbiota, which stimulates several metabolic disturbances, including uremic toxicity, inflammation, and immunosuppression, ultimately leading to the progressive kidney failure and CKD. Therefore, increased dietary fibers could be a nutritional therapeutic option for CKD, by helping to reestablish the microbial symbiosis, neutralizing the bacterial endotoxins, and inhibiting the absorption of colonic-derived uremic toxins (Ramezani and Raj 2014; Salmean et al. 2015). It has been shown that increase of the fiber consumption from 17 g to 27 g/day reduces the circulating p-cresol, colonic-derived uremic toxin, by 20% Salmean et al. 2015).

6.4.5 Cancer

6.4.5.1 Colorectal Cancer

Colorectal cancer (CRC) accounts for almost 10% of all cancers. Its incidence is steadily increased, which strongly attributed to the Western lifestyle (Brenner et al. 2014). There is convincing evidence indicating the protective effects of fiber-rich diets against the risk of CRC. Conversely, reduced dietary fiber intake followed by an increase in incidence of CRC (Brenner et al. 2014). As expected, the Food and Drug Administration (FDA), World Cancer Research Fund (WCRF), American Institute of Cancer Research (AICR) recommend consuming grain products rich in fibers, vegetables, and fruits for decreasing the risk of development or progression of CRC. Adherence to vegetarian and Mediterranean diets significantly lowers the risk of CRC (Orlich et al. 2015; Schwingshackl et al. 2017). The protective effects of fibers on CRC depend on their properties such as lowering the colonic pH to prevent the carcinogens formation; better weight control; promoting healthy gut microbiota; and increasing the production of SCFAs, supporting the normal colonic mucosa; exerting anticarcinogenic and anti-inflammatory activities (Gomes et al. 2019).

A meta-analysis of 16 prospective studies has declared that each 10 g/day intake of total fiber approximately lowers CRC risk by 10% and each 90 g/day of whole grain causes a 17% reduction of cancer risk (Aune et al. 2011). The EPIC cohort study over 11 years follow-up period of 477,312 individuals with mean age of 51 years found a 13% reduction of CRC incidence with 10 g/day rise in fiber intake (Murphy et al. 2012). In case–control studies, each daily 10 g increase of total or cereal fibers is estimated to reduce CRC risk by 9% and 30%, respectively (Ben et al.

2014). Three large US-based randomized controlled trials (2007 US Polyp Prevention Trial, 2002 US Wheat Bran Fiber Trial, and 2000 US Polyp Prevention Trial) did not found any significant association between fiber-rich foods consumption and risk of CRC recurrence (Lanza et al. 2007). However, the US Phoenix Colon Cancer Prevention Physicians' Network including 1429 subjects reported a significant reduction in risk of adenoma recurrence by 12% in the wheat bran-rich vs low-wheat bran fiber diets (Alberts et al. 2000). In addition, a pooled analysis of two large RCTs (Wheat Bran Fiber Trial and 2000 Polyp Prevention Trial) showed that dietary fibers remarkably lowered the risk of recurrent CRC by 19% in men, while no significant association was found for women explaining the disparity in RCT results (Gomes et al. 2019). The US Prostate, Lung, Colorectal, and Ovarian Cancer Screening Trial on 57,774 adults with mean age of 62 years found that consumption of $\geq$12.8 g fiber/1000 kcal is protective against the development of rectal or distal colorectal adenoma, reducing their risk by 24% compared with the intake of <9.9 g fiber/1000 kcal (Kunzmann et al. 2015). Increased total fiber intake did not reduce the risk of colorectal cancer, whereas the association was statistically significant for distal colon cancer risk, interestingly cereal and fruits fiber sources were the most effective in this regard (Ahmed et al. 2006).

6.4.5.2 Breast Cancer

Breast cancer (BC) is the most common cancer ended to death among women (Fitzmaurice et al. 2017). There is controversial data on the link between dietary fibers and the risk of BC. Several mechanisms proposed to underlie the anti-breast cancer effects of fibers including reduction of abdominal fat gain, elevated circulating estrogen levels, C-reactive protein, and insulin resistance (Dieli-Conwright et al. 2016). The 2017 Continuous Update of the World Cancer Research Fund and American Institute for Cancer Research (WCRF and AICR) Continuous Update Project Report found such association to be inconclusive (For 2017). Its older version in 2014 suggested a limited data on inverse association of the fiber-rich foods with risk of breast cancer and all-cause mortality (Research 2016). Pooling seven cohort studies, there was a minor decrease in breast cancer risk of 6% receiving Mediterranean diet (Schwingshackl et al. 2017). Several meta-analyses of prospective studies have shown a mild but significant decrease in breast cancer risk of 5–7% for each 10 g/day augmentation in the fiber intake (Dong et al. 2011; Aune et al. 2012). A meta-analysis of two large RCTs and a cohort study showed that in patients diagnosed with BC, fiber-rich diet significantly lessened the mortality risk by 17% and recurrence risk by 23% (Xing et al. 2014). An EPIC prospective study included 334,849 women with mean age of 50 years reported that BC risk is reduced by 5% when received a diet enriched by total fiber and by 10% for fiber from non-starchy vegetables, with a notable relation; for estrogen- and progesterone-receptors negative tumors. However, fibers from fruits, cereals, or legumes do not have any significant protective effects against BC (Ferrari et al. 2013). The Nurses' Health Study II on 90,534 premenopausal women declared that risk of BC reduced by 25% at higher compared with lower intake of fiber during adolescence and early adulthood (Farvid et al. 2016). Prospective cohort meta-analyses have estimated a lower BC risk by 4–7% for each 10 g/day fiber intake in all women. Meanwhile, 15 g/day fiber intake leads to decreased BC risk by 7–12% in postmenopausal women (Dong et al. 2011; Aune et al. 2012; Chen et al. 2016).

6.4.5.3 Other Cancers

Prostate and gastric cancers may also be affected by the adequate fiber intake (World Cancer Research Fund International/American Institute for Cancer Research Continuous Update Project Report 2014). A meta-analysis of 19 case-control and 2 cohort studies included 580,064 subjects demonstrated that dietary fibers have a negative dose–response association with gastric cancer risk as each 10 g/day increase in fiber results in 44% reduction of gastric cancer (Zhang et al. 2013a, b). The Physicians' Health Study found that fiber enriched foods lowers all-cause mortality, while Western diet contributes to higher prostate cancer-specific and all-cause mortality (Yang et al. 2015a). However, the 2014 WCRF and AICR continuous update report did not confirm the direct protective effect of fiber intake against the risk of prostate cancer (World Cancer Research Fund International/ American Institute for Cancer Research Continuous Update Project Report 2014).

6.4.6 Dietary Fiber and the Immune System

The gastrointestinal (GI) system is the largest immune organ. The lymphoid tissue of GI composed of Peyer's patches and other non-aggregated lymphocytes that all contain 60% of body lymphocytes. Diet composition and prebiotics intake regulate the function of the gut immune system. Fibers especially inulin and other oligofructoses are found to stimulate the growth of health-promoting bacteria in the colon such as *Bifidobacteria* and lactobacilli that generate SCFAs and activate the immune system. Few human studies show that inulin consumption exerts favorable effects on the types and amounts of circulating lymphocytes by increasing the fecal contents of *Bifidobacteria*. These fibers are not digested by GI enzymes and remain intact when reach the colon, where are completely fermented by microbiota resulting in SCFA production and *Bifidobacteria* growth. *Bifidobacteria* is shown to protect from intestinal infection; decrease the intestinal pH to form acids from carbohydrates fermentation; produce different antioxidants and vitamins; activate intestinal function and eases digestion and absorption; prevent constipation; stimulate the immune system; and reduce the colorectal cancer risk (Mohiuddin 2019).

Intake of prebiotic fiber mixture has the following benefits for neonates: enhancing immune development followed by decreased risk of respiratory infections and atopic dermatitis; boosting the bowel function. Prebiotic fibers is shown to alleviate gut inflammation in animal models, thus is suggested to be beneficial to treat IBD. Prebiotic fibers consumption is also found to significantly decrease infection risk in liver transplant patients. Moreover, preliminary human studies have suggested their favorable effects in UC, CD, or pouchitis (Anderson et al. 2009).

6.4.7 Allergic Disorders

Fiber-rich diet and probiotics are able to modulate the microbiota composition, regulate the tolerogenic tone of the mucosal immune system, balance the activities of Th1 and Th2 cells, and affect the bacterial–epithelium-immune interactions, thereby, could protect against the allergic disorders (Daien et al. 2017). There is controversial data on the positive influence of fibers on eczema, hay fever, atopic dermatitis, and food allergy (Devereux 2006; Cochrane et al. 2009). Interestingly, dietary fiber supplementation is associated with the reduced risk of allergic asthma development in mice and recurrent wheeze in human infants, when mother supplemented with fibers (Hogenkamp et al. 2015; Gray et al. 2017). The recent elevation of the prevalence of allergic diseases in developed countries attributed to the Western lifestyle and reduced consumption of dietary fiber (Daien et al. 2017). The byproducts (SCFAs including propionate, acetate, and butyrate) of fiber-fermented products by gut bacteria bind to the G protein-coupled receptors such as GPR109a, GPR41, and GPR43 expressing by the immune and non-immune cells (Macia et al. 2015; McKenzie et al. 2015; Daien et al. 2017). They also inhibited the histone deacetylases (HDACs) activity in immune cells, promoting the activity of the anti-inflammatory cells (Daien et al. 2017). SCFAs support the epithelial integrity by increasing its resistance, boosting mucus and prostaglandin secretions and the epithelium repair ability via direct activation of NLR Family Pyrin Domain Containing 3 (NLRP3) inflammasome, leading to release of IL-18 (Daien et al. 2017).

Moreover, SCFAs alter dendritic cells (DCs) precursor generation in the bone marrow. Subsequently the DCs have less ability to boost Th2 effector cells in the lungs (Trompette et al. 2014). They also increase the ability of DCs to generate tolerogenic regulatory T cells (Treg), and promote the transformation of naïve T to Treg, Th1, and Th17; while inhibiting Th2 differentiation (Daien et al. 2017). Fibers appear to modulate antibody responses; both systemically and locally in the gastro-intestinal tract by increased T follicular helper response, IgA^+ B cells in the small intestine, and promoting the plasma cell differentiation (Kim et al. 2016; Tan et al. 2016). Interestingly, dietary fibers affect leukocyte migration by identification of GPR43, the SCFAs receptor, as a neutrophil chemoattractant (Vinolo et al. 2011). Dietary fibers (especially polysaccharides and oligosaccharides) and SCFAs also suppress the mast cells activation through the inhibition of Mitogen-activated protein kinase (MAPK) signaling pathway (Folkerts et al. 2018). Fiber-rich diets prevent the sensitization of mast cells; either by inhibiting digestion of allergens or by activating the galactin-9 expression, which leads to the blockade of the IgE–antigen complexes formation (Folkerts et al. 2018). Accordingly, the 2015 World Allergy Organization evidence-based guideline panel has recommended the consumption of sufficient amount of prebiotic supplements to prevent the sensitization and development of allergic disorders in non-exclusively breastfed infants, but not in exclusively breastfed infants. However, the panel has not suggested the use of prebiotics during pregnancy or breastfeeding due to the lack of experimental or observational studies on this topic (Simons et al. 2015).

6.5 Conclusion

Many trials showed that dietary interventions cause remarkable positive effects on aging-associated diseases such as reducing the risk of diabetes, promoting the reversion of metabolic syndrome and its complications, decreasing the risk of CVD and the incidence of CVD mortality, ameliorating the neurodegenerative impairments, and soothing the allergic disorders. Recent data from preclinical and clinical studies indicate that dietary fibers play pervasive role in controlling the mechanisms of aging and its associated diseases. More attempts are necessary to design predictive biomarkers that are applicable in randomized clinical trials to examine the efficacy of designated dietary interventions on risk of aging-related neurodegenerative and chronic diseases, and the rate of biological aging. In addition, researches should move towards the discovery of the interactions between the dietary fiber intake, the gut microbiome, and persuasive molecular and cellular effectors in modulating the key mechanisms involved in maintenance of cellular, tissue, and organ function during aging.

References

Abdullah MM, Gyles CL et al (2015) Cost-of-illness analysis reveals potential healthcare savings with reductions in type 2 diabetes and cardiovascular disease following recommended intakes of dietary fiber in Canada. Front Pharmacol 6:167

Adams JB, Johansen LJ et al (2011) Gastrointestinal flora and gastrointestinal status in children with autism--comparisons to typical children and correlation with autism severity. BMC Gastroenterol 11:22

Adeva MM, Souto G (2011) Diet-induced metabolic acidosis. Clin Nutr 30(4):416–421

Ahmed RL, Schmitz KH et al (2006) The metabolic syndrome and risk of incident colorectal cancer. Cancer 107(1):28–36

Alberts DS, Martinez ME et al (2000) Lack of effect of a high-fiber cereal supplement on the recurrence of colorectal adenomas. Phoenix Colon Cancer Prevention Physicians' Network. N Engl J Med 342(16):1156–1162

Aleixandre A, Miguel M (2016) Dietary fiber and blood pressure control. Food Funct 7 (4):1864–1871

Aljuraiban GS, Griep LM et al (2015) Total, insoluble and soluble dietary fibre intake in relation to blood pressure: the INTERMAP Study. Br J Nutr 114(9):1480–1486

Altman M, Holland JC et al (2015) Reduction in food away from home is associated with improved child relative weight and body composition outcomes and this relation is mediated by changes in diet quality. J Acad Nutr Diet 115(9):1400–1407

Ananthakrishnan AN, Khalili H et al (2013) A prospective study of long-term intake of dietary fiber and risk of Crohn's disease and ulcerative colitis. Gastroenterology 145(5):970–977

Anderson JW, Baird P et al (2009) Health benefits of dietary fiber. Nutr Rev 67(4):188–205

Assmann KE, Lassale C et al (2015) A healthy dietary pattern at midlife, combined with a regulated energy intake, is related to increased odds for healthy aging. J Nutr 145(9):2139–2145

Aune D, Chan DS et al (2011) Dietary fibre, whole grains, and risk of colorectal cancer: systematic review and dose-response meta-analysis of prospective studies. BMJ 343:d6617

Aune D, Chan D et al (2012) Dietary fiber and breast cancer risk: a systematic review and meta-analysis of prospective studies. Ann Oncol 23(6):1394–1402

Bahadoran Z, Mirmiran P et al (2019) Circulating markers of nitric oxide homeostasis and cardiometabolic diseases: insights from population-based studies. Free Radic Res 53:359–376

Bai W, Sinclair M et al (2018) Automated cardiovascular magnetic resonance image analysis with fully convolutional networks. J Cardiovasc Magn Reson 20(1):65
Ben Q, Sun Y et al (2014) Dietary fiber intake reduces risk for colorectal adenoma: a meta-analysis. Gastroenterology 146(3):689–699.e686
Bhupathiraju SN, Tobias DK et al (2014) Glycemic index, glycemic load, and risk of type 2 diabetes: results from 3 large US cohorts and an updated meta-analysis. Am J Clin Nutr 100 (1):218–232
Boban M, Bulj N et al (2019) Nutritional considerations of cardiovascular diseases and treatments. Nutr Metab Insights 12:1178638819833705
Boulange CL, Neves AL et al (2016) Impact of the gut microbiota on inflammation, obesity, and metabolic disease. Genome Med 8(1):42
Brenner H, Kloor M et al (2014) Colorectal cancer. Lancet 383(9927):1490–1502
Brouns F, Theuwissen E et al (2012) Cholesterol-lowering properties of different pectin types in mildly hyper-cholesterolemic men and women. Eur J Clin Nutr 66(5):591
Brussow H (2013) Microbiota and healthy ageing: observational and nutritional intervention studies. Microb Biotechnol 6(4):326–334
Buil-Cosiales P, Zazpe I et al (2014) Fiber intake and all-cause mortality in the Prevención con Dieta Mediterránea (PREDIMED) study. Am J Clin Nutr 100(6):1498–1507
Cani PD, Dewever C et al (2004) Inulin-type fructans modulate gastrointestinal peptides involved in appetite regulation (glucagon-like peptide-1 and ghrelin) in rats. Br J Nutr 92(3):521–526
Cani PD, Neyrinck AM et al (2005) Oligofructose promotes satiety in rats fed a high-fat diet: involvement of glucagon-like peptide-1. Obes Res 13(6):1000–1007
Casiglia E, Tikhonoff V et al (2013) High dietary fiber intake prevents stroke at a population level. Clin Nutr 32(5):811–818
Cassidy A, De Vivo I et al (2010) Associations between diet, lifestyle factors, and telomere length in women. Am J Clin Nutr 91(5):1273–1280
Chan C, Lee P (2016) Association between dietary fibre intake with cancer and all-cause mortality among 15740 adults: the N ational H ealth and N utrition E xamination S urvey III. J Hum Nutr Diet 29(5):633–642
Chan AO, Leung G et al (2007) Increasing dietary fiber intake in terms of kiwifruit improves constipation in Chinese patients. World J Gastroenterol 13(35):4771–4775
Chang CC, Lin YT et al (2010) Kiwifruit improves bowel function in patients with irritable bowel syndrome with constipation. Asia Pac J Clin Nutr 19(4):451–457
Cheah MC, McCullough AJ et al (2019) Dietary manipulations for nonalcoholic fatty liver disease (NAFLD). In: Bioactive food as dietary interventions for diabetes. Elsevier, London, pp 69–88
Chen J, He J et al (2006) A randomized controlled trial of dietary fiber intake on serum lipids. Eur J Clin Nutr 60(1):62–68
Chen G, Lv D et al (2013) Dietary fiber intake and stroke risk: a meta-analysis of prospective cohort studies. Eur J Clin Nutr 67(1):96
Chen S, Chen Y et al (2016) Dietary fibre intake and risk of breast cancer: a systematic review and meta-analysis of epidemiological studies. Oncotarget 7(49):80980–80989
Chen J-P, Chen G-C et al (2017) Dietary fiber and metabolic syndrome: a meta-analysis and review of related mechanisms. Nutrients 10(1):24
Chew KY, Brownlee IA (2018) The impact of supplementation with dietary fibers on weight loss: a systematic review of randomised controlled trials. Bioact Carbohydr Diet Fibre 14:9–19
Chiba M, Abe T et al (2010) Lifestyle-related disease in Crohn's disease: relapse prevention by a semi-vegetarian diet. World J Gastroenterol 16(20):2484–2495
Chiba M, Tsuji T et al (2015) High amount of dietary fiber not harmful but favorable for Crohn disease. Perm J 19(1):58
Chuang S-C, Norat T et al (2012) Fiber intake and total and cause-specific mortality in the European Prospective Investigation into Cancer and Nutrition cohort. Am J Clin Nutr 96(1):164–174
Claesson MJ, Jeffery IB et al (2012) Gut microbiota composition correlates with diet and health in the elderly. Nature 488(7410):178–184

Clemente JC, Ursell LK et al (2012) The impact of the gut microbiota on human health: an integrative view. Cell 148(6):1258–1270
Cochrane S, Beyer K et al (2009) Factors influencing the incidence and prevalence of food allergy. Allergy 64(9):1246–1255
Conlon MA, Bird AR (2014) The impact of diet and lifestyle on gut microbiota and human health. Nutrients 7(1):17–44
Copaci I, Chira C et al (2000) Maintenance of remission of ulcerative colitis (UC): mesalamine, dietary fiber, S. boulardi. Dig Liver Dis 32:A24
Cosco TD, Prina AM et al (2014) Operational definitions of successful aging: a systematic review. Int Psychogeriatr 26(3):373–381
Crowe FL, Appleby PN et al (2011) Diet and risk of diverticular disease in Oxford cohort of European Prospective Investigation into Cancer and Nutrition (EPIC): prospective study of British vegetarians and non-vegetarians. BMJ 343:d4131
Crowe FL, Balkwill A et al (2014) Source of dietary fibre and diverticular disease incidence: a prospective study of UK women. Gut 63(9):1450–1456
Cugnet-Anceau C, Nazare J-A et al (2010) A controlled study of consumption of β-glucan-enriched soups for 2 months by type 2 diabetic free-living subjects. Br J Nutr 103(3):422–428
Dahl WJ, Stewart ML (2015) Position of the academy of nutrition and dietetics: health implications of dietary fiber. J Acad Nutr Diet 115(11):1861–1870
Daien CI, Pinget GV et al (2017) Detrimental impact of microbiota-accessible carbohydrate-deprived diet on gut and immune homeostasis: an overview. Front Immunol 8:548
Davis JN, Hodges VA et al (2006) Normal-weight adults consume more fiber and fruit than their age- and height-matched overweight/obese counterparts. J Am Diet Assoc 106(6):833–840
Davis JN, Alexander KE et al (2009) Inverse relation between dietary fiber intake and visceral adiposity in overweight Latino youth. Am J Clin Nutr 90(5):1160–1166
de Mello VDF, Pulkkinen L et al (2014) DNA methylation in obesity and type 2 diabetes. Ann Med 46(3):103–113
Deehan EC, Walter J (2016) The fiber gap and the disappearing gut microbiome: implications for human nutrition. Trends Endocrinol Metab 27(5):239–242
Devereux G (2006) The increase in the prevalence of asthma and allergy: food for thought. Nat Rev Immunol 6(11):869–874
Dhingra D, Michael M et al (2012) Dietary fibre in foods: a review. J Food Sci Technol 49 (3):255–266
Dieli-Conwright CM, Lee K et al (2016) Reducing the risk of breast cancer recurrence: an evaluation of the effects and mechanisms of diet and exercise. Curr Breast Cancer Rep 8 (3):139–150
Dietary Guidelines Advisory Committee (2015a) Advisory report to the Secretary of Health and Human Services and the Secretary of Agriculture. Part D. Chapter 2: dietary patterns, foods and nutrients and health outcomes. USDA, Washington, DC
Dietary Guidelines Advisory Committee (2015b) Scientific report of the 2015 dietary guidelines advisory committee: advisory report to the secretary of health and human services and the secretary of agriculture. A. R. S. U.S. Department of Agriculture. Washington, DC, pp 1–35
Dietary Guidelines Advisory Committee (2015c) Scientific report of the 2015 dietary guidelines advisory committee: advisory report to the secretary of health and human services and the secretary of agriculture. A. R. S. U.S. Department of Agriculture. Washington, DC, pp 1–78
Dong J-Y, He K et al (2011) Dietary fiber intake and risk of breast cancer: a meta-analysis of prospective cohort studies. Am J Clin Nutr 94(3):900–905
Dreher ML (2018a) Dietary patterns in aging and disease. In: Bendich CWBA (ed) Dietary patterns and whole plant foods in aging and disease. Springer, Cham
Dreher ML (2018b) Fiber and diverticular disease. In: Adrianne Bendich CWB (ed) Dietary fiber in health and disease. Springer, New York
Dreher ML (2018c) Fiber in type 2 diabetes prevention and management. In: Dietary fiber in health and disease. Springer, New York

Dreher ML (2018d) Fiber and coronary heart disease. In: Dietary fiber in health and disease. Springer, New York, pp 273–289
Dreher ML (2018e) Fiber-rich diets in chronic kidney disease. In: Dietary fiber in health and disease. Springer, New York
Drossman DA (2016) Functional gastrointestinal disorders: history, pathophysiology, clinical features, and Rome IV. Gastroenterology 150(6):1262–1279. e1262
EFSA Panel on Dietetic Products, Nutrition and Allergies (2010) Scientific opinion on dietary reference values for carbohydrates and dietary fibre. EFSA J 8(3):1462
El-Ansary AK, Ben Bacha A et al (2012) Etiology of autistic features: the persisting neurotoxic effects of propionic acid. J Neuroinflammation 9:74
Esposito K, Marfella R et al (2004) Effect of a mediterranean-style diet on endothelial dysfunction and markers of vascular inflammation in the metabolic syndrome: a randomized trial. JAMA 292(12):1440–1446
Evans CE, Greenwood DC et al (2015) Effects of dietary fibre type on blood pressure: a systematic review and meta-analysis of randomized controlled trials of healthy individuals. J Hypertens 33 (5):897–911
Evenepoel P, Meijers BK (2012) Dietary fiber and protein: nutritional therapy in chronic kidney disease and beyond. Kidney Int 81(3):227–229
Evert AB, Boucher JL et al (2014) Nutrition therapy recommendations for the management of adults with diabetes. Diabetes Care 37(Supplement 1):S120–S143
Faghfoori Z, Shakerhosseini R et al (2014) Effects of an oral supplementation of germinated barley foodstuff on serum CRP level and clinical signs in patients with ulcerative colitis. Health Promot Perspect 4(1):116–121
Fåk F, Jakobsdottir G et al (2015) The physico-chemical properties of dietary fibre determine metabolic responses, short-chain fatty acid profiles and gut microbiota composition in rats fed low-and high-fat diets. PLoS One 10(5):e0127252
Farooqui AA, Farooqui T (2018) Effects of mediterranean diet components on neurodegenerative diseases. In: Role of the Mediterranean diet in the brain and neurodegenerative diseases. Elsevier, London, pp 1–16
Farvid MS, Eliassen AH et al (2016) Dietary fiber intake in young adults and breast cancer risk. Pediatrics 137(3):e20151226
Fatani S, Pickavance LC et al (2007) Differential vascular dysfunction in response to diets of differing macronutrient composition: a phenomenonological study. Nutr Metab 4(1):15
Fernández-Vallinas S, Miguel M et al (2016) Long-term antihypertensive effect of a soluble cocoa fiber product in spontaneously hypertensive rats. Food Nutr Res 60(1):29418
Ferrari P, Rinaldi S et al (2013) Dietary fiber intake and risk of hormonal receptor-defined breast cancer in the European Prospective Investigation into Cancer and Nutrition study. Am J Clin Nutr 97(2):344–353
Finegold SM, Dowd SE et al (2010) Pyrosequencing study of fecal microflora of autistic and control children. Anaerobe 16(4):444–453
Fitzmaurice C, Allen C et al (2017) Global, regional, and national cancer incidence, mortality, years of life lost, years lived with disability, and disability-adjusted life-years for 32 cancer groups, 1990 to 2015: a systematic analysis for the global burden of disease study. JAMA Oncol 3 (4):524–548
Fogelholm M, Anderssen S et al (2012) Dietary macronutrients and food consumption as determinants of long-term weight change in adult populations: a systematic literature review. Food Nutr Res 56(1):19103
Folkerts J, Stadhouders R et al (2018) Effect of dietary fiber and metabolites on mast cell activation and mast cell-associated diseases. Front Immunol 9:1067
For WCRFAI (2017) Breast cancer. How diet, nutrition and physical activity affect breast cancer risk
Ford AC, Moayyedi P et al (2014) American College of Gastroenterology monograph on the management of irritable bowel syndrome and chronic idiopathic constipation. Am J Gastroenterol 109(Suppl 1):S2–S26; quiz S27

Fuhrman J, Beetham R et al (2016) Nutrient-dense, plant-rich dietary intervention effective at reducing cardiovascular disease risk factors for worksites: a pilot study. Altern Ther Health Med 22(5):32
Fung TC, Olson CA et al (2017) Interactions between the microbiota, immune and nervous systems in health and disease. Nat Neurosci 20(2):145–155
Galisteo M, Duarte J et al (2008) Effects of dietary fibers on disturbances clustered in the metabolic syndrome. J Nutr Biochem 19(2):71–84
Gallegos-Orozco JF, Foxx-Orenstein AE et al (2012) Chronic constipation in the elderly. Am J Gastroenterol 107(1):18–25; quiz 26
Gänzle M, Follador R (2012) Metabolism of oligosaccharides and starch in lactobacilli: a review. Front Microbiol 3:340
Gao A, Yan X et al (2015) Physicochemical and bioactive properties of soluble dietary fibers from Blasting Extrusion Processing (BEP)-extruded carrot residues. Food Bioprocess Technol 8 (10):2036–2046
Gibb RD, McRorie JW Jr et al (2015) Psyllium fiber improves glycemic control proportional to loss of glycemic control: a meta-analysis of data in euglycemic subjects, patients at risk of type 2 diabetes mellitus, and patients being treated for type 2 diabetes mellitus. Am J Clin Nutr 102 (6):1604–1614
Giglia TM, Witmer CM et al (2019) Thrombosis and modern anticoagulation options for the adult with congenital heart disease. In: Intensive care of the adult with congenital heart disease. Springer, Cham, pp 535–564
Gilhooly CH, Das SK et al (2008) Use of cereal fiber to facilitate adherence to a human caloric restriction program. Aging Clin Exp Res 20(6):513–520
Gomes S, Oliveira C et al (2019) The role of diet-related short-chain fatty acids in colorectal cancer metabolism and survival: prevention and therapeutic implications. Curr Med Chem 25. https://doi.org/10.2174/0929867325666180530102050
Gopinath B, Flood VM et al (2016) Association between carbohydrate nutrition and successful aging over 10 years. J Gerontol Ser A Biol Sci Med Sci 71(10):1335–1340
Gower BA, Bergman R et al (2016) Baseline insulin sensitivity affects response to high-amylose maize resistant starch in women: a randomized, controlled trial. Nutr Metab 13(1):2
Gray LE, O'Hely M et al (2017) The maternal diet, gut bacteria, and bacterial metabolites during pregnancy influence offspring asthma. Front Immunol 8:365
Grundy SM (2008) Metabolic syndrome pandemic. Arterioscler Thromb Vasc Biol 28(4):629–636
Hajishafiee M, Saneei P et al (2016) Cereal fibre intake and risk of mortality from all causes, CVD, cancer and inflammatory diseases: a systematic review and meta-analysis of prospective cohort studies. Br J Nutr 116(2):343–352
Halmos EP, Gibson PR (2015) Dietary management of IBD—insights and advice. Nat Rev Gastroenterol Hepatol 12(3):133–146
Hartley L, May MD et al (2016) Dietary fibre for the primary prevention of cardiovascular disease. Cochrane Database Syst Rev (1):Cd011472
Hategan A, Hirsch CH (2018) Physiology and pathology of aging. In: Hategan A, Bourgeois JA, Hirsch CH, Giroux C (eds) Geriatric psychiatry: a case-based textbook. Springer, Cham
Heidemann C, Schulze MB et al (2008) Dietary patterns and risk of mortality from cardiovascular disease, cancer, and all causes in a prospective cohort of women. Circulation 118(3):230–237
Hogenkamp A, Thijssen S et al (2015) Supplementing pregnant mice with a specific mixture of nondigestible oligosaccharides reduces symptoms of allergic asthma in male offspring. J Nutr 145(3):640–646
Hu FB, Manson JE et al (2001) Diet, lifestyle, and the risk of type 2 diabetes mellitus in women. N Engl J Med 345(11):790–797
Huang T-b, Ding P-p et al (2014) Dietary fiber intake and risk of renal cell carcinoma: evidence from a meta-analysis. Med Oncol 31(8):125
InterAct Consortium (2015) Dietary fibre and incidence of type 2 diabetes in eight European countries: the EPIC-InterAct Study and a meta-analysis of prospective studies. Diabetologia 58(7):1394–1408

Ishikawa H, Matsumoto S et al (2011) Beneficial effects of probiotic bifidobacterium and galacto-oligosaccharide in patients with ulcerative colitis: a randomized controlled study. Digestion 84 (2):128–133

Janssen AW, Kersten S (2015) The role of the gut microbiota in metabolic health. FASEB J 29 (8):3111–3123

Jeffery IB, Lynch DB et al (2016) Composition and temporal stability of the gut microbiota in older persons. ISME J 10(1):170–182

Jenkins DJ, Kendall CW et al (2000) Viscous fibers, health claims, and strategies to reduce cardiovascular disease risk. Oxford University Press, New York

Jenkins DJ, Kendall CW et al (2008) Effect of a low-glycemic index or a high-cereal fiber diet on type 2 diabetes: a randomized trial. JAMA 300(23):2742–2753

Jiao J, Xu J-Y et al (2015) Effect of dietary fiber on circulating C-reactive protein in overweight and obese adults: a meta-analysis of randomized controlled trials. Int J Food Sci Nutr 66(1):114–119

Jones JR, Lineback DM et al (2006) Dietary reference intakes: implications for fiber labeling and consumption: a summary of the International Life Sciences Institute North America Fiber Workshop, June 1–2, 2004, Washington, DC. Nutr Rev 64(1):31–38

Kang DW, Park JG et al (2013) Reduced incidence of Prevotella and other fermenters in intestinal microflora of autistic children. PLoS One 8(7):e68322

Karlsen MC, Rogers G et al (2019) Theoretical food and nutrient composition of whole-food plant-based and vegan diets compared to current dietary recommendations. Nutrients 11(3):625

Kaulmann A, Bohn T et al (2016) Bioactivity of polyphenols: preventive and adjuvant strategies toward reducing inflammatory bowel diseases—promises, perspectives, and pitfalls. Oxidative Med Cell Longev 2016:1

Kaye EK, Valencia A et al (2010) Tooth loss and periodontal disease predict poor cognitive function in older men. J Am Geriatr Soc 58(4):713–718

Keenan MJ, Marco ML et al (2015) Improving healthspan via changes in gut microbiota and fermentation. Age 37(5):98

Khan NA, Raine LB et al (2015) Dietary fiber is positively associated with cognitive control among prepubertal children. J Nutr 145(1):143–149

Khan K, Jovanovski E et al (2018) The effect of viscous soluble fiber on blood pressure: a systematic review and meta-analysis of randomized controlled trials. Nutr Metab Cardiovasc Dis 28(1):3–13

Kieffer DA, Martin RJ et al (2016) Impact of dietary fibers on nutrient management and detoxification organs: gut, liver, and kidneys. Adv Nutr 7(6):1111–1121

Kim Y, Je Y (2014) Dietary fiber intake and total mortality: a meta-analysis of prospective cohort studies. Am J Epidemiol 180(6):565–573

Kim Y, Je Y (2016) Dietary fibre intake and mortality from cardiovascular disease and all cancers: a meta-analysis of prospective cohort studies. Arch Cardiovasc Dis 109(1):39–54

Kim H, Stote KS et al (2009) Glucose and insulin responses to whole grain breakfasts varying in soluble fiber, beta-glucan: a dose response study in obese women with increased risk for insulin resistance. Eur J Nutr 48(3):170–175

Kim M, Qie Y et al (2016) Gut microbial metabolites fuel host antibody responses. Cell Host Microbe 20(2):202–214

King DE (2005) Dietary fiber, inflammation, and cardiovascular disease. Mol Nutr Food Res 49 (6):594–600

Kirkwood TBL (2017) Why and how are we living longer? Exp Physiol 102(9):1067–1074

Kondo K, Ishikado A et al (2014) A high-fiber, low-fat diet improves periodontal disease markers in high-risk subjects: a pilot study. Nutr Res 34(6):491–498

Krishnamurthy VM, Wei G et al (2012) High dietary fiber intake is associated with decreased inflammation and all-cause mortality in patients with chronic kidney disease. Kidney Int 81 (3):300–306

Kulkarni G, Shen B (2019) Maintenance of a healthy pouch. In: Pouchitis and ileal pouch disorders. Elsevier, London, pp 313–333

Kunzmann AT, Coleman HG et al (2015) Dietary fiber intake and risk of colorectal cancer and incident and recurrent adenoma in the Prostate, Lung, Colorectal, and Ovarian Cancer Screening Trial. Am J Clin Nutr 102(4):881–890
Lairon D (2007) Dietary fiber and control of body weight. Nutr Metab Cardiovasc Dis 17(1):1–5
Lane ER, Lee D et al (2017) Dietary therapies in pediatric inflammatory bowel disease: an evolving inflammatory bowel disease paradigm. Gastroenterol Clin N Am 46(4):731–744
Lanza E, Yu B et al (2007) The polyp prevention trial continued follow-up study: no effect of a low-fat, high-fiber, high-fruit, and -vegetable diet on adenoma recurrence eight years after randomization. Cancer Epidemiol Biomark Prev 16(9):1745–1752
Larsson SC, Wolk A (2014) Dietary fiber intake is inversely associated with stroke incidence in healthy Swedish adults. J Nutr 144(12):1952–1955
Lattimer JM, Haub MD (2010) Effects of dietary fiber and its components on metabolic health. Nutrients 2(12):1266–1289
Liljeberg H, Björck I (2000) Effects of a low-glycaemic index spaghetti meal on glucose tolerance and lipaemia at a subsequent meal in healthy subjects. Eur J Clin Nutr 54(1):24
Lindström J, Peltonen M et al (2006) High-fibre, low-fat diet predicts long-term weight loss and decreased type 2 diabetes risk: the Finnish Diabetes Prevention Study. Diabetologia 49 (5):912–920
Liu S, Willett WC et al (2003) Relation between changes in intakes of dietary fiber and grain products and changes in weight and development of obesity among middle-aged women. Am J Clin Nutr 78(5):920–927
Liu L, Wang S et al (2015a) Fiber consumption and all-cause, cardiovascular, and cancer mortalities: a systematic review and meta-analysis of cohort studies. Mol Nutr Food Res 59 (1):139–146
Liu X, Wu Y et al (2015b) Dietary fiber intake reduces risk of inflammatory bowel disease: result from a meta-analysis. Nutr Res 35(9):753–758
Loddo I, Romano C (2015) Inflammatory bowel disease: genetics, epigenetics, and pathogenesis. Front Immunol 6:551
López-Otín C, Blasco MA et al (2013) The hallmarks of aging. Cell 153(6):1194–1217
Loveman E, Colquitt J et al (2016) Cochrane corner: does increasing intake of dietary fibre help prevent cardiovascular disease? Heart 102(20):1607
Ma Y, Olendzki BC et al (2015) Single-component versus multicomponent dietary goals for the metabolic syndrome: a randomized trial. Ann Intern Med 162(4):248–257
Macia L, Tan J et al (2015) Metabolite-sensing receptors GPR43 and GPR109A facilitate dietary fibre-induced gut homeostasis through regulation of the inflammasome. Nat Commun 6:6734
Maćkowiak K, Torlińska-Walkowiak N et al (2016) Dietary fibre as an important constituent of the diet. Adv Hyg Exp Med/Postepy Hig Med Dosw 70:104
Mahalle N, Garg MK et al (2016) Association of dietary factors with severity of coronary artery disease. Clin Nutr ESPEN 15:75–79
Mandaliya D, Patel S et al (2018) Fiber in our diet and its role in health and disease. In: Functional food and human health. Springer, Singapore, pp 247–255
Mariat D, Firmesse O et al (2009) The Firmicutes/Bacteroidetes ratio of the human microbiota changes with age. BMC Microbiol 9:123
Marlett JA, McBurney MI et al (2002) Position of the American Dietetic Association: health implications of dietary fiber. J Am Diet Assoc 102(7):993–1000
Martinez I, Lattimer JM et al (2013) Gut microbiome composition is linked to whole grain-induced immunological improvements. ISME J 7(2):269–280
Mathers JC (2015) Impact of nutrition on the ageing process. Br J Nutr 113(S1):S18–S22
Mazidi M, Rezaie P et al (2016) Gut microbiome and metabolic syndrome. Diabetes Metab Syndr 10(2 Suppl 1):S150–S157
McCullough K, Sharma P et al (2012) Measuring the population burden of chronic kidney disease: a systematic literature review of the estimated prevalence of impaired kidney function. Nephrol Dial Transplant 27(5):1812–1821

McEvoy CT, Guyer H et al (2017) Neuroprotective diets are associated with better cognitive function: the health and retirement study. J Am Geriatr Soc 65(8):1857–1862
McKenzie CI, Mackay CR et al (2015) GPR43 - a prototypic metabolite sensor linking metabolic and inflammatory diseases. Trends Endocrinol Metab 26(10):511–512
McRae MP (2017) Dietary fiber is beneficial for the prevention of cardiovascular disease: an umbrella review of meta-analyses. J Chiropr Med 16(4):289–299
McRorie JW Jr (2015) Evidence-based approach to fiber supplements and clinically meaningful health benefits, part 1: what to look for and how to recommend an effective fiber therapy. Nutr Today 50(2):82
Messaoudi M, Lalonde R et al (2011) Assessment of psychotropic-like properties of a probiotic formulation (Lactobacillus helveticus R0052 and Bifidobacterium longum R0175) in rats and human subjects. Br J Nutr 105(5):755–764
Milani C, Ferrario C et al (2016) The human gut microbiota and its interactive connections to diet. J Hum Nutr Diet 29(5):539–546
Miyamoto S (2019) Autophagy and cardiac aging. Cell Death Differ 26:653
Moayyedi P, Quigley EM et al (2014) The effect of fiber supplementation on irritable bowel syndrome: a systematic review and meta-analysis. Am J Gastroenterol 109(9):1367–1374
Mohiuddin A (2019) Complimentary treatments for minor GI disorders. Adv Pharmacol 4 (1):000153
Mollard RC, Sénéchal M et al (2014) Dietary determinants of hepatic steatosis and visceral adiposity in overweight and obese youth at risk of type 2 diabetes. Am J Clin Nutr 99 (4):804–812
Morris MC, Tangney CC et al (2015) MIND diet slows cognitive decline with aging. Alzheimers Dement 11(9):1015–1022
Mottillo S, Filion KB et al (2010) The metabolic syndrome and cardiovascular risk a systematic review and meta-analysis. J Am Coll Cardiol 56(14):1113–1132
Mudgil D, Barak S (2013) Composition, properties and health benefits of indigestible carbohydrate polymers as dietary fiber: a review. Int J Biol Macromol 61:1–6
Murphy N, Norat T et al (2012) Dietary fibre intake and risks of cancers of the colon and rectum in the European prospective investigation into cancer and nutrition (EPIC). PLoS One 7(6):e39361
Nagarajan N, Morden A et al (2015) The role of fiber supplementation in the treatment of irritable bowel syndrome: a systematic review and meta-analysis. Eur J Gastroenterol Hepatol 27 (9):1002–1010
Nagpal R, Kumar M et al (2016) Gut microbiota in health and disease: an overview focused on metabolic inflammation. Benefic Microbes 7(2):181–194
Navarro S, Neuhouser M et al (2016) The interaction between dietary fiber and fat and risk of colorectal cancer in the women's health initiative. Nutrients 8(12):779
Niv E, Halak A et al (2016) Randomized clinical study: partially hydrolyzed guar gum (PHGG) versus placebo in the treatment of patients with irritable bowel syndrome. Nutr Metab (Lond) 13:10
O'Callaghan NJ, Toden S et al (2012) Colonocyte telomere shortening is greater with dietary red meat than white meat and is attenuated by resistant starch. Clin Nutr 31(1):60–64
O'Toole PW, Jeffery IB (2015) Gut microbiota and aging. Science 350(6265):1214–1215
Ofori-Asenso R, Garcia D et al (2016) Cardiovascular diseases in Ghana within the context of globalization. Cardiovasc Diagn Ther 6(1):67
Orlich MJ, Singh PN et al (2015) Vegetarian dietary patterns and the risk of colorectal cancers. JAMA Intern Med 175(5):767–776
Pal S, Khossousi A et al (2011) The effect of a fibre supplement compared to a healthy diet on body composition, lipids, glucose, insulin and other metabolic syndrome risk factors in overweight and obese individuals. Br J Nutr 105(1):90–100
Palaniswamy S, Piltonen T et al (2019) The association between blood copper concentration and biomarkers related to cardiovascular disease risk–analysis of 206 individuals in the Northern Finland Birth Cohort 1966. J Trace Elem Med Biol 51:12–18

Parekh PJ, Balart LA et al (2015) The influence of the gut microbiome on obesity, metabolic syndrome and gastrointestinal disease. Clin Transl Gastroenterol 6:e91
Parikh S, Pollock NK et al (2012) Adolescent fiber consumption is associated with visceral fat and inflammatory markers. J Clin Endocrinol Metab 97(8):E1451–E1457
Parisi G, Zilli M et al (2002) High-fiber diet supplementation in patients with irritable bowel syndrome (IBS): a multicenter, randomized, open trial comparison between wheat bran diet and partially hydrolyzed guar gum (PHGG). Dig Dis Sci 47(8):1697–1704
Park Y, Subar AF et al (2011) Dietary fiber intake and mortality in the NIH-AARP diet and health study. Arch Intern Med 171(12):1061–1068
Pokusaeva K, Fitzgerald GF et al (2011) Carbohydrate metabolism in Bifidobacteria. Genes Nutr 6 (3):285
Post RE, Mainous AG et al (2012) Dietary fiber for the treatment of type 2 diabetes mellitus: a meta-analysis. J Am Board Fam Med 25(1):16–23
Racine A, Carbonnel F et al (2016) Dietary patterns and risk of inflammatory bowel disease in Europe: results from the EPIC study. Inflamm Bowel Dis 22(2):345–354
Ramezani A, Raj DS (2014) The gut microbiome, kidney disease, and targeted interventions. J Am Soc Nephrol 25(4):657–670
Redgwell RJ, Fischer M (2005) Dietary fiber as a versatile food component: an industrial perspective. Mol Nutr Food Res 49(6):521–535
Research WCRF (2016) Continuous update project report. Continuous update project report: diet, nutrition, physical activity, and breast cancer survivors
Rivière A, Selak M et al (2016) Bifidobacteria and butyrate-producing colon bacteria: importance and strategies for their stimulation in the human gut. Front Microbiol 7:979
Ross AB, Bruce SJ et al (2011) A whole-grain cereal-rich diet increases plasma betaine, and tends to decrease total and LDL-cholesterol compared with a refined-grain diet in healthy subjects. Br J Nutr 105(10):1492–1502
Russo L, Andreozzi P et al (2015) Partially hydrolyzed guar gum in the treatment of irritable bowel syndrome with constipation: effects of gender, age, and body mass index. Saudi J Gastroenterol 21(2):104–110
Salmean YA, Segal MS et al (2015) Fiber supplementation lowers plasma p-cresol in chronic kidney disease patients. J Ren Nutr 25(3):316–320
Sánchez-Muniz FJ (2012) Dietary fibre and cardiovascular health. Nutr Hosp 27(1):31
Schmier JK, Miller PE et al (2014) Cost savings of reduced constipation rates attributed to increased dietary fiber intakes: a decision-analytic model. BMC Public Health 14:374
Schwartz N, Kaye EK et al (2012) High-fiber foods reduce periodontal disease progression in men aged 65 and older: the Veterans Affairs normative aging study/Dental Longitudinal Study. J Am Geriatr Soc 60(4):676–683
Schwingshackl L, Schwedhelm C et al (2017) Adherence to Mediterranean diet and risk of cancer: an updated systematic review and meta-analysis. Nutrients 9(10):1063
Shay CM, Van Horn L et al (2012) Food and nutrient intakes and their associations with lower BMI in middle-aged US adults: the International Study of Macro-/Micronutrients and Blood Pressure (INTERMAP). Am J Clin Nutr 96(3):483–491
Sheng T, Shen R-l et al (2015) No association between fiber intake and prostate cancer risk: a meta-analysis of epidemiological studies. World J Surg Oncol 13(1):264
Shlisky J, Bloom DE et al (2017) Nutritional considerations for healthy aging and reduction in age-related chronic disease. Adv Nutr 8(1):17
Shu H-S, Tai Y-Y et al (2015) Plasma high-sensitivity C-reactive protein level is associated with impaired estimated glomerular filtration rate in hypertensives. Acta Cardiol Sin 31(2):91
Silva FM, Kramer CK et al (2013) Fiber intake and glycemic control in patients with type 2 diabetes mellitus: a systematic review with meta-analysis of randomized controlled trials. Nutr Rev 71 (12):790–801
Simons FE, Ebisawa M et al (2015) 2015 Update of the evidence base: World Allergy Organization anaphylaxis guidelines. World Allergy Organ J 8(1):32

Singh A, Bishayee A et al (2018) Targeting histone deacetylases with natural and synthetic agents: an emerging anticancer strategy. Nutrients 10(6):731
Siyame EWP (2018) Dietary practices of hypertensive patients attending hypertensive clinics in Malawi. North Carolina State University
Slavin JL (2005) Dietary fiber and body weight. Nutrition 21(3):411–418
Slavin JL (2008) Position of the American Dietetic Association: health implications of dietary fiber. J Am Diet Assoc 108(10):1716–1731
Slavin JL, Lloyd B (2012) Health benefits of fruits and vegetables. Adv Nutr 3(4):506–516
Smith A, Sutherland D et al (2015) An investigation of the acute effects of oligofructose-enriched inulin on subjective wellbeing, mood and cognitive performance. Nutrients 7(11):8887–8896
Sonnenburg ED, Sonnenburg JL (2014) Starving our microbial self: the deleterious consequences of a diet deficient in microbiota-accessible carbohydrates. Cell Metab 20(5):779–786
Strate LL, Keeley BR et al (2017) Western dietary pattern increases, and prudent dietary pattern decreases, risk of incident diverticulitis in a prospective cohort study. Gastroenterology 152 (5):1023–1030.e1022
Streppel MT, Arends LR et al (2005) Dietary fiber and blood pressure: a meta-analysis of randomized placebo-controlled trials. Arch Intern Med 165(2):150–156
Streppel MT, Ocké MC et al (2008) Dietary fiber intake in relation to coronary heart disease and all-cause mortality over 40 y: the Zutphen Study. Am J Clin Nutr 88(4):1119–1125
Sun L, Zhang Z et al (2017) Dietary fiber intake reduces risk for Barrett's esophagus and esophageal cancer. Crit Rev Food Sci Nutr 57(13):2749–2757
Tan J, McKenzie C et al (2016) Dietary fiber and bacterial SCFA enhance oral tolerance and protect against food allergy through diverse cellular pathways. Cell Rep 15(12):2809–2824
Tetens I (2011) EFSA Panel on Dietetic Products, Nutrition and Allergies (NDA); Scientific Opinion Part II on the substantiation of health claims related to various food (s)/food constituent (s) not supported by pertinent human data (ID 406, 462, 472, 543, 659, 678, 696, 858, 1381, 1403, 1437, 1438, 1513, 1536, 1537, 1538, 1539, 1540, 1543, 1613, 1627, 1855, 1860, 1981, 2126, 2514, 3127, 4038, 4501, 4672, 4712, 4718) pursuant to Article 13 (1) of Regulation (EC) No 1924/2006
Thies F, Masson LF et al (2014) Oats and CVD risk markers: a systematic literature review. Br J Nutr 112(S2):S19–S30
Threapleton DE, Greenwood DC et al (2013a) Dietary fiber intake and risk of first stroke: a systematic review and meta-analysis. Stroke 44(5):1360–1368
Threapleton DE, Greenwood DC et al (2013b) Dietary fibre intake and risk of cardiovascular disease: systematic review and meta-analysis. BMJ 347:f6879
Triff K, Kim E et al (2015) Chemoprotective epigenetic mechanisms in a colorectal cancer model: modulation by n-3 PUFA in combination with fermentable fiber. Curr Pharmacol Rep 1 (1):11–20
Trindade M, Oigman W et al (2017) Potential role of endothelin in early vascular aging. Curr Hypertens Rev 13(1):33–40
Trompette A, Gollwitzer ES et al (2014) Gut microbiota metabolism of dietary fiber influences allergic airway disease and hematopoiesis. Nat Med 20(2):159–166
van Bennekum AM, Nguyen DV et al (2005) Mechanisms of cholesterol-lowering effects of dietary insoluble fibres: relationships with intestinal and hepatic cholesterol parameters. Br J Nutr 94 (3):331–337
Vercambre M-N, Boutron-Ruault M-C et al (2009) Long-term association of food and nutrient intakes with cognitive and functional decline: a 13-year follow-up study of elderly French women. Br J Nutr 102(3):419–427
Verma S, Jagtap U et al (2018) A novel gene-diet pair modulates C. elegans aging. PLoS Genet 14 (8):e1007608
Veronese N, Solmi M et al (2018) Dietary fiber and health outcomes: an umbrella review of systematic reviews and meta-analyses. Am J Clin Nutr 107(3):436–444

Vetrani C, Costabile G et al (2016) Effects of whole-grain cereal foods on plasma short chain fatty acid concentrations in individuals with the metabolic syndrome. Nutrition 32(2):217–221
Vieytes CAM, Taha HM et al (2019) Carbohydrate nutrition and the risk of cancer. Curr Nutr Rep 8:230–239
Vinolo MA, Ferguson GJ et al (2011) SCFAs induce mouse neutrophil chemotaxis through the GPR43 receptor. PLoS One 6(6):e21205
Vuksan V, Jenkins AL et al (2011) Viscosity rather than quantity of dietary fibre predicts cholesterol-lowering effect in healthy individuals. Br J Nutr 106(9):1349–1352
Wanders AJ, van den Borne JJ et al (2011) Effects of dietary fibre on subjective appetite, energy intake and body weight: a systematic review of randomized controlled trials. Obes Rev 12 (9):724–739
Wang F, Yu T et al (2015) Gut microbiota community and its assembly associated with age and diet in chinese centenarians. J Microbiol Biotechnol 25(8):1195–1204
Wan-Loy C, Siew-Moi P (2016) Marine algae as a potential source for anti-obesity agents. Mar Drugs 14(12):222
Wei ZH, Wang H et al (2009) Time- and dose-dependent effect of psyllium on serum lipids in mild-to-moderate hypercholesterolemia: a meta-analysis of controlled clinical trials. Eur J Clin Nutr 63(7):821–827
Wei B, Liu Y et al (2018) Dietary fiber intake and risk of metabolic syndrome: a meta-analysis of observational studies. Clin Nutr 37(6):1935–1942
Weickert M, Mohlig M et al (2005) Impact of cereal fibre on glucose-regulating factors. Diabetologia 48(11):2343–2353
Weickert MO, Hattersley JG et al (2018) Effects of supplemented isoenergetic diets varying in cereal fiber and protein content on the bile acid metabolic signature and relation to insulin resistance. Nutr Diabetes 8(1):11
Weinhold KR, Miller CK et al (2015) A randomized controlled trial translating the diabetes prevention program to a University Worksite, Ohio, 2012-2014. Prev Chronic Dis 12:E210
Whelton SP, Hyre AD et al (2005) Effect of dietary fiber intake on blood pressure: a meta-analysis of randomized, controlled clinical trials. J Hypertens 23:475
Whitehead A, Beck EJ et al (2014) Cholesterol-lowering effects of oat β-glucan: a meta-analysis of randomized controlled trials. Am J Clin Nutr 100(6):1413–1421
Wilkins T, Pepitone C et al (2012) Diagnosis and management of IBS in adults. Am Fam Physician 86(5):419–426
Wolfram T, Ismail-Beigi F (2010) Efficacy of high-fiber diets in the management of type 2 diabetes mellitus. Endocr Pract 17(1):132–142
World Cancer Research Fund International/American Institute for Cancer Research Continuous Update Project Report (2014) Diet, nutrition, physical activity, and prostate cancer. World Cancer Research Fund International
Wu K, Bowman R et al (2007) Apolipoprotein E polymorphisms, dietary fat and fibre, and serum lipids: the EPIC Norfolk study. Eur Heart J 28(23):2930–2936
Wu Y, Qian Y et al (2015) Association between dietary fiber intake and risk of coronary heart disease: a meta-analysis. Clin Nutr 34(4):603–611
Xing MY, Xu SZ et al (2014) Effect of low-fat diet on breast cancer survival: a meta-analysis. Asian Pac J Cancer Prev 15(3):1141–1144
Xu H, Huang X et al (2014) Dietary fiber, kidney function, inflammation, and mortality risk. Clin J Am Soc Nephrol 9(12):2104–2110
Xu L, Yu W et al (2015) Efficacy of pectin in the treatment of diarrhea predominant irritable bowel syndrome. Zhonghua Wei Chang Wai Ke Za Zhi 18(3):267–271
Yang M, Kenfield SA et al (2015a) Dietary patterns after prostate cancer diagnosis in relation to disease-specific and total mortality. Cancer Prev Res 8(6):545–551
Yang Y, Zhao L-G et al (2015b) Association between dietary fiber and lower risk of all-cause mortality: a meta-analysis of cohort studies. Am J Epidemiol 181(2):83–91

Yao B, Fang H et al (2014) Dietary fiber intake and risk of type 2 diabetes: a dose-response analysis of prospective studies. Eur J Epidemiol 29(2):79–88
Zapata HJ, Quagliarello VJ (2015) The microbiota and microbiome in aging: potential implications in health and age-related diseases. J Am Geriatr Soc 63(4):776–781
Zeng L, Hu S et al (2017) Macronutrient intake and risk of Crohn's disease: systematic review and dose-response meta-analysis of epidemiological studies. Nutrients 9(5):500
Zhang Z, Xu G et al (2013a) Dietary fiber consumption and risk of stroke. Eur J Epidemiol 28 (2):119–130
Zhang Z, Xu G et al (2013b) Dietary fiber intake reduces risk for gastric cancer: a meta-analysis. Gastroenterology 145(1):113–120. e113
Zhang L, Pagoto S et al (2018) A nonrestrictive, weight loss diet focused on fiber and lean protein increase. Nutrition 54:12–18
Zhu X, Sun X et al (2015) Quantitative assessment of the effects of beta-glucan consumption on serum lipid profile and glucose level in hypercholesterolemic subjects. Nutr Metab Cardiovasc Dis 25(8):714–723
Zimmer J, Lange B et al (2012) A vegan or vegetarian diet substantially alters the human colonic faecal microbiota. Eur J Clin Nutr 66(1):53–60
Zou J, Chassaing B et al (2018) Fiber-mediated nourishment of gut microbiota protects against diet-induced obesity by restoring IL-22-mediated colonic health. Cell Host Microbe 23(1):41–53. e44

Dietary Polyphenols for Active and Healthy Ageing

7

L. Testai and V. Calderone

Abstract

The prolongation of lifespan is a desired condition, strictly depending from several factors, including genetics and environmental factors. In this context, nutrition can be a powerful tool, useful to modulate ageing. In particular, polyphenols constituents, widely distributed in edible vegetables and fruits, can act on specific intracellular markers, among which the enzyme sirtuin-1 (SIRT1) and several kinases, mainly AMPK, mTOR and MAPK/ERK and then contribute to reduce proinflammatory and fibrotic processes typical of the ageing; therefore inhibiting the progression of age-related pathologies, including neurodegenerative, cardiovascular and cancer. Indeed polyphenol family is very big and includes numerous subgroups of chemical constituents, ranging from rather elementary substances, such as phenolic acids and stilbenes, to complex polymerized molecules, such as tannins. They are different for distribution in the natural king and for bio-pharmacological profile. In this chapter the role played from polyphenols in the senescence has been evaluated.

Keywords

Ageing · Polyphenols · Lifespan · Health · Diet

L. Testai (✉) · V. Calderone
Department of Pharmacy, University of Pisa, Pisa, Italy

Interdepartmental Centre of Research "Nutraceutical and Healthy Food", Pisa, Italy

Interdepartmental Centre of Research "Biology and Pathology of the Ageing", Pisa, Italy
e-mail: lara.testai@unipi.it

S. M. Nabavi et al. (eds.), *Nutrients and Nutraceuticals for Active & Healthy Ageing*,
https://doi.org/10.1007/978-981-15-3552-9_7

7.1 Introduction

The increase in life expectancy, in industrialized as well as developing countries, raises the number of age-related pathology incidences, such as cardiovascular, metabolic and neurodegenerative diseases, to a point beyond social and economic sustainability, making ageing an important and worldwide topic (Dilberger et al. 2019; Pray 2017; Tzioras et al. 2018). Several studies have demonstrated that healthy ageing depends on several factors, such as genetics and environmental factors; therefore, nutrition can be a powerful tool to modulate ageing (Mico et al. 2017).

Senescence is a process in which cells stop proliferating and become dysfunctional and secrete proinflammatory molecules, reactive oxygen species (ROS) and extracellular matrix components that cause inflammation in the surrounding tissues, responsible for a low-grade systemic inflammation and oxidative stress (Franceschi et al. 2018). Indeed, it is well known that an accumulation of cellular damage and a reduction in protective stress response pathways are at the basis of age-related diseases. In particular, the accumulation of genetic damage, including nuclear mutations, telomere attrition and epigenetic changes, can accelerate the shortening of the lifespan (López-Otín et al. 2013).

In addition, NF-κB is implicated as a key transcription factor in the development of chronic inflammation; in fact, NF-κB is implicated in the initiation of tissue-specific stress responses throughout the body (Aggarwal et al. 2004; Sethi et al. 2008). NF-kB is the principal supporter of the senescence-associated secretory phenotype (SASP), which is characterized by the production and secretion of proinflammatory cytokines, thereby sustaining the chronic inflammation typical of ageing (Queen and Tollefsbol 2010).

Another intrinsic age-related state observed in numerous organisms is the accumulation of toxins at the level of cells and organs; in neurodegenerative conditions, such as Alzheimer's disease (AD), β and γ secretases are critical for the production of insoluble amyloid proteins, known to play a role in Alzheimer's disease (Glukhov et al. 2008; Sebai et al. 2009).

Moreover, another event associated with ageing and the development of age-related diseases is mitochondrial dysfunction (Wagner et al. 2009; Modrick et al. 2009). A decline in mitochondrial function, and the commensurate loss of a sufficient energy production, supposedly plays a key role during ageing. Indeed, the free radical theory proposes that an age-related progressive loss of mitochondrial function is associated with ROS production (Green et al. 2011). However, the reduced efficiency of mitochondrial bioenergetics with ageing can result from reduced biogenesis of organelles, in which the transcriptional cofactor PGC1α, regulated through SIRT1, is crucial (Sahin and DePinho 2012). Therefore, developing ways to prevent mitochondrial dysfunction can prove a potent strategy to counteract adverse effects associated with ageing (Ikonomovic et al. 2009; Anisimov et al. 2009; Weinreb et al. 2004; Levites et al. 2003).

7.2 Targets Responsible for the Anti-Ageing Effects of Polyphenols

Polyphenols are natural compounds that are widely produced by plants, endowed with antibiotic and antifungal properties (Leiro et al. 2004) and moreover are well known for their antioxidant activity. However, in addition to promoting specific antioxidant and anti-inflammatory activities, polyphenols have garnered considerable interest for their role in the modulation of a plethora of targets.

New findings point to the possibility that polyphenols have suitable properties for treatment of several disease states in humans, and a combination of these may hold therapeutic benefits not yet realized (Cheynier 2005).

7.2.1 Sirtuin-Isoform 1 (SIRT1)

The role of sirtuin in the protecting against cellular senescence has mainly been investigated with mammalian sirtuin isoform 1 (SIRT1), which is a species-conserved NAD^+-dependent protein deacetylase that targets acetylated lysine residues in substrate proteins. It plays a crucial role in many cell signalling pathways (Heger et al. 2019); indeed, its overexpression has been shown to extend the lifespan of lower eukaryotes, such as yeast and worms, and to reduce senescence in several cell types, particularly when exposed to oxidants (Sasaki et al. 2006). In addition, the reduction of SIRT1 promotes premature senescence-like phenotypes in endothelial cells (Ota et al. 2007).

Taken together, these results support the idea that SIRT1 plays a role in cellular senescence. In vitro evidence shows that SIRT1 is capable of deacetylating a wide variety of proteins, including AMP-activated protein kinase (AMPK), p53, mechanistic target of rapamycin (mTOR) NF-κB, HSF-1, FOXO1, FOXO3, and FOXO4, and PGC-1α by directly activating them (Chaudhary and Pfluger 2009; Saunders and Verdin 2009), highlighting the complex network involved in longevity (Lee et al. 2019).

7.2.2 AMPK

AMPK is a pivotal energy sensor that alleviates or delays the process of fibrogenesis. Therefore, considering that fibrosis is a common process characterized by excessive extracellular matrix accumulation after inflammatory injury, AMPK is considered a crucial player that orchestrates ageing and the main diseases of the heart, liver, kidney and lung (Jiang et al. 2017; Hardie 2007; Steinberg and Kemp 2009; Mihaylova and Shaw 2011; Reznick et al. 2007).

Notably, Reznick et al. demonstrated that the ageing process is associated with a decline in AMPK, suggesting a reduced capability of organisms to respond to age-related stress (Reznick et al. 2007). Accordingly, several studies have highlighted that upregulated AMPK expression may be associated with a prolonged lifespan in lower organisms (Ulgherait et al. 2014; Stenesen et al. 2013).

7.2.3 mTOR

mTOR is a serine-threonine kinase that senses and integrates diverse environmental and intracellular signals, such as those initiated by growth factors and nutrients, to direct cellular and organismal responses (Saxton and Sabatini 2017). It is evolutionarily conserved, its name is derived from the first inhibitor rapamycin found to affect this signal transduction pathway, which is identified in the 1970s (Pazoki-Toroudi et al. 2016).

mTOR has been recognized as a regulator of lifespan of stem cells in the nematode *Caenorhabditis elegans* (Vellai et al. 2003), in the fruit fly *Drosophila melanogaster* (Kapahi et al. 2004) and in the yeast strain *Saccharomyces cerevisiae* (Kaeberlein et al. 2005). The inhibition of mTOR by rapamycin doubles the lifespan of these simple organisms and in more developed animal species, including rodents; suggesting that also mTOR network can be considered as a regulator of ageing and lifespan (Harrison et al. 2009; Selman et al. 2009).

Moreover, recently, a main role of mTOR in the promotion of SASP has been suggested, leading to hypotheses implicating its role in inflammation and in the increase in mitochondrial mass and markers of mitochondrial activity (Weichhart 2018).

7.2.4 MAPK/ERK

The mitogen-activated protein kinase (MAPK) signalling pathway is shared by four distinct cascades, which have been named accordingly to their components: extracellular related kinase (ERK1/2), Jun amino terminal kinase (JNK1/2/3), p38 MAPK and ERK5. These enzymes are activated through a sequential phosphorylation cascade that amplifies the signals transduced from the cell membrane to the nucleus. Depending on the duration and magnitude of its expression and its subcellular localization, ERK regulates various cell responses, such as proliferation, migration, differentiation and death (Ramos 2008). In particular, the activation of the ERK pathway can induce senescence; indeed, an increase in β-galactosidase activity and an induction of classical senescence-associated genes, including p16 and p21, have been reported (Denoyelle et al. 2006). Recent evidence has revealed that numerous bioflavonoids, obtained from a variety of dietary fruits, plants and medicinal herbal sources, exhibit protective functions against the development of neurodegenerative diseases, mainly through the modulation of different compartments of the ERK signalling pathway (Farzaei et al. 2018).

7.2.5 Classification of Polyphenols

Polyphenols are one of the most abundant and extensively studied family of chemical entities, naturally produced by the plant kingdom (Bravo 1998). Polyphenols can be found in fruits, vegetables, nuts, seeds, flowers and tree bark. These components are involved in the attraction of pollinators, execution of structural functions, defence against ultraviolet radiation and protection against microbial invasion and herbivores (Manach et al. 2004). A wide number of substances containing numerous hydroxyl moieties on aromatic rings are included in this family. Furthermore, this class of compounds is highly diversified and comprises several subgroups, ranging from rather elementary substances, such as phenolic acids and stilbenes, to complex polymerized molecules, such as tannins (Cheynier et al. 2017). Natural polyphenols are present in nature in conjugated form, with one or more sugar residues associated with a hydroxyl group, although the direct linkage of a sugar unit to an aromatic carbon atom can also occur. Monosaccharides, disaccharides and oligosaccharides are also attached (Bravo 1998). Polyphenols are usually classified on the basis of the number of phenol rings and by the structural components that bind these rings together; therefore, phenolic acids, flavonoids, stilbenes and lignans are naturally produced (Fig. 7.1). Moreover, flavonoids can be further organized into distinctive subgroups, including anthocyanins, flavan-3-ols, flavones, flavanones and flavonols (Fig. 7.2) (Bravo 1998; Manach et al. 2004; Tsao 2010).

With regard to the physical properties of polyphenols, they played a critical role in determining the sensory and nutritional characteristics of foods and, in addition, to pigmentation. Although some volatile polyphenols, such as vanillin and eugenol, are extremely potent odorants; the most common flavour perceptions stimulated by polyphenols are astringency and bitterness, primarily elicited by flavonol polymers (proanthocyanidins or condensed tannins) (Bravo 1998; Lesschaeve and Noble 2005).

7.2.6 Phenolic Derivatives

Oleuropein, a polyphenol typically found in extra-virgin olive oil, in cultured neuroblastoma and in a mouse model of amyloid beta (Aβ) deposition, triggers autophagy through a rapid release of Ca^{2+} from stores, which in turn activates the phosphorylation and activation of AMPK. The link between AMPK activation and mTOR inhibition has also been observed in animal models, supporting the idea that autophagy activation by oleuropein proceeds through mTOR inhibition (Rigacci et al. 2015).

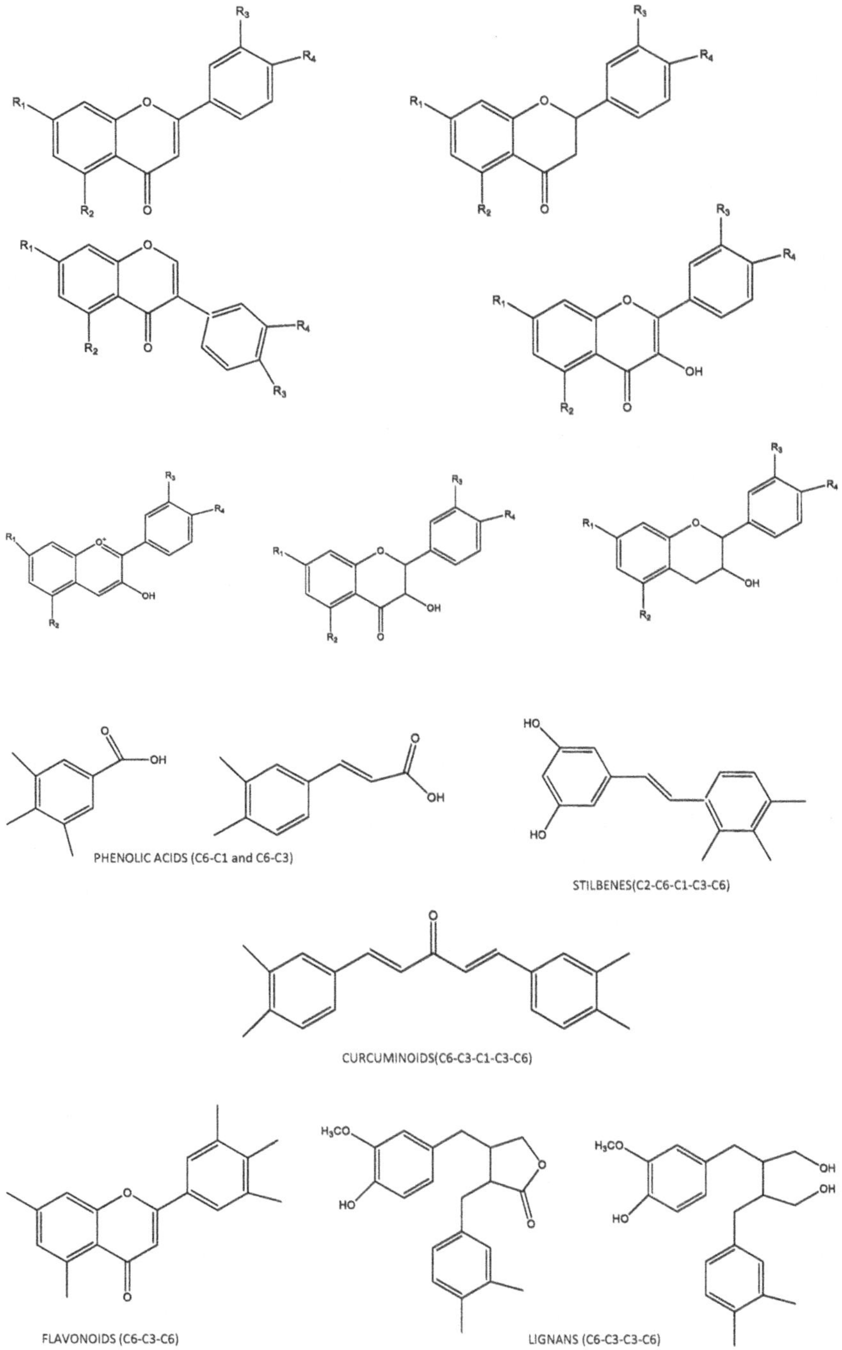

Fig. 7.1 Chemical structures of polyphenols classes

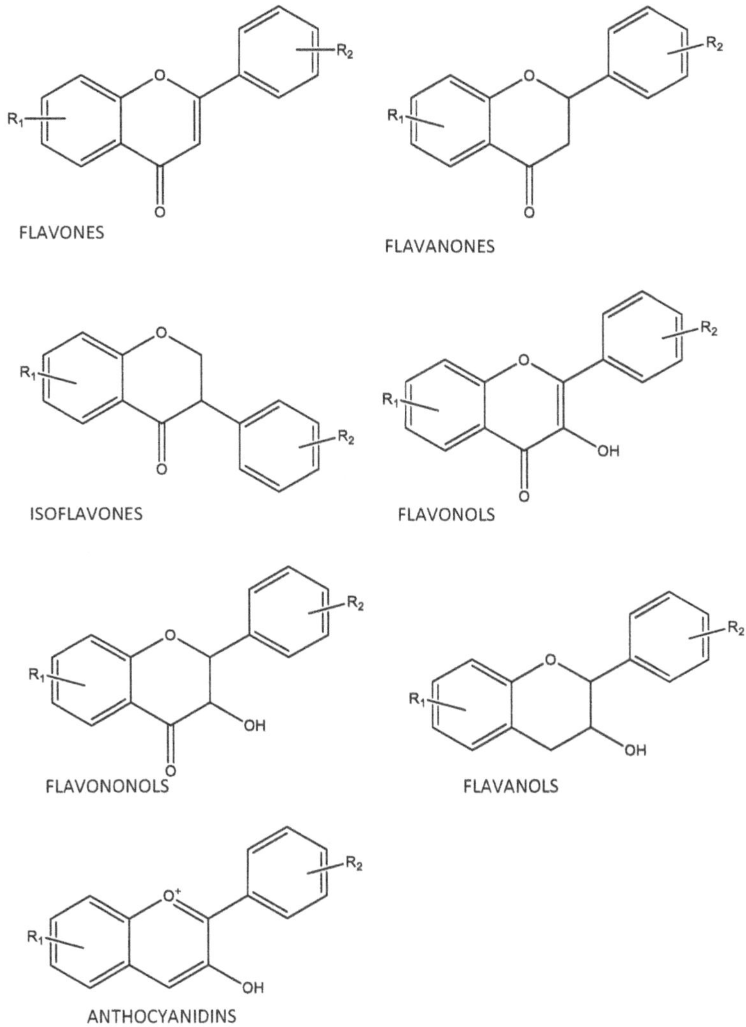

Fig. 7.2 Chemical structures of flavonoids sub-classes

7.2.7 Resveratrol and Other Stilbene Derivatives

A wide body of evidence suggests the positive effects of resveratrol (RSV), a well-known stilbene derivative, for enhancing health through multiple signalling pathways. First, RSV is recognized as a potent activator of the SIRT1 enzyme (Abolaji et al. 2018).

Interestingly, Smith et al. have shown that an innovative formulation of RSV, called SRT501, has a molecular profile similar to that observed under calorie-restricted conditions in both genetically and diet-induced obese mice. Recently, other polyphenols structurally similar to RSV can activate SIRT1, as demonstrated in in vivo investigations, producing effects very similar to those of calorie restriction (Smith et al. 2009). Furthermore, RSV is able to induce AMPK, even in SIRT1-knockout mice, ensuring additional effective protection against ageing (Cheng et al. 2019). In zebrafish, retinal is involved in the same signalling pathways, and it has been hypothesized that RSV inhibits mitochondrial dysfunction through this pathway (Wang et al. 2019).

It has been demonstrated that RSV can modulate several mechanisms involved in cognitive decline, including antioxidant, anti-inflammatory and anti-apoptotic processes and autophagy regulation; however, it also increases blood flow and enhances the plasticity of synaptic pathways, suggesting that these functions can be, at least in part, the means through which RSV can be leveraged to support healthy ageing (Fontana 2009). Moreover, RSV activates the transmembrane protein α-secretase, which is associated with the formation of a soluble, non-amyloidogenic—non plaque-forming—protein from the amyloid precursor protein (APP) located in the membrane of neuronal cells. Interestingly, when soluble APP is produced, any neuritic plaque, a hallmark feature of AD, is formed. Indeed, APP is modified through a pathway that involves two additional enzymes, β- and γ-secretase, which sequentially process the APP protein, leading to the formation of insoluble, amyloidogenic oligomers or fibrils. Under these conditions, neuritic plaques are formed, suggesting the crucial role of RSV for understanding the aetiology of AD (Adlard et al. 2009; Mandel et al. 2008; Rao et al. 2020).

In addition, Benitez et al. have suggested that RSV may exert anti-proliferative and apoptotic effects by mediating the inhibition of NF-κB, as observed during in vitro studies of human prostate cancer cells (Benitez et al. 2009).

Moreover, RSV, in fish gut, reverses senescence-associated β-galactosidase activity, downregulates the levels of proinflammatory cytokines, IL-8 and TNFα, and upregulates the expression of the anti-inflammatory cytokine IL-10. Furthermore, RSV increases SIRT1 expression and inhibits NF-κB by decreasing RelA/p65, Ac-RelA/p65 and p-IκBα levels and by increasing the interaction between SIRT1 and RelA/p65. Moreover, it reverses the decline in intestinal epithelial cells (IECs) and intestinal stem cells (ISCs) in fish gut caused by ageing (Liu et al. 2018).

Interestingly, RSV alone or in combination with exercise significantly increases the expression of phosphorylated AMPK and SIRT1, decreases the expression of acetyl P53 and the Bax/Bcl-2 ratio in aged rats, showing significant improvement in gastrocnemius muscle morphology and ultrastructure and having possible positive effects in sarcopenia, an age-related syndrome characterized by progressive loss of muscle mass and function (Liao et al. 2017). Recently, RSV has been correlated with an improvement in the osteogenic differentiation of mesenchymal stem cells in bone through the activation of the AMPK signalling pathway (Zhou et al. 2019).

7.2.8 Curcumin

Curcumin (CRM) is a ferulic derivative obtained from the rhizome of *Curcuma longa*, commonly used as a spice (curry and turmeric) and yellow food dye (E100). It has a pleiotropic profile due to its ability to interact simultaneously with many receptors, growth factors, kinases, transcription factors, enzymes, adhesion molecules, apoptotic regulators, proinflammatory cytokines and other compounds (Bielak-Zmijewska et al. 2019). Furthermore, CRM upregulates and downregulates different kinds of miRNA and participates in epigenetic changes, such as the regulation of histone acetyltransferases and histone deacetylases (Gupta et al. 2013; Boyanapalli and Kong 2015; Remely et al. 2015; Reuter et al. 2011). Several in vitro and in vivo preclinical studies have demonstrated the potential therapeutic value of CRM against ageing-associated disorders, including atherosclerosis, hypertension, diabetes, neurodegenerative diseases, osteoporosis and cancer. In fact, it is the object of clinical trials for the treatment of these disorders (Kumar et al. 2018).

In particular, CRM reduces amyloid burden, rescues neuronal damage and restores normal cognitive and sensory motor functions in different animal models of neurodegenerative diseases (Maiti and Dunbar 2018). In this regard, CRM binds Aβ plaques, reducing their neurotoxicity and initiating their degradation; CRM injected intraventricularly in Aβ1–42-expressing rats reduces the cognitive decline and promotes hippocampal regeneration (Voulgaropoulou et al. 2019). At the vascular level, CRM increases the level of sirtuins and AMPK in vascular smooth muscle cells undergoing replicative senescence (Grabowska et al. 2016). Finally, mice and rats receiving CRM supplements showed enhanced effects from exercise in terms of time until exhaustion and prevention of fatigue, most likely because of an increased level of AMPK and SIRT1 expression and/or activation in muscles (Huang et al. 2015; Ray Hamidie et al. 2015).

7.2.9 Classification of Flavonoids

Flavonoids are abundantly present in plants as secondary metabolites. The basic chemical structure of flavonoids is represented by two benzene rings (A and C) connected by a pyran ring B (Fig. 7.2). One of the benzene rings (A) is fused with the pyran ring, while the other benzene ring (C) is attached as a substituent to the pyran ring. Various derivatives of flavonoids are produced, depending on the pattern of substitution on the benzene rings and of the oxidation and saturation status of the pyran ring.

Therefore, isoflavones present a benzene ring (C) attached to position 3 of the pyran ring and are typical of various natural products, mainly soybean (Wang and Murphy 1994).

Neoflavonoids have a benzene ring (C) attached to position 4 of the pyran ring (Donnelly and Boland 1995).

Flavones contain a double bond in the pyran ring between positions 2 and 3 and hydroxyl substituents in both aromatic rings (Fukui et al. 1968).

Flavonols differ from the flavones in the hydroxyl group at position 3 of the pyran ring; indeed they are the alcoholic derivatives of flavones and generally, they are known as 3-hydroxyflavones.

Flavanones, saturated flavones, are also known as dihydroflavones. They differ from flavones and flavanones for the absence of a double bond between positions 2 and 3. Flavanones are characteristic, but not exclusive, to the *Citrus* genus.

The flavanonols are 3-hydroxy flavanones and are also called dihydroflavonols. They present a saturated pyran ring, a hydroxyl group at position 3 and a carbonyl group at position 4.

Finally, flavanols, also called flavan-3-ol, are lacking of a carbonyl group at position 4. The pyran ring is saturated and disubstituted at positions 2 and 3. This chemical characteristic leads to four possible diastereomers of the flavanol. In flavanols, the benzene ring (C) is attached to position 2, while the hydroxyl groups are attached at position 3 of the pyran ring (Ayaz et al. 2019).

7.2.10 Evidence of Anti-Ageing Properties of Flavonoids

Several flavonoids have been demonstrated to inhibit the progression of age-related neurodegenerative pathologies and to alleviate cognitive deficits in numerous normal and transgenic preclinical animal models. Indeed, flavonoids are recognized as inhibitors of cholinesterases, including acetylcholinesterase and butyrylcholinesterase; β-secretase; and free radicals and considered possible modulators of the signalling pathways implicated in cognitive and neuroprotective functions. Moreover, flavonoids can to interact with various signalling protein pathways, including ERK and PI3-kinase/Akt, to modulate their actions, thereby leading to beneficial neuroprotective effects. Finally, they enhance vascular blood flow and stimulate neurogenesis, particularly in the hippocampus (Macready et al. 2009; Spencer 2010; Zhang et al. 2019a).

7.2.10.1 Anthocyanidins

Berry fruits, which contain high amounts of flavonoids, especially anthocyanidins, have received particular attention for preventing age-related cognitive decline. In particular, blueberry supplements has been shown to improve memory and learning. Moreover, a long-term prospective study on neurologically healthy elderly people emphasized that increased intake of berry anthocyanidins is associated with a slower rate of cognitive decline and delays in the onset of deficits by approximately 2.5 years (Bakoyiannis et al. 2019).

7.2.10.2 Flavanols

A number of preclinical and clinical evidence points to the beneficial effects of flavanols in age-related disorders, particularly neurodegenerative and cardiovascular disorders. Cocoa and green tea contain high amounts of flavanols, among which are epicatechin, catechin and their derivatives (Ottaviani et al. 2018).

Evidence suggests that cocoa flavanols can benefit brain function via mechanisms that include enhanced neuronal plasticity and cerebrovascular function. For example, epicatechin has been found to enhance the retention of spatial memory in male C57BL/6 mice (8–10 weeks old), particularly in combination with exercise, and catechins of green tea administered to 14-month-old female mice for 6 months prevented spatial learning and memory decline (Yevchak et al. 2008; Mastroiacovo et al. 2015; Alonso-Alonso 2015). An increase in hippocampal brain-derived neurotrophic factor has also been observed in adult C57BL/6 mice treated with 4 mg/day of epicatechin (Stringer et al. 2015). Cocoa flavanol intake for 2 weeks improved flow-mediated dilatation in both the young and elderly patients included in the clinical study, where it enhanced arteriolar and microvascular vasodilator capacity and decreased systolic blood pressure (Heiss et al. 2015). More recently, cocoa flavanol consumption was found to improve the endothelial functional integrity in healthy humans, thereby decreasing endothelial microparticle levels, a marker inversely correlated with flow-mediated dilation (Gröne et al. 2019).

7.2.10.3 Citrus Flavanones

Recently, the citrus flavonoids naringenin (NAR) and hesperetin (HSP) have been reported to prevent senescence. In particular, Da Pozzo and colleagues demonstrated that a juice obtained from *Citrus bergamia*, Rizzo, called bergamot, (BJ) was able to induce antioxidant effects and to inhibit the expression of senescence markers (β-galactosidase, p16 and p21) in cardiomyoblasts subjected to damage following treatment with doxorubicin or hydroxyperoxide. Moreover, BJ upregulated SIRT1, Nrf2 and FOXO3 expression in 12-month-old mice that had received it as a supplement for 3 months (Da Pozzo et al. 2018).

NAR is well known for the anti-ischemic cardioprotective profile it confers to adult rats, which is mediated by stimulation caused by large-conductance calcium-activated potassium channels in mitochondria (Martelli et al. 2013). Later, Testai et al. confirmed that this profile is maintained in older animals (12 months), suggesting that NAR could be considered a valid supplement that benefits elderly patients (Testai et al. 2017).

Interestingly, Testai and colleagues recently demonstrated that NAR is endowed with anti-senescence activity, which it confers through the activation of the SIRT1 enzyme. Indeed, NAR presented a similar profile to that presented by BJ with respect to senescent doxorubicin-treated cardiomyoblasts, showing a reduction in β-galactosidase and in p21 and p16; moreover, these researchers demonstrated that NAR administered to 12-month-old mice for 6 months inhibited oxidative stress, chronic inflammation and fibrosis processes at the myocardial level (Da Pozzo et al. 2017). It is noteworthy that NAR slowed the progression of degenerative processes in a retinitis pigmentosa model (Piano et al. 2019).

HSP has antioxidant properties that are not limited to its radical scavenging activity; in fact, HSP exhibits a pivotal role in senescence processes by enhancing cellular antioxidant defences via the ERK/Nrf2 signalling pathway (Parhiz et al. 2015; Kim et al. 2006). Of note, HSP-glycoside (hesperidin) can inhibit ROS production in *Saccharomyces cerevisiae*, as demonstrated by the reduction in SOD gene expression. This effect seems to be correlated with an increase in Sir2 activity (homologue of the mammalian SIRT1 enzyme) (Sun et al. 2012). HSP also significantly reduces the secretion of inflammatory cytokines, including IL-1β and IL-6, and downregulates the phosphorylation of ERK/MAPK pathway components during LPS-induced neuroinflammation involving BV2 cells (Jo et al. 2019).

7.2.10.4 Flavonols

Quercetin (QRC) is the main flavonol in the daily human diet, and most of the quercetin in plants is in the glycoside form. QRC has been demonstrated to modulate sirtuins and to protect against several chronic diseases; in particular, this flavonol stimulates the SIRT1 enzyme both in vitro and in vivo models through direct and indirect mechanisms (Howitz et al. 2003; Trevino-Saldana and Garcia-Rivas 2017). Notably, the role of QRC is not restricted to SIRT1; in fact, it also upregulates SIRT2 expression in vivo (Peredo-Escarcega et al. 2015), and similar to kaempferol, it can inhibit SIRT6 (Rahnasto-Rilla et al. 2018). In this regard, through molecular docking approaches the mechanisms involved in the modulation of the SIRT6 enzyme by QRC and its derivatives have been investigated. The authors of these studies observed that diquercetin preferred a binding site in the nicotinamide (NAM) moiety, whereas 2-chloro-1,4-naphthoquinone-quercetin preferred to dock at a substrate binding site, leading to speculation that different interaction sites are involved in QRC activity; indeed, diquercetin competes with nicotinamide adenine dinucleotide (NAD^+), whereas 2-chloro-1,4-naphthoquinone-quercetin competes with the acetylated substrate in the catalytic site of SIRT6 (Papaevgeniou and Chondrogianni 2018).

A short treatment with QRC *plus* caloric restriction (but not alone) is effective to counteract t age-related accumulation of oxidative macromolecular damage, including the decline in SOD and catalase activity, counteract, in 21-month-old Wistar rats (Alugoju and Periyasamy 2018). An interesting reaction in which antioxidant defences were ameliorated has also been observed in mice with D-galactose-induced neurotoxicity. The administration of QRC at 20 and 50 mg/kg for 8 weeks improved mouse learning and memory compared to the memory and learning observed in the control mice. QRC also prevents changes in neuronal cell morphology and apoptosis rate in the hippocampus, and it increases the expression of Nrf2, HO-1 and SOD in D-galactose-treated mice. Finally, the abolition of these effects in the presence of a Nrf2 inhibitor corroborates the involvement of the Nrf2-ARE signalling pathway in the beneficial action of QRC (Dong et al. 2017).

Sarubbo et al. reported in vivo effects of long-term QRC administration in 18-month-old Sprague-Dawley rats when administered in combination with silymarin and naringenin (20 mg/kg/day i.p., 4 weeks). The restorative effects of QRC on cognition and motor coordination were consistent with the biochemical and molecular results. In addition, polyphenols increased SIRT1 levels and decreased NF-κB levels in the hippocampus, confirming it is a valuable potential therapeutic for attenuating inflammation and brain function decline (Sarubbo et al. 2018).

Fisetin is another flavonol with a promising senotherapeutic profile; indeed, an acute or intermittent fisetin treatment of progeroid and aged mice reduced senescence markers in multiple tissues in a dose-dependent manner (Yousefzadeh et al. 2018).

Of note, fisetin suppresses markers of senescence in vivo in transgenic mice characterized by accelerated accumulation of senescent cells but not in wild-type mice, suggesting that this natural compound intriguingly can confer specific health benefits to elderly patients (Yousefzadeh et al. 2018).

Furthermore, Wistar rats that received fisetin supplements had modulated membrane transporters, namely, calcium-ATPase, sodium potassium-ATPase and sodium hydrogen exchanger, during senescence-induced by D-galactose and during natural ageing. Fisetin also protected against oxidative modifications in rat ageing (Singh et al. 2019).

7.2.10.5 Flavones

Chrysin, an abundant flavone present in honey and propolis, at a dose of 10 mg/kg significantly protects against age-related memory decline, attenuating the increase in ROS levels and the inhibition of SOD, CAT and GPx action in aged mice. Moreover, chrysin markedly mitigates the decrease in brain-derived neurotrophic factor (BDNF) levels in prefrontal and hippocampal regions (Souza et al. 2015). Generally, chrysin supplementation for 8 weeks, protected against oxidative damage from with the exposure to D-galactose, as confirmed by histopathological evaluation (Anand et al. 2012).

More recently, Farkhondeh et al. observed that chrysin was effective in attenuating age-related lipid abnormalities, glucose elevation and inflammation, in young (2-month-old) and old (20-month-old) rats (Farkhondeh et al. 2019).

7.2.10.6 Isoflavones

Anti-senescence activity has also been demonstrated by genistein, an isoflavone, in human umbilical vein endothelial cells. This effect seems to be associated with induced autophagy through the involvement of the SIRT1/AMPK pathway (Zhang et al. 2019b). Similarly, formononetin (another isoflavone) has been associated with cardioprotective effects in a model of ischaemia/reperfusion injury in the hearts of aged rats, which was also induced through enhanced autophagy (Huang et al. 2018).

7.2.11 Bioavailability of Polyphenols: A Limit to or an Advantage for Clinical Efficacy?

A body of evidence on human disease models demonstrates beneficial effects of polyphenolic compounds, however, the development of new and effective treatments greatly depends on obtaining a greater understanding of the bioavailability and metabolism of polyphenols (Cheynier et al. 2017). After oral intake, polyphenols are rapidly hydrolysed into the aglycones by lactase-phlorizin hydrolase or cytosolic β-glucosidase in the small intestine or by bacterial glycosidases in the colon. The plasma concentrations in humans after intake of polyphenol-rich food can reach to a maximum on the order of several μM (Murota et al. 2007; Mullen et al. 2006). The aglycones are further metabolized through phase reaction into glucuronides by UDP-glucuronosyltransferase and/or sulphates by sulfotransferase, and those with the catechol moiety can also be methylated by catechol-O-methyltransferase (COMT) (Nemeth et al. 2003; Murota and Terao 2003; Kawai 2018). Indeed, phase reactions are generally recognized as responsible for production of more polar, hydrophilic, and biologically inactive molecules that are more readily excreted in urine.

Most recently, the interactions between polyphenols and intestinal microbiota have been considered crucial to gaining greater knowledge of the metabolic destiny of these natural compounds. Indeed, recent evidence suggests the existence of reciprocal relationship between polyphenols and microbiota. On the one hand, microbiota, in the gastrointestinal tract, can influence the absorption and bioavailability of polyphenols in ingested food, and on the other hand, the microbiota can also influence the metabolism of polyphenols, leading to the production of propionic and acetic derivatives, also known as short chain fatty acids (SCFAs). Interestingly, SCFAs are thought to greatly modulate the local immune system and intestinal permeability and are critical to the microbiota composition. Therefore, SCFAs have been suggested to have the ability to stimulate mitochondrial biogenesis and metabolism; specifically, it has been hypothesized that, despite the low bioavailability of polyphenols, significant beneficial effects can be mediated by their microbiota-produced metabolites. In light of this hypothesis, a substantial impact on the potential benefits of polyphenols has been correlated with the intestinal microbiota population (Kumar Singh et al. 2019). Further work is required to fully understand the complex interactions between polyphenols and the microbiota and to determine the exact nature of their subsequent local and systemic health benefits, particularly in humans.

References

Abolaji AO, Adedara AO, Adie MA, Vicente-Crespo M, Farombi EO (2018) Resveratrol prolongs lifespan and improves 1-methyl-4-phenyl-1,2,3,6-tetrahydropyridine-induced oxidative damage and behavioural deficits in Drosophila melanogaster. Biochem Biophys Res Commun 503:1042–1048

Adlard PA, James SA, Bush AI, Masters CL (2009) Beta-amyloid as a molecular therapeutic target in Alzheimer's disease. Drugs Today (Barc) 45:293–304

Aggarwal BB, Bhardwaj A, Aggarwal RS, Seeram NP, Shishodia S, Takada Y (2004) Role of resveratrol in prevention and therapy of cancer: preclinical and clinical studies. Anticancer Res 24:2783–2840

Alonso-Alonso M (2015) Cocoa flavanols and cognition: regaining chocolate in old age? Am J Clin Nutr 101(3):423–424. https://doi.org/10.3945/ajcn.114.106146. Epub 2015 Jan 28

Alugoju P, Periyasamy L (2018) Effect of short-term quercetin, caloric restriction and combined treatment on age-related oxidative stress markers in the rat cerebral cortex. CNS Neurol Disord Drug Targets 17:119–131
Anand KV, Mohamed Jaabir MS, Thomas PA, Geraldine P (2012) Protective role of chrysin against oxidative stress in d-galactose-induced aging in an experimental rat model. Geriatr Gerontol Int 12:741–750
Anisimov VN, Sikora E, Pawelec G (2009) Relationships between cancer and aging: a multilevel approach. Biogerontology 10:323–338
Ayaz M, Sadiq A, Junaid M, Ullah F, Ovais M, Ullah I, Ahmed J, Shahid M (2019) Flavonoids as prospective neuroprotectants and their therapeutic propensity in aging associated neurological disorders. Front Aging Neurosci 11:155
Bakoyiannis I, Daskalopoulou A, Pergialiotis V, Perrea D (2019) Phytochemicals and cognitive health: are flavonoids doing the trick? Biomed Pharmacother 109:1488–1497
Benitez DA, Hermoso MA, Pozo-Guisado E, Fernández-Salguero PM, Castellón EA (2009) Regulation of cell survival by resveratrol involves inhibition of NF kappa B-regulated gene expression in prostate cancer cells. Prostate 69:1045–1054
Bielak-Zmijewska A, Grabowska W, Ciolko A, Bojko A, Mosieniak G, Bijoch Ł, Sikora E (2019) The role of curcumin in the modulation of ageing. Int J Mol Sci 20(5):1239
Boyanapalli SSS, Kong ANT (2015) Curcumin, the king of spices: epigenetic regulatory mechanisms in the prevention of cancer, neurological, and inflammatory diseases. Curr Pharmacol Rep 1:129–139
Bravo L (1998) Polyphenols: chemistry, dietary sources, metabolism, and nutritional significance. Nutr Rev 56(11):317–333
Chaudhary N, Pfluger PT (2009) Metabolic benefits from Sirt1 and Sirt1 activators. Curr Opin Clin Nutr Metab Care 12:431–437
Cheng CK, Luo JY, Lau CW, Chen ZY, Tian XY, Huang Y (2019) Pharmacological basis and new insights of resveratrol action in the cardiovascular system. Br J Pharmacol 177:1258. https://doi.org/10.1111/bph.14801
Cheynier V (2005) Polyphenols in foods are more complex than often thought. Am J Clin Nutr 81:223S–229S
Cheynier V, Halbwirth H, Stich K, Martens S (2017) Foreword focus on polyphenols. Planta 246:183
Da Pozzo E, Costa B, Cavallini C, Testai L, Martelli A, Calderone V, Martini C (2017) The citrus flavanone naringenin protects myocardial cells against age-associated damage. Oxidative Med Cell Longev 2017:9536148
Da Pozzo E, De Leo M, Faraone I, Milella L, Cavallini C, Piragine E, Testai L, Calderone V, Pistelli L, Braca A, Martini C (2018) Antioxidant and antisenescence effects of bergamot juice. Oxidative Med Cell Longev 2018:9395804
Denoyelle C, Abou-Rjaily G, Bezrookove V, Verhaegen M, Johnson TM, Fullen DR, Pointer JN, Gruber SB, Su LD, Nikiforov MA, Kaufman RJ, Bastian BC, Soengas MS (2006) Anti-oncogenic role of the endoplasmic reticulum differentially activated by mutations in the MAPK pathway. Nat Cell Biol 8:1053–1063
Dilberger B, Passon M, Asseburg H, Silaidos CV, Schmitt F, Schmiedl T, Schieber A, Eckert GP (2019) Polyphenols and metabolites enhance survival in rodents and nematodes-impact of mitochondria. Nutrients 11(8):E1886. https://doi.org/10.3390/nu11081886
Dong F, Wang S, Wang Y, Yang X, Jiang J, Wu D, Qu X, Fan H, Yao R (2017) Quercetin ameliorates learning and memory via the Nrf2-ARE signaling pathway in d-galactose-induced neurotoxicity in mice. Biochem Biophys Res Commun 491:636–641
Donnelly D, Boland G (1995) Isoflavonoids and neoflavonoids: naturally occurring O-heterocycles. Nat Prod Rep 12:321–338
Farkhondeh T, Abedi F, Samarghandian S (2019) Chrysin attenuates inflammatory and metabolic disorder indices in aged male rat. Biomed Pharmacother 109:1120–1125

Farzaei MH, Tewari D, Momtaz S, Argüelles S, Nabavi SM (2018) Targeting ERK signaling pathway by polyphenols as novel therapeutic strategy for neurodegeneration. Food Chem Toxicol 120:183–195
Fontana L (2009) The scientific basis of caloric restriction leading to longer life. Curr Opin Gastroenterol 25:144–150
Franceschi C, Zaikin A, Gordleeva S, Ivanchenko M, Bonifazi F, Storci G, Bonafè M (2018) Inflammaging 2018: an update and a model. Semin Immunol 40:1–5
Fukui K, Matsumoto T, Nakamura S, Nakayama M, Horie T (1968) Synthetic studies of the flavone derivatives: VII. The synthesis of jaceidin. Bull Chem Soc Jpn 41:1413–1417
Glukhov IL, Sirota NP, Kuznetsova EA (2008) DNA damage in human mononuclear cells induced by bacterial endotoxin. Bull Exp Biol Med 146:301–303
Grabowska W, Suszek M, Wnuk M, Lewinska A, Wasiak E, Sikora E, Bielak-Zmijewska A (2016) Curcumin elevates sirtuin level but does not postpone in vitro senescence of human cells building the vasculature. Oncotarget 7:19201–19213
Green DR, Galluzzi L, Kroemer G (2011) Mitochondria and the autophagy-inflammation-cell death axis in organismal aging. Science 333(6046):1109–1112
Gröne M, Sansone R, Höffken P, Horn P, Rodriguez-Mateos A, Schroeter H, Kelm M, Heiss C (2019) Cocoa flavanols improve endothelial functional integrity in healthy young and elderly subjects. J Agric Food Chem 68:1871
Gupta SC, Kismali G, Aggarwal BB (2013) Curcumin, a component of turmeric: from farm to pharmacy. Biofactors 39:2–13
Hardie DG (2007) AMPK and SNF1: snuffing out stress. Cell Metab 6(5):339–340
Harrison DE, Strong R, Sharp ZD, Nelson JF, Astle CM, Flurkey K et al (2009) Rapamycin fed late in life extends lifespan in genetically heterogeneous mice. Nature 460:392–395
Heger V, Tyni J, Hunyadi A, Horáková L, Lahtela-Kakkonen M, Rahnasto-Rilla M (2019) Quercetin based derivatives as sirtuin inhibitors. Biomed Pharmacother 111:1326–1333
Heiss C, Sansone R, Karimi H, Krabbe M, Schuler D, Rodriguez-Mateos A, Kraemer T, Cortese-Krott MM, Kuhnle GG, Spencer JP, Schroeter H, Merx MW, Kelm M, FLAVIOLA Consortium, European Union 7th Framework Program (2015) Impact of cocoa flavanol intake on age-dependent vascular stiffness in healthy men: a randomized, controlled, double-masked trial. Age (Dordr) 37(3):9794
Howitz KT, Bitterman KJ, Cohen HY, Lamming DW, Lavu S, Wood JG, Zipkin RE, Chung P, Kisielewski A, Zhang L-L, Scherer B, Sinclair DA (2003) Small molecule activators of sirtuins extend Saccharomyces cervisiae lifespan. Nature 425:191–196
Huang WC, Chiu WC, Chuang HL, Tang DW, Lee ZM, Li W, Chen FA, Huang CC (2015) Effect of curcumin supplementation on physiological fatigue and physical performance in mice. Nutrients 7:905–921
Huang Z, Liu Y, Huang X (2018) Formononetin may protect aged hearts from ischemia/reperfusion damage by enhancing autophagic degradation. Mol Med Rep 18:4821–4830
Ikonomovic MD, Wecker L, Abrahamson EE et al (2009) Cortical alpha 7 nicotinic acetylcholine receptorand beta-amyloid levels in early Alzheimer disease. Arch Neurol 66:646–651
Jiang S, Li T, Yang Z, Yi W, Di S, Sun Y, Wang D, Yang Y (2017) AMPK orchestrates an elaborate cascade protecting tissue from fibrosis and aging. Ageing Res Rev 38:18–27
Jo SH, Kim ME, Cho JH, Lee Y, Lee J, Park YD, Lee JS (2019) Hesperetin inhibits neuroinflammation on microglia by suppressing inflammatory cytokines and MAPK pathways. Arch Pharm Res 42:695–703
Kaeberlein M, Powers RW, Steffen KK, Westman EA, Hu D, Dang N et al (2005) Regulation of yeast replicative life span by TOR and Sch9 in response to nutrients. Science 310:1193–1196
Kapahi P, Zid BM, Harper T, Koslover D, Sapin V, Benzer S (2004) Regulation of lifespan in Drosophila by modulation of genes in the TOR signaling pathway. Curr Biol 14:885–890
Kawai Y (2018) Understanding metabolic conversions and molecular actions of flavonoids in vivo: toward new strategies for effective utilization of natural polyphenols in human health. J Med Investig 65(3.4):162–165

Kim JY, Jung KJ, Choi JS, Chung HY (2006) Modulation of the age-related nuclear factor-kappaB (NF-kappaB) pathway by hesperetin. Aging Cell 5:401–411
Kumar Singh A, Cabral C, Kumar R, Ganguly R, Kumar Rana H, Gupta A, Rosaria Lauro M, Carbone C, Reis F, Pandey AK (2019) Beneficial effects of dietary polyphenols on gut microbiota and strategies to improve delivery efficiency. Nutrients 11:2216
Kumar SD, Houreld NN, Abrahamse H (2018) Therapeutic potential and recent advances of curcumin in the treatment of aging-associated diseases. Molecules 23(4):835
Lee SH, Lee JH, Lee HY, Min KJ (2019) Sirtuin signaling in cellular senescence and aging. BMB Rep 52(1):24–34
Leiro J, Cano E, Ubeira FM, Orallo F, Sanmartín ML (2004) In vitro effects of resveratrol on the viability and infectivity of the microsporidian Encephalitozoon cuniculi. Antimicrob Agents Chemother 48:2497–2501
Lesschaeve I, Noble AC (2005) Polyphenols: factors influencing their sensory properties and their effects on food and beverage preferences. Am J Clin Nutr 81:330S–335S
Levites Y, Amit T, Mandel S, Youdim MB (2003) Neuroprotection and neurorescue against Abeta toxicity and PKC-dependent release of nonamyloidogenic soluble precursor protein by green tea polyphenol (−)-epigallocatechin-3-gallate. FASEB J 17:952–954
Liao ZY, Chen JL, Xiao MH, Sun Y, Zhao YX, Pu D, Lv AK, Wang ML, Zhou J, Zhu SY, Zhao KX, Xiao Q (2017) The effect of exercise, resveratrol or their combination on Sarcopenia in aged rats via regulation of AMPK/Sirt1 pathway. Exp Gerontol 98:177–183
Liu S, Zheng Z, Ji S, Liu T, Hou Y, Li S, Li G (2018) Resveratrol reduces senescence-associated secretory phenotype by SIRT1/NF-κB pathway in gut of the annual fish Nothobranchius guentheri. Fish Shellfish Immunol 80:473–479
López-Otín C, Blasco MA, Partridge L, Serrano M, Kroemer G (2013) The hallmarks of aging. Cell 153(6):1194–1217
Macready AL, Kennedy OB, Ellis JA, Williams CM, Spencer JP, Butler LT (2009) Flavonoids and cognitive function: a review of human randomized controlled trial studies and recommendations for future studies. Genes Nutr 4:227–242
Maiti P, Dunbar GL (2018) Use of curcumin, a natural polyphenol for targeting molecular pathways in treating age-related neurodegenerative diseases. Int J Mol Sci 19(6):1637
Manach C, Scalbert A, Morand C, Rémésy C, Jiménez L (2004) Polyphenols: food sources and bioavailability. Am J Clin Nutr 79:727–747
Mandel SA, Amit T, Weinreb O, Reznichenko L, Youdim MB (2008) Simultaneous manipulation of multiple brain targets by green tea catechins: a potential neuroprotective strategy for Alzheimer and Parkinson diseases. CNS Neurosci Ther 14:352–365
Mastroiacovo D, Kwik-Uribe C, Grassi D, Necozione S, Raffaele A, Pistacchio L, Righetti R, Bocale R, Lechiara MC, Marini C, Ferri C, Desideri G (2015) Cocoa flavanol consumption improves cognitive function, blood pressure control, and metabolic profile in elderly subjects: the Cocoa, Cognition, and Aging (CoCoA) Study—a randomized controlled trial. Am J Clin Nutr 101:538–548
Mico V, Berninches L, Tapia J, Daimiel L (2017) NutrimiRAging: micromanaging nutrient sensing pathways through nutrition to promote healthy aging. Int J Mol Sci 18:E915
Mihaylova MM, Shaw RJ (2011) The AMPK signalling pathway coordinates cell growth, autophagy and metabolism. Nat Cell Biol 13(9):1016–1023
Modrick ML, Didion SP, Sigmund CD, Faraci FM (2009) Role of oxidative stress and AT (1) receptors in cerebral vascular dysfunction with aging. Am J Physiol Heart Circ Physiol 296:H1914–H1919
Mullen W, Edwards CA, Crozier A (2006) Absorption, excretion and metabolite profiling of methyl- glucuronyl-, glucosyl- and sulpho-conjugates of quercetin in human plasma and urine after ingestion of onions. Br J Nutr 96:107–116
Murota K, Terao J (2003) Antioxidative flavonoid quercetin : implication of its intestinal absorption and metabolism. Arch Biochem Biophys 417:12–17

Murota K, Hotta A, Ido H, Kawai Y, Moon JH, Sekido K, Hayashi H, Inakuma T, Terao J (2007) Antioxidant capacity of albumin-bound quercetin metabolites after onion consumption in humans. J Med Investig 54:370–374

Nemeth K, Plumb GW, Berrin JG, Juge N, Jacob R, Naim HY, Williamson G, Swallow DM, Kroon PA (2003) Deglycosylation by small intestinal epithelial cell beta-glucosidases is a critical step in the absorption and metabolism of dietary flavonoid glycosides in humans. Eur J Nutr 42:29–42

Ota H, Akishita M, Eto M, Iijima K, Kaneki M, Ouchi Y (2007) Sirt1 modulates premature senescence-like phenotype in human endothelial cells. J Mol Cell Cardiol 43:571–579

Ottaviani JI, Heiss C, Spencer JPE, Kelm M, Schroeter H (2018) Recommending flavanols and procyanidins for cardiovascular health: revisited. Mol Asp Med 61:63–75

Papaevgeniou N, Chondrogianni N (2018) Anti-aging and anti-aggregation properties of polyphenolic compounds in C. elegans. Curr Pharm Des 24:2107–2120

Parhiz H, Roohbakhsh A, Soltani F, Rezaee R, Iranshahi M (2015) Antioxidant and anti-inflammatory properties of the citrus flavonoids hesperidin and hesperetin: an updated review of their molecular mechanisms and experimental models. Phytother Res 29:323–331

Pazoki-Toroudi H, Amani H, Ajami M, Nabavi SF, Braidy N, Kasi PD, Nabavi SM (2016) Targeting mTOR signaling by polyphenols: a new therapeutic target for ageing. Ageing Res Rev 31:55–66

Peredo-Escarcega AE, Guarner-Lans V, Perez-Torres I, Ortega-Ocampo S, Carreon-Torres E, Castrejon-Tellez V, Diaz-Diaz E, Rubio-Ruiz ME (2015) The combination of resveratrol and quercetin attenuates metabolic syndrome in rats by modifying the serum fatty acid composition and by upregulating SIRT 1 and SIRT 2 expression in white adipose tissue. Evid Based Complement Alternat Med 2015:474032

Piano I, D'Antongiovanni V, Testai L, Calderone V, Gargini C (2019) A nutraceutical strategy to slowing down the progression of cone death in an animal model of retinitis pigmentosa. Front Neurosci 13:461

Pray L (2017) Nutrition across the lifespan for healthy aging. In Proceedings of the National Academies of Sciences Engineering and Medicine's Food Forum, Washington DC, USA, 12 May 2017

Queen BL, Tollefsbol TO (2010) Polyphenols and aging. Curr Aging Sci 3(1):34–42

Rahnasto-Rilla M, Tyni J, Huovinen M, Jarho E, Kulikowicz T, Ravichandran S, Bohr VA, Ferucci L, Lahtela-Kakkonen M, Moaddel R (2018) Natural polyphenols as sirtuin 6 modulators. Sci Rep 8:4163

Ramos JW (2008) The regulation of extracellular signal-regulated kinase (ERK) in mammalian cells. Int J Biochem Cell Biol 40:2707–2719

Rao YL, Ganaraja B, Joy T, Pai MM, Ullal SD, Murlimanju BV (2020) Neuroprotective effects of resveratrol in Alzheimer's disease. Front Biosci (Elite Ed) 12:139–149

Ray Hamidie R, Yamada T, Ishizawa R, Saito Y, Masuda K (2015) Curcumin treatment enhances the effect of exercise on mitochondrial biogenesis in skeletal muscle by increasing cAMP levels. Metabolism 64:1334–1347

Remely M, Lovrecic L, de la Garza AL, Migliore L, Peterlin B, Milagro F, Martinez A, Haslberger A (2015) Therapeutic perspectives of epigenetically active nutrients. Br J Pharmacol 172:2756–2768

Reuter S, Gupta SC, Park B, Goel A, Aggarwal BB (2011) Epigenetic changes induced by curcumin and other natural compounds. Genes Nutr 6:93–108

Reznick RM, Zong H, Li J, Morino K, Moore IK, Yu HJ, Liu ZX, Dong J, Mustard KJ, Hawley SA, Befroy D, Pypaert M, Hardie DG, Young LH, Shulman GI (2007) Aging-associated reductions in AMP-activated protein kinase activity and mitochondrial biogenesis. Cell Metab 5 (2):151–156

Rigacci S, Miceli C, Nediani C, Berti A, Cascella R, Pantano D, Nardiello P, Luccarini I, Casamenti F, Stefani M (2015) Oleuropein aglycone induces autophagy via the AMPK/mTOR signalling pathway: a mechanistic insight. Oncotarget 6:35344–35357

Sahin E, DePinho RA (2012) Axis of ageing: telomeres, p53 and mitochondria. Nat Rev Mol Cell Biol 13(6):397–404
Sarubbo F, Ramis MR, Kienzer C, Aparicio S, Esteban S, Miralles A, Moranta D (2018) Chronic silymarin, quercetin and naringenin treatments increase monoamines synthesis and hippocampal Sirt1 levels improving cognition in aged rats. J Neuroimmune Pharmacol 13:24–38
Sasaki T, Maier B, Bartke A, Scrable H (2006) Progressive loss of SIRT1 with cell cycle withdrawal. Aging Cell 5:413–422
Saunders LR, Verdin E (2009) Stress response and aging. Science 323:1021–1022
Saxton RA, Sabatini DM (2017) mTOR signaling in growth, metabolism, and disease. Cell 169 (2):361–371
Sebai H, Ben-Attia M, Sani M, Aouani E, Ghanem-Boughanmi N (2009) Protective effect of resveratrol in endotoxemia-induced acute phase response in rats. Arch Toxicol 83:335–340
Selman C, Tullet JMA, Wieser D, Irvine E, Lingard SJ, Choudhury AI et al (2009) Ribosomal protein S6 kinase 1 signaling regulates mammalian life span. Science 326:140–144
Sethi G, Sung B, Aggarwal BB (2008) Nuclear factor-kB activation: from bench to bedside. Exp Biol Med 233:21–31
Singh S, Garg G, Singh AK, Tripathi SS, Rizvi SI (2019) Fisetin, a potential caloric restriction mimetic, modulates ionic homeostasis in senescence induced and naturally aged rats. Arch Physiol Biochem 9:1–8
Smith JJ, Kenney RD, Gagne DJ et al (2009) Small molecule activators of SIRT1 replicate signaling pathways triggered by calorie restriction in vivo. BMC Syst Biol 3:31
Souza LC, Antunes MS, Filho CB, Del Fabbro L, de Gomes MG, Goes AT, Donato F, Prigol M, Boeira SP, Jesse CR (2015) Flavonoid Chrysin prevents age-related cognitive decline via attenuation of oxidative stress and modulation of BDNF levels in aged mouse brain. Pharmacol Biochem Behav 134:22–30
Spencer JP (2010) The impact of fruit flavonoids on memory and cognition. Br J Nutr 104:S40–S47
Steinberg GR, Kemp BE (2009) AMPK in health and disease. Physiol Rev 89(3):1025–1078
Stenesen D, Suh JM, Seo J, Yu K, Lee KS, Kim JS, Min KJ, Graff JM (2013) Adenosine nucleotide biosynthesis and AMPK regulate adult life span and mediate the longevity benefit of caloric restriction in flies. Cell Metab 17(1):101–112
Stringer TP, Guerrieri D, Vivar C, van Praag H (2015) Plant-derived flavanol (−)epicatechin mitigates anxiety in association with elevated hippocampal monoamine and BDNF levels, but does not influence pattern separation in mice. Transl Psychiatry 5:e493
Sun K, Xiang L, Ishihara S, Matsuura A, Sakagami Y, Qi J (2012) Anti-aging effects of hesperidin on Saccharomyces cerevisiae via inhibition of reactive oxygen species and UTH1 gene expression. Biosci Biotechnol Biochem 76:640–645
Testai L, Martelli A, Marino A, D'Antongiovanni V, Ciregia F, Giusti L, Lucacchini A, Chericoni S, Breschi MC, Calderone V (2013) The activation of mitochondrial BK potassium channels contributes to the protective effects of naringenin against myocardial ischemia/reperfusion injury. Biochem Pharmacol 85:1634–1643
Testai L, Da Pozzo E, Piano I, Pistelli L, Gargini C, Breschi MC, Braca A, Martini C, Martelli A, Calderone V (2017) The citrus flavanone naringenin produces cardioprotective effects in hearts from 1 year old rat, through activation of mitoBK channels. Front Pharmacol 8:71
Trevino-Saldana N, Garcia-Rivas G (2017) Regulation of sirtuin-mediated protein deacetylation by cardioprotective phytochemicals. Oxidative Med Cell Longev 2017:1750306
Tsao R (2010) Chemistry and biochemistry of dietary polyphenols. Nutrients 2:1231–1246
Tzioras M, Davies C, Newman A, Jackson R, Spires-Jones T (2018) APOE at the interface of inflammation, neurodegeneration and pathological protein spread in Alzheimer's disease. Neuropathol Appl Neurobiol 45:327–346
Ulgherait M, Rana A, Rera M, Graniel J, Walker DW (2014) AMPK modulates tissue and organismal aging in a non-cell-autonomous manner. Cell Rep 8(6):1767–1780
Vellai T, Takacs-Vellai K, Zhang Y, Kovacs AL, Orosz L, Müller F (2003) Genetics: influence of TOR kinase on lifespan in C. elegans. Nature 426:620

Voulgaropoulou SD, van Amelsvoort TAMJ, Prickaerts J, Vingerhoets C (2019) The effect of curcumin on cognition in Alzheimer's disease and healthy aging: a systematic review of pre-clinical and clinical studies. Brain Res 1725:146476

Wagner W, Bork S, Horn P, Krunic D, Walenda T, Diehlmann A, Benes V, Blake J, Huber FX, Eckstein V, Boukamp P, Ho AD (2009) Aging and replicative senescence have related effects on human stem and progenitor cells. PLoS One 4(6):e5846

Wang H-J, Murphy PA (1994) Isoflavone content in commercial soybean foods. J Agric Food Chem 42:1666–1673

Wang N, Luo Z, Jin M, Sheng W, Wang HT, Long X, Wu Y, Hu P, Xu H, Zhang X (2019) Exploration of age-related mitochondrial dysfunction and the anti-aging effects of resveratrol in zebrafish retina. Aging (Albany NY) 11:3117–3137

Weichhart T (2018) mTOR as regulator of lifespan, aging and cellular senescence. Gerontology 64:127–134

Weinreb O, Mandel S, Amit T, Moussa BHY (2004) Nuerological mechanisms of green tea polyphenol's in Alzheimer's and Parkinson's disease. J Nutr Biochem 15:506–516

Yevchak AM, Loeb SJ, Fick DM (2008) Promoting cognitive health and vitality: a review of clinical implications. Geriatr Nurs 29:302–310

Yousefzadeh MJ, Zhu Y, McGowan SJ, Angelini L, Fuhrmann-Stroissnigg H, Xu M, Ling YY, Melos KI, Pirtskhalava T, Inman CL, McGuckian C, Wade EA, Kato JI, Grassi D, Wentworth M, Burd CE, Arriaga EA, Ladiges WL, Tchkonia T, Kirkland JL, Robbins PD, Niedernhofer LJ (2018) Fisetin is a senotherapeutic that extends health and lifespan. EBioMedicine 36:18–28

Zhang J, Wu J, Liu F, Tong L, Chen Z, Chen J, He H, Xu R, Ma Y, Huang C (2019a) Neuroprotective effects of anthocyanins and its major component cyanidin-3-O-glucoside (C3G) in the central nervous system: an outlined review. Eur J Pharmacol 858:172500

Zhang H, Yang X, Pang X, Zhao Z, Yu H, Zhou H (2019b) Genistein protects against ox-LDL-induced senescence through enhancing SIRT1/LKB1/AMPK-mediated autophagy flux in HUVECs. Mol Cell Biochem 455:127–134

Zhou T, Yan Y, Zhao C, Xu Y, Wang Q, Xu N (2019) Resveratrol improves osteogenic differentiation of senescent bone mesenchymal stem cells through inhibiting endogenous reactive oxygen species production via AMPK activation. Redox Rep 24(1):62–69

Anti-Inflammatory Nutrients and Nutraceuticals for Active and Healthy Aging

8

Sepideh Goudarzi and Mohammad Abdollahi

Abstract

Inflammaging is the continuation of a mild chronic asymptomatic inflammation over a long time in the body, followed by clinical symptoms of aging as the result. This process underlies many diseases associated with aging, including cancers, diabetes, and diseases with cognitive decline. There is a variety of anti-inflammatory agents to help prevent or ameliorate the process. Best and most frequent natural sources that contain anti-inflammatory agents are the food and herbs, out of which there are some dietary supplements manufactured as well.

Major anti-inflammatory components in food include certain amino acids, vitamins, and minerals, and most importantly, polyunsaturated fatty acids. As it is hard to precisely separate food from edible herbs—since some herbs are a part of the daily diet—the great family of polyphenols is discussed as the major anti-inflammatory in herbs, despite their existence in what is called food, as well. The molecular mechanisms of the most frequent and effective agents and their role in the prevention of certain diseases related to inflammaging are discussed. And finally, briefly categorized tables are composed, to compare the sources of the compounds in more detail.

Keywords

Inflammaging · Neuroinflammation · Polyunsaturated fatty acids · Polyphenols · Flavonoids

S. Goudarzi · M. Abdollahi (✉)
Department of Toxicology and Pharmacology, School of Pharmacy, Tehran University of Medical Sciences, Tehran, Iran

Toxicology and Diseases Group, Pharmaceutical Sciences Research Center (PSRC), The Institute of Pharmaceutical Sciences (TIPS), Tehran University of Medical Sciences, Tehran, Iran
e-mail: s-goudarzi@student.tums.ac.ir; Mohammad@tums.ac.ir

S. M. Nabavi et al. (eds.), *Nutrients and Nutraceuticals for Active & Healthy Ageing*,
https://doi.org/10.1007/978-981-15-3552-9_8

8.1 Inflammation

Inflammation is a process occurring in response to an infection or injury in the body. It involves many specific cell pathways and cytokine release that cope with the unusual cells; and it may come to production of cells or factors that might cause pathological conditions themselves, this is associated with the incapability of the immune system to distinguish self from non-self (Fasano 2012). This is where the inflammation can be potentially harmful. If the underlying cause of inflammation is not removed, the inflammatory processes will continue until they fall into a chronic inflammatory state.

Inflammaging refers to the condition in which a prolonged, low-grade inflammation is triggered and continued for a part of a life span, which is in this term, the aging span. It is at first subclinical, with no notable symptoms, but it will eventually show up with diseases, the predominant ones being cancers, and diseases related to cognitive decline as well as diabetes and other metabolic disorders. Table 8.1 overviews the common diseases and symptoms of inflammation. Franceschi et al. have described and discussed the pathophysiology associated with inflammaging in their review (Franceschi and Campisi 2014). As proposed by them, there are six main sources to inflammaging:

1. the damaged cells and macromolecules (which should have been eliminated but accumulated); these could mimic a possible inflammatory state as they appear pathogenic to the immune system.

Table 8.1 Common diseases associated with inflammaging

Neurologic and cognitive disorders	Alzheimer's Disease Parkinson's Disease Multiple Sclerosis Bipolar Mood Disorder Major Depression Disorder General Anxiety Disorder
Metabolic disorders	Diabetes Dyslipidemia
Cancers	Gastric cancer Colorectal cancer Pancreatic cancer Hepatocellular carcinoma Bladder cancer Prostate cancer
Eye and vision	Cataract Macular degeneration
Joint and skeletal	Arthritis
Cardiovascular diseases	Hypertension Ischemic heart diseases and atherosclerosis
Pulmonary diseases	Infection Lung cancer

2. products of the bacteria of the normal flora. Along with aging, the ability of the intestinal cells to prevent leakage to the inner tissues decreases, which allows more pathogenic molecules to be introduced to the immune system.
3. persistently senescent cells, because they produce high amounts of proinflammatory cytokines
4. Increased activation of the coagulation system
5. Immunosenescence; As adaptive immunity declines with aging, the innate immune system increases its activity, which does not possess the specificity of the adaptive immunity.
6. Dysregulation in the complement pathway, a cause for many degenerative diseases associated with aging

One of the most significant changes in chronic inflammation is the imbalanced ratio of proinflammatory to anti-inflammatory cytokines. Cytokines are the dominant factors in inducing or suppressing the inflammation. The mechanisms involved are generation of molecules with the ability to eliminate the undesired, such as ROS, or the iNOS protein; producer of NO, which plays an active role in inflammation in the neurons, joints, and lung (Sharma et al. 2007). Abundant ROS can cause DNA damage and genome instability, thereby inducing cancer. It also promotes tumor proliferation, angiogenesis, and survival (Wu et al. 2014). NF-κB is a protein activated by proinflammatory stimuli such as viruses, cytokines, oncogenes, toll-like receptors, etc. This protein is also tumorigenic as it mediates cell proliferation and angiogenesis (Wu et al. 2014). Moreover, the ROS, NO, cytokines, and prostaglandins are the main actors in inducing inflammation in the CNS, following different proteinopathies, which stand for the neurodegenerative state caused by over activated microglia (Glass et al. 2010).

During the subclinical period of *inflammaging*, lifestyle intervention, including exercise and healthy diet, contribute to prevent or delay the progression of inflammation.

Healthy diet, such as Mediterranean pattern, has been recognized as one of the best approaches to ameliorate the inflammatory status, due to the high amount of the bioactive compounds that have been found to be efficient in treatment of inflammation; some of them are also used as supplements or nutraceuticals. Many vitamins and minerals fall into this category; however, they are mainly taken by the normal daily diet. It is only of importance that necessary considerations be taken regarding the changes in the need for vitamins and minerals, especially those with an anti-inflammatory role, in the elderly.

8.2 Food

Poor nutrition contributes to many chronic aging diseases, including cardiovascular diseases, diabetes and most importantly, cancer. It is known what a poor diet might consist of; however, what exactly a "rich" or healthy diet might mean, varies upon individuals, for the beneficial and detrimental sides of the constituents might be difficult to describe. Subsequent studies at the end of the twentieth century focused on diets rich in plants and fruits, as they showed a promise to prevent cancer. These studies had focused on the radical scavenging and antioxidant characteristics of the natural constituents, including polyphenols, flavonoids, and tannins.

Dominant contents that a diet covers, which could influence the inflammatory state are vitamins, minerals, amino acids, and beneficial fats, i.e. long chain fatty acids.

8.2.1 Vitamins and Minerals

Vitamins and minerals are essential to the diet at any age; however, their amount of intake changes during aging. Some of them, such as Zinc, Folate, vitamin B12, Calcium, and vitamin D have an anti-inflammatory activity. It is only recommended that vitamins and minerals should be taken within the normal dosage range as they are normally found in a rich and healthy diet. Main mechanisms are charted in Table 8.2.

8.2.2 Amino Acids

Amino acids have a variety of functions in the body. Some have antioxidant and anti-inflammatory effects. Most effective ones are *Arginine*, *Glutamine*, *L-carnitine*, *Tryptophan*, *Creatine,* and *Taurine*. A number of in vivo studies evaluated their efficacy as a supplement/nutraceutical in reducing inflammation. Main mechanisms are charted in Table 8.3, in vivo studies in Table 8.4 as well.

Table 8.2 Main anti-inflammatory mechanisms of the vitamins/minerals

Vitamin/mineral	Anti-inflammatory mechanism	Ref.
Vitamin B12	May inhibit the inflammatory mediators	Hosseinzadeh et al. (2012)
Folate	Through folate receptor beta on activated macrophages	Poh et al. (2017)
Calcium	Major role is through calcium sensing receptors, which promote inflammatory response in deficiency of calcium	Hendy and Canaff (2016)
Zinc	Modulation of NF-κB, regulation of inflammatory cytokines	Gammoh and Rink (2017)
Vitamin D	Might reduce inflammatory markers specially NF-κB, modulates lymphocytes and immunoglobulins	de Carvalho and Ribeiro (2017)

Table 8.3 Main anti-inflammatory mechanisms of the amino acids

Amino acid	Anti-inflammatory mechanism	Ref.
Arginine	Suppresses T cell function	Raber et al. (2012)
Glutamine	Activates MKP-1 and inhibits NF-κB and MAPK	Kim et al. (2015)
L-carnitine	Inhibits NF-κB	Koc et al. (2011)
Tryptophan	Suppresses T cell function	Santhanam et al. (2016)
Creatine	Reduces inflammatory mediators, including TLRs and NF-κB	Riesberg et al. (2016)

8.2.3 Omega-3 Fatty Acids

Omega-3 fatty acids are polyunsaturated fatty acids (PUFA), having their last pi bond between C3 and C4. Most active biological omega-3 acids include eicosapentaenoic acid (EPA, 20:5n-3), docosapentaenoic acid (DPA, 22:5n-3), and docosahexaenoic acid (DHA, 22:6n-3). Omega 6 fatty acids are characterized by a double bond between the C6–7. These fatty acids have a major role in preventing the inflammation.

It is thought that PUFAs will integrate into the membrane of many cells, including immune cells, e.g. macrophages, lymphocytes, and neutrophils. This integration stands for the fatty acid composition, when responding to the inflammatory stimuli, i.e. the more (PUFA) in the membrane of the cells, the more they will be involved in the eicosanoid cascade (Calder 2015). This hypothesis has been proved in several studies (Healy et al. 2000; Faber et al. 2011; Thies et al. 2001; Kew et al. 2003). By an inflammatory trigger, phospholipase A2 will release the ARA from the membrane; however, it is clear that by a decreased amount of ARA, the omega-3 fatty acids will substitute the ARA, and stand as substrates for the COX and LOX enzymes. This has been further detailed below.

The cascade of biological inflammatory response involves two main pathways. The N-3 pathway results in the production of PGE3 (anti-inflammatory), the n-6 produces Dihomo-gamma-linolenic acid, which, if interrupted by the environmental factors, will be the substrate for the production of PGE1 (anti-inflammatory). But when continued, it produces Arachidonic acid, the primary substrate for COX, LOX, and CYP450 enzymes, responsible for PGE2 and leukotriene production (inflammatory). The two pathways are linked to same enzymes (mostly elongase and desaturase); but what determines the predominant pathway, will be the availability of the nutrients needed for each equation, as well as the metabolic environment (Mahan et al. 2017). Hence, a well-thought diet will have to make sure of adequate intake of n-3 fatty acids, and the supplements and minerals supporting the PGE1 and PGE3 production Table 8.5.

Table 8.4 Studies on the anti-inflammatory effects of amino acids

Target group	Amino acid	Study	Dosage	Duration	Outcome	Ref.
Animals	Taurine	Rats with liver injury	200 mg/kg/day	Single dose	Decreased inflammatory factors and liver transaminases significantly	Liu et al. (2017)
	L-carnitine	Albino mice	100 or 300 mg/kg/day	5 days	Reduced microglial activation and increased BDNF concentration	Kazak and Yarim (2017)
Humans		Healthy obese women	2 g/day	8 weeks	Therapy alone did not change the inflammatory markers	Rafraf et al. (2015)
		Patients with coronary artery diseases	1000 mg/day	12 weeks	Levels of inflammation markers were significantly reduced	Lee et al. (2015)
	Glutamine	Patients with abdominal malignancies treated with radiotherapy	30 g/day	From 3 days before radiotherapy until end of treatment	Decreased the inflammation caused by radiation	de Urbina et al. (2017)
	Creatine	Young athletes	0.3 mg/kg	7 days	Inhibited the increase of inflammatory markers after exercise	Deminice et al. (2013)

Table 8.5 Common herbs to prevent or fight inflammation

Herb	Content	Indications and dosage	Safety profile
Agrimony (*Agrimonia eupatoria*)	Catechin tannins Flavonoids (quercetin, Kaempferol, apigenin, luteolin) Procyanidins	No pharmacologic data available on the dosage (European medicine agency, committee on herbal medicinal products EMA/HMPC/680595/2013 n.d.) Cannot be recommended as a nutraceutical	No pharmacologic data available on the safety profile of systemic chronic and subchronic use (European medicine agency, committee on herbal medicinal products EMA/HMPC/680595/2013 n.d.) Cannot be recommended as a nutraceutical
Arnica (*Arnica montana*)	Sesquiterpene lactones Short-chained fatty acids: Acetic acid and isobutyric acid Flavone and flavonol glycosides	Approved by the commission E for rheumatism No oral dosage range	Classified by the US Food and Drug Administration as unsafe because of its toxicity It should not be taken orally
Brewer's yeast (*Saccharomyces cerevisiae*)	Mannans, glucans Group B vitamins	No indications reported regarding inflammaging A study of 500 mg daily for 12 weeks on rhinitis, has reported improvement, with no side effects (Moyad et al. 2009)	Classified as GRAS
English plantain (*Plantago lanceolata*)	Iridoids: Rhinantin, catalpol Flavonoids: Apigenin, luteolin Tannins: Aesculetin Salicylic acid	A polyherbal formulation containing plantago psyllium reduced blood glucose and serum lipids in 30 DM2 patients (Zarvandi et al. 2017). However, there isn't enough data on dosage and period of use for diabetes or hyperlipidemia (Ota and Ulrih 2017) A study on 18 PD patients proved benefits on levodopa pharmacokinetic profile, administering 3.5 g TDS of plantago husk (Fernandez-Martinez et al. 2014)	It contains allergenic substances and antigens, thus can result in flu-like symptoms and anaphylactoid reactions
Fenugreek (*Trigonella foenum-graecum*)	Saponins, known as graecunins (glycosides of diosgenin) Flavones: Vitexin, vitexin glycoside, quercetin Amino acids: Isoleucine,	Cholesterol lowering effect (with no significant change on HDL profile) (Valette et al. 1984; Singhal et al. 1982; Hannan et al. 2003; Stark and Madar 1993; Thompson Coon and Ernst 2003; Yadav et al. 2004; Sauvaire et al. 1991; Gupta et al. 2001)	No special adverse reactions, except for a probable hypoglycemia

(continued)

Table 8.5 (continued)

Herb	Content	Indications and dosage	Safety profile
	4-hydroxyisoleucin Sapinogens: Diosgenin, smilagenin, sarsasapogenin, tigogenin, neotigogenin, gitogenin, neogitogenin	Antidiabetic effect, seen especially with the amino acid 4-hydroxyisoleucine. The mechanism is through *rising plasma insulin, either secretion, or synthesis* The anti-inflammatory effects are proved on animals (Sur et al. 2001). In a 21-day trial on rats, the concentration of inflammatory mediators was decreased on fenugreek mucilage; the effects were concluded to be higher than that of indomethacin (Sindhu et al. 2018). Thus, the plant carries a promising anti-arthritic property. However, literature lacks clinical data The plant exerts its anti-inflammatory effects On neurological diseases as a neuroprotective, antidepressant, antianxiety as well as a modulator on cognitive functions (Sharma et al. 2017; Zameer et al. 2017)	
Flax (*Linum usitatissimum*)	Unsaturated fatty acids mainly ALA and linoleic acid Lignans: Secoisolarciresinol diglucoside	It has shown cholesterol lowering effect (Lucas et al. 2002; Kaul et al. 2008; Edel et al. 2015; Torkan et al. 2015) By the EMA, it is only approved for its mucilage effects and constipation relief	The US Food and Drug Administration has NOT classified flaxseed [oil] as GRAS. However, it does allow up to 12% (weight) flaxseed in food
German chamomile (*Matricaria recutita*)	Volatile oil: (−)-alpha-bisabolol (levomenol), bisabolol oxide A, bisabolol oxide B, bisabololone oxide A, chamazulenespathulenol Flavonoids: Apigenin, luteolin, chrysoeriol, apigenin-7-O-glucoside, apigenin glucoside acetate, quercetin, isorhamnetin, patuletin	On scopolamine-induced memory impaired rats, the drug has improved the neuroinflammation through modulation of the cholinergic system and antioxidant activity in the hippocampus (Ionita et al. 2018) A study on type-2 DM patients, has reported reduction in glycemic index with the extract (Zemestani et al. 2016) A clinical trial has proved that the extract can be a safe and effective treatment for moderate cyclic mastalgia (Saghafi et al. 2018)	No special side effects reported with the studies mentioned (Zemestani et al. 2016; Saghafi et al. 2018; Mao et al. 2016)

	Hydroxycoumarins: Umbelliferone, herniarin Mucilages: Rhamanogalacturonane	A study has proved long-term use of chamomile extract to improve moderate to severe GAD symptoms, but not the rate of relapse (Mao et al. 2016) The study consisted of two phases. Phase 1 was defined as 12 weeks of open-label treatment with chamomile extract 500 mg TDS for 12 weeks; and phase 2 double-blind randomized trial, in which patients were divided into two groups, receiving either the extract or placebo for 26 weeks in substitution No established pharmaceutical use by EMA by 2011 (EMA/HMPC/55843/2011 n.d.)	
Oats (*Avena sativa*)	Fiber (beta-gluacan), manganese, selenium, phosphorous, tryptophan, thiamine, and vitamin E (mainly as alpha-tocopherol) Unsaturated triglycerides enzymes: Lipase, lipoxygenase, superoxide dismutase	A study has shown reduction in need for laxatives and increased bioavailability of vitamin B12, as well as maintaining body weight in the elderly at nursing homes who received oat bran in addition to food (Sturtzel et al. 2010) Many, but not all, trials show reductions in LDL-cholesterol using beta-glucan (Cugnet-Anceau et al. 2010; Liatis et al. 2009b; Hallikainen et al. 2006; Chen et al. 2006; Queenan et al. 2007). Oat fiber can decrease cholesterol levels modestly. Eight grams per day (Chen et al. 2006) of the soluble fiber or 3 g/day of beta-glucan (Liatis et al. 2009b; Maki et al. 2010) has been recommended in the studies. The study reports 5 mg/dL decrease in LDL upon addition of 3 g/day of beta-glucan to American diet (Maki et al. 2010)	Classified as GRAS by US Food and Drug Administration (FDA) when taken as food. Safety and efficacy of dosages higher than those found in food are unproven
St. John's wort (*Hypericum perforatum*)	Flavonoids: Isoquercetin, quercetin, phenols	It shows anti-inflammatory effects by inhibition of arachidonic acid release and inflammatory-cytokine production (Panossian et al. 1996) SJW extracts exhibit neuroprotection, improve cognitive function as well as decreasing Amyloidβ plaques (Hofrichter et al. 2013; Cao et al. 2017) Mild antidepressive effects in elderly (Ng et al. 2017;	May result in photosensitivity and allergic reactions Its vast interactions profile with drugs may limit its consumption

(continued)

Table 8.5 (continued)

Herb	Content	Indications and dosage	Safety profile
		Apaydin et al. 2016) Has shown antiaging effects on mice receiving 135 mg/kg/day for 4 weeks of the lyophilized extract (Mohammadirad et al. 2013)	
California peppertree (*Schinus molle*)	Volatile oil: Alpha-phellandrene, beta-phellandrene, limonene Flavonoids: Kaempferol, myricetin, quercetin Resins Mucilages	Because of its anti-inflammatory effects, might be effective in rheumatism, however, pharmacological data lack	No pharmacological data available on the use as a nutraceutical A study on rats reported safety in dosages used as an insecticide (Ferreroa et al. 2007)
Cape Aloe (*Aloe ferox*)		Purgative and laxative (van Wyk and Smith 2008; Hutchings et al. 1996; Grace et al. 2008; Chen et al. 2012; Cock 2015) Rheumatoid arthritis Topically used to treat erythema Available in forms of capsule, cream, gel	No pharmacological data available
Sage (*Salvia officinalis*)	Essential oil Flavones Phenolic acids Phenylpropanoid glycosides (martynoside) Terpenes and terpenoids: Camphor, thujone	It has shown memory enhancement in animals and human clinical trials. (doses of 300 and 600 mg) (Kennedy et al. 2006; Scholey et al. 2008; Kennedy and Scholey 2006) Might be used to treat Alzheimer's disease (333 mg) (Akhondzadeh and Abbasi 2006; Akhondzadeh et al. 2003) Topically used to treat erythema	Its thujone and camphor are recognized as neurotoxic when taken at higher doses Lima et al. (2004)
Strawberry (*Fragaria vesca*)	Tannins Flavonoids: Anthocyanin Quercetin	Reduces NF-κB activity, upregulates endothelial NOS activity (Basu et al. 2014) There are no enough data on established clinical use of the herb (Basu et al. 2014; EMA/HMPC/432276/2015 n.d.)	Might cause allergic reactions due to allergens Strawberries raise the plasma level of potassium, which might result in renal stones in prone patients Great care must be given in patients with heart failure

Turmeric (*Curcuma longa)*	Curcumin Monoterpenes Sesquiterpenes	Anti-hyperlipidemia Antitumor	GRAS Contraindicated in patients with bile duct obstruction, liver disease, gallstones and any bile-related disease
Ginseng (*Panax ginseng)*	Terpenoids: Sapogenins, oleanolic acid Group B vitamins	Corticosteroidal activity Clinical data shows: Enhanced cognitive performance, hypoglycemic activity, cancer prevention, enhanced quality of life, immunomodulatory effects	The long-term safety has not been established, but it is better not to be used in long term. Scientifically unproven data suggest preventing the use of Ginseng when under hormone therapy or receiving stimulants like coffee or antipsychotics, as well as ADHD patients and those with anxiety or mania

Not all the contents of a plant are mentioned, only the ones related to an anti-inflammatory effect
Not all the medicinal and other uses are mentioned, only those related to inflammaging
The contents are taken from Gruenwald et al. (2007)

Linoleic acid (18:2n-6)
↓
gamma Linolenic acid (18:3n-6)
↓
Dihomo-gamma-linolenic acid (20:3n-6)
↓
Arachidonic acid (20:4n-6)
↓
Adrenic acid

alpha Linolenic acid (18:3-n3)
↓
Stearidonic acid (18:4n-3)
↓
Eicosatetraenoic acid (20:4n-3)
↓
Eicosapentaenoic acid (EPA, 20:5n-3)
↓
Docosapentaenoic acid (DPA, 22:5n-3)
↓
Docosahexaenoic acid (DHA, 22:6n-3)

Apart from the role of Omega-3 fatty acids in enhancing the lipid profile, their role in the prevention of Arachidonic acid-derived products will induce the hypothesis of preventing inflammation caused by 2-series prostanoids, such as rheumatoid arthritis (RA). Some clinical trials have reported anti-inflammatory effects of omega-3 rich diets in the RA (Kremer et al. 1985, 1987; Cleland et al. 1988).

8.2.3.1 Dosage Considerations

There seems to be a threshold for the anti-inflammatory responses of n-3 fatty acids (Calder 2015). Based on a trial, an intake of more than 2 grams per day of DHA + EPA is needed to trigger the effects (Calder 2015; Rees et al. 2006). On the other hand, it is also important not to go beyond a maximum border of daily intake, this will suppress the arachidonic acid production, affecting the balance of the two pathways (Mahan et al. 2017).

8.2.3.2 Omega3 Fatty Acids in RA

A number of clinical trials supported omega-3 supplementation alongside rheumatoid medications. A trial on almost 700 patients with early RA suggests that taking higher amounts of omega 3 a year before the initiation of *disease modifying antirheumatic drug* (DMARD), might yield to better outcomes of the drug therapy (Lourdudoss et al. 2017). Level of omega-3 is also associated with the inflammatory/refractory pain of early RA patients receiving methotrexate for 3 months (Lourdudoss et al. 2018). A study evaluated the rate of concomitant need to analgesics after consumption of 3.6 g of EPA and 4.2 g of DHA daily for 12 weeks (Rajaei et al. 2016). They reported an improved disease state as well as a reduction in need for analgesics.

8.2.3.3 Omega3 Fatty Acids in Cardiovascular Diseases

Omega-3 can impact the cardiovascular system through several mechanisms. They are capable of decreasing blood cholesterol and triglyceride, preventing vessel stiffness and atherosclerosis. Omega-3 can relax the smooth muscle of the veins, which will enhance vasodilation. They also promote endothelial function. With their antioxidant activity besides their anti-inflammatory effects, they are considered key nutrients in cardiovascular complications of aging. Several studies have evaluated their role as nutrients in cardiovascular system (Pase et al. 2015; Casanova et al. 2017; Colussi et al. 2017).

8.2.3.4 Omega 3 Fatty Acids and Neuroinflammation

There are plenty of data on the role of omega-3 in preventing cognitive decline or maintaining metabolic status in neurodegenerative diseases. Both anti-inflammatory and antioxidant activity of omega-3 are the main mechanisms for these effects. Moreover, long-term metabolic imbalances like dyslipidemia are risk factors for the Alzheimer's disease, which could be modulated with omega-3 supplementation.

The major neuroprotective role of omega-3 is their lipid balancing property that is discussed in studies, rather than an antioxidant activity. As amyloid precursor protein (APP) and the secretases are integrated into the lipid bilayer of the cells, lipid balance will be of importance (Grimm et al. 2017). Although trials suggest positive changes in memory function and cognition, there is no definite conclusion (Grimm et al. 2017; Shinto et al. 2014; Samieri et al. 2017; Külzow et al. 2016). Yet, as omega-3 is generally regarded as safe (GRAS), a usual and safe daily intake could be recommended, both as prevention or supplementation of an Alzheimer's or Parkinson's disease (AD or PD) patients (Taghizadeh et al. 2017).

8.2.4 Immunomodulating Polysaccharides

There are many natural compounds who influence the immune system, especially the innate one. Beta-glucans are polysaccharides based on glucose monomers found in the cell walls of bacteria and fungi. They can be taken orally as dietary fiber or supplement, and are well-known for their cholesterol lowering effects and immunomodulation (Liatis et al. 2009a; Behall et al. 2004). A low dose of 3 g/day of beta-glucan (available as supplements) seems to be effective in mildly reducing cholesterol (Bashir and Choi 2017). Their anti-inflammatory effects are widely used in respiratory conditions like asthma and respiratory tract infections (RTI), which is not necessarily related to aging (Bashir and Choi 2017; Sarinho et al. 2009; Talbott and Talbott 2009, 2012).

8.2.5 Chondroitin and Glucosamine

Glucosamine and chondroitin are dietary supplements used in osteoarthritis because of their cartilage building property. As osteoarthritis is a condition with which mostly elderly cope, these compounds could play a role in healthy aging. However, two reviews over their efficacy have reported no significant benefit from consumption of the supplements in treating osteoarthritis (Henrotin et al. 2012; Runhaar et al. 2017). However, as they report that more data is needed to confirm the (in) effectiveness, they could still be recommended due to the low adverse effect profile.

Their anti-inflammatory effects are carried out through inhibiting NF-κB. A trial of 18 healthy overweight individuals measured some inflammatory biomarkers, including C-reactive protein (CRP), after 30 days of 1500 mg glucosamine and 1200 mg chondroitin consumption per day (Navarro et al. 2015). 23% decrease in serum CRP was reported, and the authors concluded that this supplementation may lower systemic inflammation. However, the data still lack in this area and more trials have to be run to further investigate the effects.

8.3 Polyphenols

There are a variety of chemical compounds found in herbs and natural sources that have an anti-inflammatory role. Phenols and polyphenols are a wide group of compounds, having an anti-inflammatory activity along with antiseptic and antioxidant activities, including the simple salicylic acid, rosmarinic acid, and the complex phenolic glycosides. Phenols contain one aromatic ring, while polyphenols have more than one. They might either be unconjugated or bound to a saccharide molecule, and less often, to lipids, amino acids, etc. These compounds have been widely discussed in Chap. 7.

8.3.1 Tannins

The term "tannin" refers to phenolic substances used in tanning (dyeing brown) leather and textiles. However, the concentration of these polyphenols is too low to make them functional for conversion into leather.

Tannins are most used to tighten up loose or damaged tissues, as in burns, varices or diarrhea (to dry excess secretions). Tannins are subdivided into two major groups. Proanthocyanidin groups or condensed tannins and ellagic acid-derived.

8.3.2 Flavonoids

Flavonoids are the most common polyphenols. They are known with a 6-3-6 (no. of the ring carbons) structure. The middle ring stands for the division of the family into chalcones, anthocyanins and proanthocyanidins, dihydrochalcones, flavonones, flavonols, flavones, flavonols, and isoflavones. Many flavonoids have anti-inflammatory effects. Their mechanisms will be discussed further in the current chapter.

Fruits have an abundant amount of flavonols and proanthocyanidins. Citrus family contains further types of flavonoids like flavonones and flavones. Vegetables are rich sources of flavonols, especially quercetin, which is a great anti-inflammatory agent. Celery and lettuce contain falvones.

8.3.3 Quercetin (Meletin/Sophretin)

Quercetin is a flavanol, belonging to the flavonoid's family. It is a well-known anti-inflammatory molecule that exerts its effects via inhibiting the NF-κB, resulting in prostanoid synthesis, cytokine production, and iNOS expression (Comalada et al. 2005; Morikawa et al. 2003). It is also thought to be a neuroprotective, preventing neuronal cells from inflammatory injuries (Chen et al. 2005). Together with its antioxidant effects, the molecule could be effective against cognitive impairment, especially in neurodegenerative diseases, where taupathy and Beta-amyloidosis are the problems, i.e. AD. A study on murine models of AD shows that quercetin can decrease Beta-amyloid and taupathy in the hippocampus and amygdala (Sabogal-Guáqueta et al. 2015). Shen et al. have proved a reduction in the rate of tau protein hyper phosphorylation induced by okadaic acid in vitro (Shen et al. 2018). Furthermore, in another AD mouse model, mitochondrial function was improved, as well as beta-amyloid plaque reduction, proved by higher levels of ROS and ATP in hippocampus mitochondria (Wang et al. 2014). Quercetin increased the AMPA protein kinase significantly, suggesting a mechanism by which it improves cognition.

The issue over this molecule is the pharmacokinetic profile. Quercetin undergoes significant first pass elimination and thus a high proportion of it is lost. Moreover, further in vivo data is needed to confirm the bioavailability of the drug to the brain, as the molecule can barely pass the blood–brain barrier (BBB) (Babaei et al. 2018).

Javadi et al. have studied the effects of 500 mg/day of quercetin for 8 weeks on women with rheumatoid arthritis. Quercetin could significantly reduce inflammatory markers and clinical symptoms, i.e. pain in the morning and after activity and the early morning stiffness (Javadi et al. 2017).

8.3.4 Isoflavonoids

Phytoestrogens are substances with estrogenic activity, though not steroids in structure, and they include isoflavones, coumarins, stilbenes, and lignans. "Isoflavonoid phytoestrogens" are extensively used and studied, due to their contribution in human health. They are commonly referred to as isoflavonoids. The most common types of isoflavonoids, Genistein (4′,5,7-trihydroxyisoflavone), daidzein (4′,7- dihydroxyisoflavone), and glycetein (4′,7-dihydroxy-6-methoxyisoflavone) are found in the soy. They are mostly referred to as antioxidants, anti-inflammatory, and antitumor agents, due to their estrogenic activity. The antioxidant characteristics are discussed in Chap. 7.

8.3.5 Actions of Isoflavonoids in Human Health

8.3.5.1 Cancer and Chronic Inflammatory Disease Prevention

Genistein is an estrogen agonist. By binding to the conventional nuclear estrogen receptor, it can mimic estrogenic effects. Moreover, there has recently been interest in its effects on G protein-coupled estrogen receptor (GPER), one of the massive study areas in breast cancer. This receptor is activated by estrogens (and probably other similar ligands) and its activation leads to phosphorylation of mitogen-activated protein kinases/extracellular signal-regulated kinases (MAPKs/ERK) pathway, involving the adenylate cyclase and phospholipase C, and will be continued through insulin-like growth factor-1 receptor (IGF-1R) and epidermal growth factor receptor (EGFR) (Kuo et al. 2010; Madeo and Maggiolini 2010; Maggiolini et al. 2004; Meyer et al. 2011; Prossnitz and Barton 2011). This can be a way to estrogen's anti-inflammatory and antitumor activity (Blasko et al. 2009; Rettew et al. 2010), and as the receptor can be found on microglia, its effects on the nervous system. This might stand for the mechanism that genistein can modulate the inflammatory response of microglia, as it is demonstrated in a recent study by Du et al. (Du et al. 2018).

By the stimulation of microglia by the antigen, interleukin (IL)-1β, tumor necrosis factor (TNF)-alpha, and IL-6 are the major mediators produced, in which MAPKs signaling pathway, i.e. ERK, p38, and Jun N-terminal kinase (JNK) are involved. They are responsible for the release of inflammatory mediators, including NO (by iNOS) and TNF-α. Du et al.'s study on BV2 microglia demonstrated that genistein could inhibit the MAPK pathway (Du et al. 2018). Microglial stimulation will also free NF-κB, via the TLR4, and results in its nuclear translocation (Reed-Geaghan et al. 2009). NF-κBs are regulators of the immune response. They are one of the first signals activated by immunogens. ROS, heavy metals, infections, interleukins, and TNF are of the mediators that can induce their response. The NF-κB will then bind to the promoter region of these inflammatory mediators. Many studies involving the NF-κB pathway use the LPS of Gram-negative bacteria for stimulation. NF-κBs are both involved in acute and chronic inflammations, including diseases like the rheumatoid arthritis, as well as in memory and neural survival.

8.3.5.2 NF-κB and Inflammation

A study on the molecular effects of genistein reported several mechanisms through which genistein can exert anti-inflammatory effects (Jeong et al. 2014). Genistein could block the nuclear translocation of the NF-κB and thereby regulate the inflammatory response, by reduction of the reactive oxygen species. It also attenuated the production of TNF-alpha and IL-1β generation in LPS-stimulated microglia and thus the neuroinflammation. The study further declares that genistein can also inhibit both Cox-2; resulting in inhibition of inflammatory prostaglandins (PGE2), and iNOS, resulting in lower NO production.

A recent study evaluated the anti-inflammatory effects of 8-odthohydroxydaidezin (8-OHD), an analogue of genistein, on the same cell line (BV2 microglia), and compared them with genistein (Wu et al. 2018). The results demonstrated better effects with the 8-OHD rather than genistein in preventing NO production, through downregulation of iNOS expression. It was also proved that 8-OHD could more effectively reduce the ROS released by LPS-activated microglia. However, in COX-2 inhibition, genistein could be slightly more effective.

These anti-inflammatory mechanisms contribute to genistein effects on preventing breast cancer and chronic inflammatory diseases, such as R.A. There are some synthetic agents found as GPER agonists, without feminizing effects; however, genistein's function on other estrogen receptors, including its effects on the reproductive system, has been approved as it is used in menopausal symptoms.

8.3.5.3 Prevention of Cognitive Decline

Overactivation of microglia in chronic neuroinflammation can result in neural damage and degeneration, the highly believed etiology of neurodegenerative disorders. Mechanisms are discussed further.

8.3.5.4 NF-κB and Memory

Based on the anti-inflammatory effects of estrogen, it is hypothesized that it is able to attenuate neurodegeneration (GDNF contributes to estrogen-mediated protection of midbrain dopaminergic neurons n.d.; The possible role of estrogen and selective estrogen receptor modulators in a rat model of Parkinson's disease n.d.; Estradiol protects dopaminergic neurons in a MPP+ Parkinson's disease model n.d.). Studies have proved the role of TLR4 and the NF-κB in the inflammation in neurodegeneration, suggesting an increase in their amount or activity, since TLR4 is able to recognize Amyloidβ (Okun et al. 2009; Liao et al. 2011; Frank et al. 2009). A study has confirmed that genistein could reverse the inflammatory effects of Amyloidβ, i.e. decreasing the production of IL-1β and iNOS in vitro (Zhou et al. 2014). In the study, cells treated with Amyloidβ showed a significant increase in NF-κB and TLR4 production, and genistein could reverse the signs by downregulating the expression of TLR4 and NF-κB (refer to NF-κB and inflammation).

Studies report that GPER activation, associated with the modulation of microglial activity, will yield to the neuroprotective effects on DA nigral neurons (Bessa et al. 2015; Bourque et al. 2012; Mendes-Oliveira et al. 2017; Zhao et al. 2016). In the study, DA neural loss by LPS-induced microglia was treated with GPER agonist, which will result in reduced NO and phagocytic activity; the pathway that is normally stimulated via the 17-β estradiol (estrogen). Moreover, blockade of GPER resulted in reduction of phagocytic activity and iNOS gene expression. Estrogen exerts its anti-inflammatory effects via the classical estrogen receptors (ER alpha and beta) too; in fact, agonizing their receptors also attenuate microglial NO production. However, the study utilized the synthetic agonists/antagonists (G1/G15), with neither feminizing effects nor difficulty passing the BBB. It has been proved that genistein can easily pass the BBB, (Mozolewski et al. 2017) but its effects on the reproductive system, might make it not so compatible as a long-term nutraceutical treatment, except for proper cases such as postmenopausal women. A recent review has proposed that polyphenols may be potentially beneficial to develop new drugs against neuroinflammation by targeting the TLR4 pathway (Rahimifard et al. 2017).

8.3.5.5 Lipid Profile and Cardiovascular Effects

Genistein can reduce the Apo lipoprotein B secretion from hepatocytes. Microsomal triglyceride transfer protein (MTP) is an essential molecule for lipoprotein assembly, which helps transfer esterified cholesterol (EC), triglyceride (TG), and phospholipid (PL) to the recently produced apolipoprotein-B (ApoB) molecules. It is thus essential for the release of ApoB. Genistein reduces both the expression and activity of MTP in the cell (Borradaile et al. 2002), meaning that in long-term use, it might be able to keep MTP at lower concentrations, resulting in lower release of VLDL and chylomicrons. It has been also demonstrated that both genistein and daidzein can decrease the rate of LDL-receptor expression and activity, which will result in higher LDL re-uptake. Genistein can also inhibit Acyl-coenzyme A (CoA): cholesterol acyltransferases (ACATs) which will prevent cholesterol from getting esterified and apolipoproteins from getting assembled. This will reduce the lipoprotein secretion, a mechanism close to 3-hydroxy-3-methyl-glutaryl-coenzyme A (HMG-CoA) reductase inhibitors.

Cardioprotective effects of estrogen, especially ERβ receptor agonists, including genistein are well-known (Kararigas et al. 2014; Hsieh et al. 2015; Fliegner et al. 2010; Pedram et al. 2013). After an ischemic heart crisis, reperfusion will cause an inflammatory condition, maintained by accumulation of the leukocytes. This is triggered by some inflammatory cytokines, including TNF-a, which will increase the expression of Intracellular adhesion molecule-1 (ICAM-1) (also known as CD54) production, a factor for the leukocytes' intercellular adhesion. Genistein can attenuate the expression of ICAM-1, by reducing TNF-α in a dose dependent manner. Thus, it is considered as a leukocyte-endothelium interaction inhibitor, resulting in the improved atherosclerotic condition (Kayisli et al. 2013).

8.3.5.6 Menopausal Symptoms and Osteoporosis

There has been a huge load of data on the estrogen-mimicking effects of genistein. It is a full agonist of ERβ, and this receptor is a negative inhibitor of ER alpha. Thus, genistein and phytoestrogens act through anti-growth properties, whereas ERalpha has pro-growth effects. Phytoestrogens are capable of ameliorating vasomotor symptoms, depressive moods, osteopenia, and vaginosis and could probably be taken in proper doses for months or years. There is also some clinical evidence that it might be effective in osteoporosis (Kayisli et al. 2013). However, data regarding vaginal atrophy still lack (Le Donne et al. 2011; Manonai et al. 2006).

8.3.5.7 Potential Adverse Effects

There might have been some concern about the feminizing effects of isoflavonoids; however, many studies do not approve reproductive problems or infertility, neither in pediatrics nor in adults. In fact, it is suggested that infants who are breastfed from mothers whose diets are soy-rich, have lower incidences of cancer, with respect to the adults with a soy-rich diet (Franke et al. 1999). A study has reported no special feminizing effects on male (Mitchell et al. 2001).

As isoflavonoids are substrates for thyroid peroxidase, high-dose isoflavonoid might be thought as an inducer of hypothyroidism. However, a recent clinical trial on almost 400 postmenopausal women who took up to 54 mg/day of genistein over 2–3 years, reported no significant decline in thyroid hormones (Bitto et al. 2010). As long as iodine storage is sufficient, no potential harm might threat thyroid conditions (Marini et al. 2012; Bruce et al. 2003).

8.3.5.8 Dietary Intake and Herbal Sources

Although there are many soy-based supplements available in the market, it is recommended to consume soy-rich diets instead, as a nutraceutical, as not to lose the other components in the plant, that would enhance the efficacy, or cause synergism. Moreover, this approach will ensure a safe amount of intake rather than supplements.

Genistein is mostly found in leguminosae plants; beans, peas, and peanuts as ones found in the normal human diet. In some fruits, flavonols are mostly found in the skin, and thus peeling will substantially reduce the intake. Other herbs containing flavonoids can be found in Table 8.3.

8.3.6 Cannabinoids

Cannabis smokers are less capable of recovering from fungal and bacterial infections. This is because the major cannabinoid found in cannabis, delta9-tetrahydrocannabinol (THC) decreases the production of proinflammatory cytokines, acting like an immunosuppressant. THC is a lipid soluble substance that will instantly distribute to different tissues, including the brain.

Cannabidiol (CBD) is a non-psychoactive cannabinoid that has anticonvulsive, sedative, anti-inflammatory, and hypnotic effects. It is thought to have a neuroprotective role in ischemic conditions. However, a systematic review has reported that it is not possible to conclude any efficacy out of it, regarding its effects on neurodegenerative disorders (Lim et al. 2017). Moreover, cannabinoids do not show any effects of note on the conditions rumored to be effective; i.e. multiple sclerosis (MS) (Sexton et al. 2014), anorexia in cancer or acquired immunodeficiency syndrome (Whiting et al. 2015), chemotherapy-induced nausea and vomiting (Whiting et al. 2015), and the pain related to cancer or rheumatism (Häuser et al. 2017).

8.4 Taken Together

Prevention of inflammation could be beneficial at any age. Many of the substances mentioned in the chapter are able to confront inflammation through several pathways, preventing cancers, diabetes dyslipidemia, atherosclerosis, etc. However, the dosage range and long-term use data for many of them still lack, and especially those thought to have effects on neurological disorders, have not proven favorable changes on the outcome. Nevertheless, most of them, including long chain PUFAs, vitamins, minerals, and isoflavonoids are regarded as safe within the normal consumption range; thus, taking a safe daily portion of them is recommended. Anti-inflammatory herbs are mostly not safe to be taken long-term daily, but per case and in courses of time. The daily intake limit for herbs shall be taken into consideration as well as their possible interactions' profile with medications, which is a part of the life of many elderlies.

References

Akhondzadeh S, Abbasi SH (2006) Herbal medicine in the treatment of Alzheimer's disease. Am J Alzheimers Dis Other Dement 21(2):113–118

Akhondzadeh S, Noroozian M, Mohammadi M, Ohadinia S, Jamshidi A, Khani M (2003) Salvia officinalis extract in the treatment of patients with mild to moderate Alzheimer's disease: a double blind, randomized and placebo-controlled trial. J Clin Pharm Ther 28(1):53–59

Apaydin EA, Maher AR, Shanman R, Booth MS, Miles JN, Sorbero ME, Hempel S (2016) A systematic review of St. John's wort for major depressive disorder. Syst Rev 5(1):148

Babaei F, Mirzababaei M, Nassiri-Asl M (2018) Quercetin in food: possible mechanisms of its effect on memory. J Food Sci 83(9):2280–2287

Bashir KM, Choi JS (2017) Clinical and physiological perspectives of β-glucans: the past, present, and future. Int J Mol Sci 18(9):1906

Basu A, Nguyen A, Betts NM, Lyons TJ (2014) Strawberry as a functional food: an evidence-based review. Crit Rev Food Sci Nutr 54(6):790–806

Behall KM, Scholfield DJ, Hallfrisch J (2004) Lipids significantly reduced by diets containing barley in moderately hypercholesterolemic men. J Am Coll Nutr 23:55–62

Bessa A, Campos FL, Videira RA, Mendes-Oliveira J, Bessa-Neto D, Baltazar G (2015) GPER: a new tool to protect dopaminergic neurons? Biochim Biophys Acta (BBA) - Mol Basis Dis 1852 (10):2035–2041

Bitto A, Polito F, Atteritano M, Altavilla D, Mazzaferro S, Marini H, Adamo EB, D'Anna R, Granese R, Corrado F, Russo S (2010) Genistein aglycone does not affect thyroid function: results from a three-year, randomized, double-blind, placebo-controlled trial. J Clin Endocrinol Metabol 95(6):3067–3072

Blasko E, Haskell CA, Leung S, Gualtieri G, Halks-Miller M, Mahmoudi M, Dennis MK, Prossnitz ER, Karpus WJ, Horuk R (2009) Beneficial role of the GPR30 agonist G-1 in an animal model of multiple sclerosis. J Neuroimmunol 214(1–2):67–77

Borradaile NM, Wilcox LJ, Edwards JY, Murray WH (2002) Soya phytoestrogens, genistein and daidzein, decrease apolipoprotein B secretion from HepG2 cells through multiple mechanisms. Biochem J 366(2):531–539

Bourque M, Dluzen DE, Di Paolo T (2012) Signaling pathways mediating the neuroprotective effects of sex steroids and SERMs in Parkinson's disease. Front Neuroendocrinol 33 (2):169–178

Bruce B, Messina M, Spiller GA (2003) Isoflavone supplements do not affect thyroid function in iodine-replete postmenopausal women. J Med Food 6(4):309–316

Calder PC (2015) Marine omega-3 fatty acids and inflammatory processes: effects, mechanisms and clinical relevance. Biochim Biophys Acta, Mol Cell Biol Lipids 1851(4):469–484

Cao Z, Wang F, Xiu C, Zhang J, Li Y (2017) Hypericumperforatum extract attenuates behavioral, biochemical, and neurochemical abnormalities in aluminum chloride-induced Alzheimer's disease rats. Biomed Pharmacother 91:931–937

Casanova MA, Medeiros F, Trindade M, Cohen C, Oigman W, Neves MF (2017) Omega-3 fatty acids supplementation improves endothelial function and arterial stiffness in hypertensive patients with hypertriglyceridemia and high cardiovascular risk. J Am Soc Hypertens 11 (1):10–19

Chen JC et al (2005) Inhibition of iNOS gene expression by quercetin is mediated by the inhibition of I(kappa)B kinase, nuclear factor-kappa B and STAT1, and depends on heme oxygenase-1 induction in mouse BV-2 microglia. Eur J Pharmacol 521(1–3):9–20

Chen J, He J, Wildman RP, Reynolds K, Streiffer RH, Whelton PK (2006) A randomized controlled trial of dietary fiber intake on serum lipids. Eur J Clin Nutr 60(1):62–68

Chen W, van Wyk B, Vermaak I, Viljoen AM (2012) Cape aloes—a review of the phytochemistry, pharmacology and commercialisation of Aloe ferox. Phytochem Lett 5:1–12

Cleland LG, French JK, Betts WH, Murphy GA, Elliott MJ (1988) Clinical and biochemical effects of dietary fish oil supplements in rheumatoid arthritis. J Rheumatol 15(10):1471–1475

Cock IE (2015) The genus aloe: phytochemistry and therapeutic uses including treatments for gastrointestinal conditions and chronic inflammation. In: Novel natural products: therapeutic effects in pain, arthritis and gastro-intestinal diseases. Springer, Basel, pp 179–235

Colussi G, Catena C, Novello M, Bertin N, Sechi LA (2017) Impact of omega-3 polyunsaturated fatty acids on vascular function and blood pressure: relevance for cardiovascular outcomes. Nutr Metab Cardiovasc Dis 27(3):191–200

Comalada M et al (2005) In vivo quercitrin anti-inflammatory effect involves release of quercetin, which inhibits inflammation through downregulation of the NF-kappaB pathway. Eur J Immunol 35(2):584–592

Cugnet-Anceau C, Nazare JA, Biorklund M et al (2010) A controlled study of consumption of beta-glucan-enriched soups for 2 months by type 2 diabetic free-living subjects. Br J Nutr 103 (3):422–428

de Carvalho CG, Ribeiro SM (2017) Aging, low-grade systemic inflammation and vitamin D: a mini-review. Eur J Clin Nutr 71(4):434

de Urbina JJ, San-Miguel B, Vidal-Casariego A, Crespo I, Sánchez DI, Mauriz JL, Culebras JM, González-Gallego J, Tuñón MJ (2017) Effects of oral glutamine on inflammatory and autophagy responses in cancer patients treated with abdominal radiotherapy: a pilot randomized trial. Int J Med Sci 14(11):1065

Deminice R, Rosa FT, Franco GS, Jordao AA, de Freitas EC (2013) Effects of creatine supplementation on oxidative stress and inflammatory markers after repeated-sprint exercise in humans. Nutrition 29(9):1127–1132

Du ZR, Feng XQ, Li N, Qu JX, Feng L, Chen L, Chen WF (2018) G protein-coupled estrogen receptor is involved in the anti-inflammatory effects of genistein in microglia. Phytomedicine 43:11–20

Edel AL, Rodriguez-Leyva D, Maddaford TG, Caligiuri SP, Austria JA, Weighell W, Guzman R, Aliani M, Pierce GN (2015) Dietary flaxseed independently lowers circulating cholesterol and lowers it beyond the effects of cholesterol-lowering medications alone in patients with peripheral artery disease. J Nutr 145(4):749–757

EMA/HMPC/432276/2015 (n.d.)

EMA/HMPC/55843/2011 (n.d.)
Estradiol protects dopaminergic neurons in a MPP+ Parkinson's disease model (n.d.)
European medicine agency, committee on herbal medicinal products EMA/HMPC/680595/2013 (n.d.)
Faber J, Berkhout M, Vos AP, Sijben JW, Calder PC, Garssen J, van Helvoort A (2011) Supplementation with a fish oil-enriched, high-protein medical food leads to rapid incorporation of EPA into white blood cells and modulates immune responses within one week in healthy men and women. J Nutr 141(5):964–970
Fasano A (2012) Leaky gut and autoimmune diseases. Clin Rev Allergy Immunol 42(1):71–78
Fernandez-Martinez MN, Hernandez-Echevarria L, Sierra-Vega M, Diez-Liebana MJ, Calle-Pardo-A, Carriedo-Ule D, Sahagún-Prieto AM, Anguera-Vila A, Garcia-Vieitez JJ (2014) A randomised clinical trial to evaluate the effects of Plantago ovata husk in Parkinson patients: changes in levodopa pharmacokinetics and biochemical parameters. BMC Complement Altern Med 14(1):296
Ferreroa A, Minettib A, Brasa C, Zanettia N (2007) Acute and subacute toxicity evaluation of ethanolic extract from fruits of Schinusmolle in rats. J Ethnopharmacol 113(3):441–447
Fliegner D, Schubert C, Penkalla A, Witt H, Kararigas G, Dworatzek E, Staub E, Martus P, Noppinger PR, Kintscher U, Gustafsson JÅ (2010) Female sex and estrogen receptor-β attenuate cardiac remodeling and apoptosis in pressure overload. Am J Phys Regul Integr Comp Phys 298 (6):R1597–R1606
Franceschi C, Campisi J (2014) Chronic inflammation (inflammation) and its potential contribution to age-associated diseases. J Gerontol A Biomed Sci Med Sci 69(Suppl_1):S4–S9
Frank S, Copanaki E, Burbach GJ, Müller UC, Deller T (2009) Differential regulation of toll-like receptor mRNAs in amyloid plaque-associated brain tissue of aged APP23 transgenic mice. Neurosci Lett 453(1):41–44
Franke AA, Yu MC, Maskarinec G, Fanti P, Zheng W, Custer LJ (1999) Phytoestrogens in human biomatrices including breast milk. Clin Chem 23:532
Gammoh N, Rink L (2017) Zinc in infection and inflammation. Nutrients 9(6):624
GDNF contributes to estrogen-mediated protection of midbrain dopaminergic neurons (n.d.)
Glass CK, Saijo K, Winner B, Marchetto MC, Gage FH (2010) Mechanisms underlying inflammation in neurodegeneration. Cell 140(6):918–934
Grace OM, Simmonds MSJ, Smith GF, van Wyk AE (2008) Documented utility and biocultural value of Aloe L. (Asphodelaceae): a review. Econ Bot 63(2):167–178
Grimm MO, Michaelson D, Hartmann T (2017) Omega-3 fatty acids, lipids and apoE lipidation in Alzheimer's disease: a rationale for multi-nutrient dementia prevention. J Lipid Res 58:2083
Gruenwald J, Brendler T, Jaenicke C (2007) PDR for herbal medicines. Thomson, Reuters
Gupta A, Gupta R, Lal B (2001) Effect of Trigonella foenum-graecum (fenugreek) seeds on glycaemic control and insulin resistance in type 2 diabetes mellitus: a double-blind placebo controlled study. J Assoc Physicians India 49:1057–1061
Hallikainen M, Toppinen L, Mykkänen H et al (2006) Interaction between cholesterol and glucose metabolism during dietary carbohydrate modification in subjects with the metabolic syndrome. Am J Clin Nutr 84(6):1385–1392
Hannan JM, Rokeya B, Faruque O et al (2003) Effect of soluble dietary fiber fraction of Trigonella foenum-graecum on glycemic, insulinemic, lipidemic and platelet aggregation status of type 2 diabetic model rats. J Ethnopharmacol 88:73–77
Häuser W, Fitzcharles MA, Radbruch L, Petzke F (2017) Cannabinoids in pain management and palliative medicine: an overview of systematic reviews and prospective observational studies. Dtsch Arztebl Int 114(38):627
Healy DA, Wallace FA, Miles EA, Calder PC, Newsholme P (2000) Effect of low-to-moderate amounts of dietary fish oil on neutrophil lipid composition and function. Lipids 35(7):763–768
Hendy GN, Canaff L (2016) Calcium-sensing receptor, proinflammatory cytokines and calcium homeostasis. In: Seminars in cell and developmental biology, vol 49. Academic Press, London, pp 37–43

Henrotin Y, Mobasheri A, Marty M (2012) Is there any scientific evidence for the use of glucosamine in the management of human osteoarthritis? Arthritis Res Ther 14(1):201

Hofrichter J, Krohn M, Schumacher T, Lange C, Feistel B, Walbroel B, Heinze HJ, Crockett S, Sharbel TF, Pahnke J (2013) Reduced Alzheimer's disease pathology by St. John's Wort treatment is independent of hyperforin and facilitated by ABCC1 and microglia activation in mice. Curr Alzheimer Res 10(10):1057–1069

Hosseinzadeh H, Moallem SA, Moshiri M, Sarnavazi MS, Etemad L (2012) Anti-nociceptive and anti-inflammatory effects of cyanocobalamin (vitamin B12) against acute and chronic pain and inflammation in mice. Arzneimittelforschung 62(07):324–329

Hsieh DJ, Kuo WW, Lai YP, Shibu MA, Shen CY, Pai P, Yeh YL, Lin JY, Viswanadha VP, Huang CY (2015) 17β-estradiol and/or estrogen receptor β attenuate the autophagic and apoptotic effects induced by prolonged hypoxia through HIF-1α-mediated BNIP3 and IGFBP-3 signaling blockage. Cell Physiol Biochem 36(1):274–284

Hutchings A, Scott AH, Lewis G, Cunningham A (1996) Zulu medicinal plants: an inventory. University of Natal Press, Pietermaritzburg

Ionita R, Postu PA, Mihasan M, Gorgan DL, Hancianu M, Cioanca O, Hritcu L (2018) Ameliorative effects of Matricariachamomilla L. hydroalcoholic extract on scopolamine-induced memory impairment in rats: a behavioral and molecular study. Phytomedicine 47:113

Javadi F, Ahmadzadeh A, Eghtesadi S, Aryaeian N, Zabihiyeganeh M, RahimiForoushani A, Jazayeri S (2017) The effect of quercetin on inflammatory factors and clinical symptoms in women with rheumatoid arthritis: a double-blind, randomized controlled trial. J Am Coll Nutr 36(1):9–15

Jeong JW, Lee HH, Han MH, Kim GY, Kim WJ, Choi YH (2014) Anti-inflammatory effects of genistein via suppression of the toll-like receptor 4-mediated signaling pathway in lipopolysaccharide-stimulated BV2 microglia. Chem Biol Interact 212:30–39

Kararigas G, Fliegner D, Forler S, Klein O, Schubert C, Gustafsson JÅ, Klose J, Regitz-Zagrosek V (2014) Comparative proteomic analysis reveals sex and estrogen receptor β effects in the pressure overloaded heart. J Proteome Res 13(12):5829–5836

Kaul N, Kreml R, Austria JA et al (2008) A comparison of fish oil, flaxseed oil and hempseed oil supplementation on selected parameters of cardiovascular health in healthy volunteers. J Am Coll Nutr 27(1):51–58

Kayisli UA, Guzeloglu-Kayisli O, Guzel E, Arici A (2013) Genistein inhibits cell proliferation and stimulates apoptosis in human coronary artery endothelial cells. Gynecol Obstet Investig 75 (4):235–242

Kazak F, Yarim GF (2017) Neuroprotective effects of acetyl-l-carnitine on lipopolysaccharide-induced neuroinflammation in mice: involvement of brain-derived neurotrophic factor. Neurosci Lett 658:32–36

Kennedy DO, Scholey AB (2006) The psychopharmacology of European herbs with cognition-enhancing properties. Curr Pharm Des 12(35):4613–4623

Kennedy DO, Pace S, Haskell C, Okello EJ, Milne A, Scholey AB (2006) Effects of cholinesterase inhibiting sage (Salvia officinalis) on mood, anxiety and performance on a psychological stressor battery. Neuropsychopharmacology 31(4):845–852

Kew S, Banerjee T, Minihane AM, Finnegan YE, Muggli R, Albers R, Williams CM, Calder PC (2003) Lack of effect of foods enriched with plant-or marine-derived n− 3 fatty acids on human immune function. Am J Clin Nutr 77(5):1287–1295

Kim DS, Shin MR, Kim YS, Bae WJ, Roh DH, Hwang YS, Kim EC (2015) Anti-inflammatory effects of glutamine on LPS-stimulated human dental pulp cells correlate with activation of MKP-1 and attenuation of the MAPK and NF-κB pathways. Int Endod J 48(3):220–228

Koc A, Ozkan T, Karabay AZ, Sunguroglu A, Aktan F (2011) Effect of L-carnitine on the synthesis of nitric oxide in RAW 264· 7 murine macrophage cell line. Cell Biochem Funct 29(8):679–685

Kremer J, Michalek A, Lininger L, Huyck C, Bigauoette J, Timchalk M, Rynes R, Zieminski J, Bartholomew L (1985) Effects of manipulation of dietary fatty acids on clinical manifestations of rheumatoid arthritis. Lancet 325(8422):184–187

Kremer JM, Jubiz W, Michalek A, Rynes RI, Bartholomew LE, Bigaouette J, Timchalk M, Beeler D, Lininger L (1987) Fish-oil fatty acid supplementation in active rheumatoid arthritis: a double-blinded, controlled, crossover study. Ann Intern Med 106(4):497–503

Külzow N, Witte AV, Kerti L, Grittner U, Schuchardt JP, Hahn A, Flöel A (2016) Impact of omega-3 fatty acid supplementation on memory functions in healthy older adults. J Alzheimers Dis 51 (3):713–725

Kuo J, Hamid N, Bondar G, Prossnitz ER, Micevych P (2010) Membrane estrogen receptors stimulate intracellular calcium release and progesterone synthesis in hypothalamic astrocytes. J Neurosci 30:12950–12957

Le Donne M, Caruso C, Mancuso A, Costa G, Iemmo R, Pizzimenti G, Cavallari V (2011) The effect of vaginally administered genistein in comparison with hyaluronic acid on atrophic epithelium in postmenopause. Arch Gynecol Obstet 283(6):1319–1323

Lee BJ, Lin JS, Lin YC, Lin PT (2015) Antiinflammatory effects of L-carnitine supplementation (1000 mg/d) in coronary artery disease patients. Nutrition 31(3):475–479

Liao Y, Guan ZZ, Ravid R (2011) Changes of nuclear factor and inflammatory chemotactic factors in brain of patients with Alzheimer's disease. Zhonghua bing li xue za zhi = Chin J Pathol 40 (9):585–589

Liatis S, Tsapogas P, Chala E, Dimosthenopoulos C, Kyriakopoulos K, Kapantais E, Katsilambros N (2009a) The consumption of bread enriched with β-glucan reduces LDL-cholesterol and improves insulin resistance in patients with type 2 diabetes. Diabetes Metab 35:115–120

Liatis S, Tsapogas P, Chala E et al (2009b) The consumption of bread enriched with betaglucan reduces LDL-cholesterol and improves insulin resistance in patients with type 2 diabetes. Diabetes Metab 35(2):115–120

Lim K, See YM, Lee J (2017) A systematic review of the effectiveness of medical cannabis for psychiatric, movement and neurodegenerative disorders. Clin Psychopharmacol Neurosci 15 (4):301

Lima CF, Carvalho F, Fernandes E et al (2004) Evaluation of toxic/protective effects of the essential oil of Salvia officinalis on freshly isolated rat hepatocytes. Toxicol in Vitro 18(4):457–465

Liu Y, Li F, Zhang L, Wu J, Wang Y, Yu H (2017) Taurine alleviates lipopolysaccharide-induced liver injury by anti-inflammation and antioxidants in rats. Mol Med Rep 16(5):6512–6517

Lourdudoss C, Wolk A, Nise L, Alfredsson L, van Vollenhoven R (2017) Are dietary vitamin D, omega-3 fatty acids and folate associated with treatment results in patients with early rheumatoid arthritis? Data from a Swedish population-based prospective study. BMJ Open 7(6): e016154

Lourdudoss C, Di Giuseppe D, Wolk A, Westerlind H, Klareskog L, Alfredsson L, van Vollenhoven RF, Lampa J (2018) Dietary intake of polyunsaturated fatty acids and pain in spite of inflammatory control among methotrexate-treated early rheumatoid arthritis patients. Arthritis Care Res 70(2):205–212

Lucas EA, Wild RD, Hammond LJ et al (2002) Flaxseed improves lipid profile without altering biomarkers of bone metabolism in postmenopausal women. J Clin Endocrinol Metab 87 (4):1527–1532

Madeo A, Maggiolini M (2010) Nuclear alternate estrogen receptor GPR30 mediates 17beta-estradiol-induced gene expression and migration in breast cancer-associated fibroblasts. Cancer Res 70:6036–6046

Maggiolini M, Vivacqua A, Fasanella G, Recchia AG, Sisci D, Pezzi V, Montanaro D, Musti AM, Picard D, Ando S (2004) The G protein-coupled receptor GPR30 mediates c-fos up-regulation by 17beta-estradiol and phytoestrogens in breast cancer cells. J Biol Chem 279:27008–27016

Mahan L, Raymond J, Krause M (2017) Krause's food & the nutrition care process. 14th ed. Elsevier, St. Louis, p 30

Maki KC, Beiseigel JM, Jonnalagadda SS et al (2010) Whole-grain ready-to-eat oat cereal, as part of a dietary program for weight loss, reduces low-density lipoprotein cholesterol in adults with overweight and obesity more than a dietary program including low-fiber control foods. J Am Diet Assoc 110(2):205–214

Manonai J, Songchitsomboon S, Chanda K, Hong JH, Komindr S (2006) The effect of a soy-rich diet on urogenital atrophy: a randomized, cross-over trial. Maturitas 54(2):135–140

Mao JJ, Xie SX, Keefe JR, Soeller I, Li QS, Amsterdam JD (2016) Long-term chamomile (Matricariachamomilla L.) treatment for generalized anxiety disorder: a randomized clinical trial. Phytomedicine 23(14):1735–1742

Marini H, Polito F, Adamo EB, Bitto A, Squadrito F, Benvenga S (2012) Update on genistein and thyroid: an overall message of safety. Front Endocrinol 3:94

Mendes-Oliveira J, Campos FL, Videira RA, Baltazar G (2017) GPER activation is effective in protecting against inflammation-induced nigral dopaminergic loss and motor function impairment. Brain Behav Immun 64:296–307

Meyer MR, Prossnitz ER, Barton M (2011) The G protein-coupled estrogen receptor GPER/GPR30 as a regulator of cardiovascular function. Vasc Pharmacol 55:17–25

Mitchell JH, Cawood E, Kinniburgh D, Provan A, Collins AR, Irvine DS (2001) Effect of a phytoestrogen food supplement on reproductive health in normal males. Clin Sci 100 (6):613–618

Mohammadirad A, Aghamohammadali-Sarraf F, Badiei S, Faraji Z, Hajiaghaee R, Baeeri M, Gholami M, Abdollahi M (2013) Anti-aging effects of some selected Iranian folk medicinal herbs-biochemical evidences. Iran J Basic Med Sci 16(11):1170

Morikawa K et al (2003) Inhibitory effect of quercetin on carrageenan-induced inflammation in rats. Life Sci 74(6):709–721

Moyad MA, Robinson LE, Kittelsrud JM et al (2009) Immunogenic yeast-based fermentation product reduces allergic rhinitis-induced nasal congestion: a randomized, double-blind, placebo-controlled trial. Adv Ther 26(8):795–804

Mozolewski P, Moskot M, Jakobkiewicz-Banecka J, Wegrzyn G, Bochenska K, Banecki B, Gabig-Ciminska M (2017) Nonsteroidal anti-inflammatory drugs modulate cellular glycosaminoglycan synthesis by affecting EGFR and PI3K signaling pathways. Sci Rep 7:43154

Navarro SL, White E, Kantor ED, Zhang Y, Rho J, Song X, Milne GL, Lampe PD, Lampe JW (2015) Randomized trial of glucosamine and chondroitin supplementation on inflammation and oxidative stress biomarkers and plasma proteomics profiles in healthy humans. PLoS One 10(2): e0117534

Ng QX, Venkatanarayanan N, Ho CY (2017) Clinical use of Hypericumperforatum (St John's wort) in depression: a meta-analysis. J Affect Disord 210:211–221

Okun E, Griffioen KJ, Lathia JD, Tang SC, Mattson MP, Arumugam TV (2009) Toll-like receptors in neurodegeneration. Brain Res Rev 59(2):278–292

Ota A, Ulrih NP (2017) An overview of herbal products and secondary metabolites used for management of type two diabetes. Front Pharmacol 8:436

Panossian AG, Gabrielian E, Manvelian V, Jurcic K, Wagner H (1996) Immunosuppressive effects of hypericin on stimulated human leukocytes: inhibition of the arachidonic acid release, leukotriene B4 and Interleukin-Iα production, and activation of nitric oxide formation. Phytomedicine 3(1):19–28

Pase MP, Grima N, Cockerell R, Stough C, Scholey A, Sali A, Pipingas A (2015) The effects of long-chain omega-3 fish oils and multivitamins on cognitive and cardiovascular function: a randomized, controlled clinical trial. J Am Coll Nutr 34(1):21–31

Pedram A, Razandi M, Korach KS, Narayanan R, Dalton JT, Levin ER (2013) ERβ selective agonist inhibits angiotensin-induced cardiovascular pathology in female mice. Endocrinology 154(11):4352–4364

Poh S, Putt KS, Low PS (2017) Folate-targeted dendrimers selectively accumulate at sites of inflammation in mouse models of ulcerative colitis and atherosclerosis. Biomacromolecules 18(10):3082–3088

Prossnitz ER, Barton M (2011) The G-protein-coupled estrogen receptor GPER in health and disease. Nat Rev Endocrinol 7:715–726

Queenan KM, Stewart ML, Smith KN, Thomas W, Fulcher RG, Slavin JL (2007) Concentrated oat beta-glucan, a fermentable fiber, lowers serum cholesterol in hypercholesterolemic adults in a randomized controlled trial. Nutr J 6:6

Raber P, Ochoa AC, Rodríguez PC (2012) Metabolism of L-arginine by myeloid-derived suppressor cells in cancer: mechanisms of T cell suppression and therapeutic perspectives. Immunol Investig 41(6–7):614–634

Rafraf M, Karimi M, Jafari A (2015) Effect of L-carnitine supplementation in comparison with moderate aerobic training on serum inflammatory parameters in healthy obese women. J Sports Med Phys Fitness 55(11):1363–1370

Rahimifard M, Maqbool F, Moeini-Nodeh S, Niaz K, Abdollahi M, Braidy N, Nabavi SM, Nabavi SF (2017) Targeting the TLR4 signaling pathway by polyphenols: a novel therapeutic strategy for neuroinflammation. Ageing Res Rev 36:11–19

Rajaei E, Mowla K, Ghorbani A, Bahadoram S, Bahadoram M, Dargahi-Malamir M (2016) The effect of omega-3 fatty acids in patients with active rheumatoid arthritis receiving DMARDs therapy: double-blind randomized controlled trial. Global J Health Sci 8(7):18

Reed-Geaghan EG, Savage JC, Hise AG, Landreth GE (2009) CD14 and toll-like receptors 2 and 4 are required for fibrillar Aβ-stimulated microglial activation. J Neurosci 29(38):11982–11992

Rees D, Miles EA, Banerjee T, Wells SJ, Roynette CE, Wahle KW, Calder PC (2006) Dose-related effects of eicosapentaenoic acid on innate immune function in healthy humans: a comparison of young and older men. Am J Clin Nutr 83(2):331–342

Rettew JA, McCall SH, Marriott I (2010) GPR30/GPER-1 mediates rapid decreases in TLR4 expression on murine macrophages. Mol Cell Endocrinol 328(1):87–92

Riesberg LA, Weed SA, McDonald TL, Eckerson JM, Drescher KM (2016) Beyond muscles: the untapped potential of creatine. Int Immunopharmacol 37:31–42

Runhaar J, Rozendaal RM, van Middelkoop M, Bijlsma HJ, Doherty M, Dziedzic KS, Lohmander LS, McAlindon T, Zhang W, Zeinstra SB (2017) Subgroup analyses of the effectiveness of oral glucosamine for knee and hip osteoarthritis: a systematic review and individual patient data meta-analysis from the OA trail bank. Ann Rheum Dis 76:1862

Sabogal-Guáqueta AM, Munoz-Manco JI, Ramírez-Pineda JR, Lamprea-Rodriguez M, Osorio E, Cardona-Gómez GP (2015) The flavonoid quercetin ameliorates Alzheimer's disease pathology and protects cognitive and emotional function in aged triple transgenic Alzheimer's disease model mice. Neuropharmacology 93:134–145

Saghafi N, Rhkhshandeh H, Pourmoghadam N, Pourali L, Ghazanfarpour M, Behrooznia A, Vafisani F (2018) Effectiveness of Matricariachamomilla (chamomile) extract on pain control of cyclic mastalgia: a double-blind randomised controlled trial. J Obstet Gynaecol 38(1):81–84

Samieri C, Morris MC, Bennett DA, Berr C, Amouyel P, Dartigues JF, Tzourio C, Chasman DI, Grodstein F (2017) Fish intake, genetic predisposition to Alzheimer disease, and decline in global cognition and memory in 5 cohorts of older persons. Am J Epidemiol 187(5):933–940

Santhanam S, Alvarado DM, Ciorba MA (2016) Therapeutic targeting of inflammation and tryptophan metabolism in colon and gastrointestinal cancer. Transl Res 167(1):67–79

Sarinho E, Medeiros D, Schor D, Silva AR, Sales V, Motta ME, Costa A, Azoubel A, Rizzo JA (2009) Production of interleukin-10 in asthmatic children after β-1-3-glucan. Allergol Immunopathol (Madr) 37:188–192

Sauvaire Y, Ribes G, Baccou JC, Loubatieeres-Mariani MM (1991) Implication of steroid saponins and sapogenins in the hypocholesterolemic effect of fenugreek. Lipids 26:191–197

Scholey AB, Tildesley NT, Ballard CG et al (2008) An extract of Salvia (sage) with anticholinesterase properties improves memory and attention in healthy older volunteers. Psychopharmacology 198(1):127–139

Sexton M, Cudaback E, Abdullah RA, Finnell J, Mischley LK, Rozga M, Lichtman AH, Stella N (2014) Cannabis use by individuals with multiple sclerosis: effects on specific immune parameters. Inflammopharmacology 22(5):295–303

Sharma JN, Al-Omran A, Parvathy SS (2007) Role of nitric oxide in inflammatory diseases. Inflammopharmacology 15(6):252–259

Sharma N, Suresh S, Debnath A, Jha S (2017) Trigonella seed extract ameliorates inflammation via regulation of the inflammasome adaptor protein, ASC. Front Biosci (Elite Ed) 9:246–257

Shen XY, Luo T, Li S, Ting OY, He F, Xu J, Wang HQ (2018) Quercetin inhibits okadaic acid-induced tau protein hyperphosphorylation through the Ca2+−calpain-p25-CDK5 pathway in HT22 cells. Int J Mol Med 41(2):1138–1146

Shinto L, Quinn J, Montine T, Dodge HH, Woodward W, Baldauf-Wagner S, Waichunas D, Bumgarner L, Bourdette D, Silbert L, Kaye J (2014) A randomized placebo-controlled pilot trial of omega-3 fatty acids and alpha lipoic acid in Alzheimer's disease. J Alzheimers Dis 38 (1):111–120

Sindhu G, Shyni GL, Pushpan CK, Nambisan B, Helen A (2018) Evaluation of anti-arthritic potential of Trigonella foenum-graecum L. (fenugreek) mucilage against rheumatoid arthritis. Prostaglandins Other Lipid Mediat 138:48–53

Singhal PC, Gupta RK, Joshi LD (1982) Hypocholesterolemic effect of Trigonella foenum-graecum (METHI). Curr Sci 51:136

Stark A, Madar Z (1993) The effect of an ethanol extract derived from fenugreek (Trigonella foenum-graecum) on bile acid absorption and cholesterol levels in rats. Br J Nutr 69:277–287

Sturtzel B, Dietrich A, Wagner KH, Gisinger C, Elmadfa I (2010) The status of vitamins B6, B12, folate, and of homocysteine in geriatric home residents receiving laxatives or dietary fiber. J Nutr Health Aging 14(3):219–223

Sur P, Das M, Gomes A et al (2001) Trigonella foenum-graecum (fenugreek) seed extract as an antineoplastic agent. Phytother Res 15:257–259

Taghizadeh M, Tamtaji OR, Dadgostar E, Kakhaki RD, Bahmani F, Abolhassani J, Aarabi MH, Kouchaki E, Memarzadeh MR, Asemi Z (2017) The effects of omega-3 fatty acids and vitamin E co-supplementation on clinical and metabolic status in patients with Parkinson's disease: a randomized, double-blind, placebo-controlled trial. Neurochem Int 108:183–189

Talbott S, Talbott J (2009) Effect of β 1,3/1,6 glucan on upper respiratory tract infection symptoms and mood state in marathon athletes. J Sports Sci Med 8:509–515

Talbott SM, Talbott JA (2012) Baker's yeast β-glucan supplement reduces upper respiratory symptoms and improved mood state in stressed women. J Am Coll Nutr 31:295–300

The possible role of estrogen and selective estrogen receptor modulators in a rat model of Parkinson's disease (n.d.)

Thies F, Nebe-von-Caron G, Powell JR, Yaqoob P, Newsholme EA, Calder PC (2001) Dietary supplementation with γ-linolenic acid or fish oil decreases T lymphocyte proliferation in healthy older humans. J Nutr 131(7):1918–1927

Thompson Coon JS, Ernst E (2003) Herbs for serum cholesterol reduction: a systematic view. J Fam Pract 52:468–478

Torkan M, Hassan Entezari M, Siavash M (2015) Effect of flaxseed on blood lipid level in hyperlipidemic patients. Rev Recent Clin Trials 10(1):61–67

Valette G, Sauvaire Y, Baccou JC, Ribes G (1984) Hypocholesterolaemic effect of fenugreek seeds in dogs. Atherosclerosis 50:105–111

van Wyk BE, Smith G (2008) Guide to the aloes of South Africa, 2nd edn. Briza Publications, Pretoria

Wang DM, Li SQ, Wu WL, Zhu XY, Wang Y, Yuan HY (2014) Effects of long-term treatment with quercetin on cognition and mitochondrial function in a mouse model of Alzheimer's disease. Neurochem Res 39(8):1533–1543

Whiting PF, Wolff RF, Deshpande S, Di Nisio M, Duffy S, Hernandez AV, Keurentjes JC, Lang S, Misso K, Ryder S, Schmidlkofer S (2015) Cannabinoids for medical use: a systematic review and meta-analysis. JAMA 313(24):2456–2473

Wu Y, Antony S, Meitzler JL, Doroshow JH (2014) Molecular mechanisms underlying chronic inflammation-associated cancers. Cancer Lett 345(2):164–173

Wu PS, Ding HY, Yen JH, Chen SF, Lee KH, Wu MJ (2018) Anti-inflammatory activity of 8-hydroxydaidzein in LPS-stimulated BV2 microglial cells via activation of Nrf2-antioxidant and attenuation of Akt/NF-κB-inflammatory signaling pathways, as well as inhibition of COX-2 activity. J Agric Food Chem 66(23):5790–5801

Yadav UC, Moorthy K, Baquer NZ (2004) Effects of sodium-orthovanadate and Trigonella foenum-graecum seeds on hepatic and renal lipogenic enzymes and lipid profile during alloxan diabetes. J Biosci 29:81–91

Zameer S, Najmi AK, Vohora D, Akhtar M (2017) A review on therapeutic potentials of Trigonella foenum-graecum (fenugreek) and its chemical constituents in neurological disorders: complementary roles to its hypolipidemic, hypoglycemic, and antioxidant potential. Nutr Neurosci 16:1–7

Zarvandi M, Rakhshandeh H, Abazari M, Shafiee-Nick R, Ghorbani A (2017) Safety and efficacy of a polyherbal formulation for the management of dyslipidemia and hyperglycemia in patients with advanced-stage of type-2 diabetes. Biomed Pharmacother 89:69–75

Zemestani M, Rafraf M, Asghari-Jafarabadi M (2016) Chamomile tea improves glycemic indices and antioxidants status in patients with type 2 diabetes mellitus. Nutrition 32(1):66–72

Zhao TZ, Ding Q, Hu J, He SM, Shi F, Ma LT (2016) GPER expressed on microglia mediates the anti-inflammatory effect of estradiol in ischemic stroke. Brain Behav 6(4):e00449

Zhou X, Yuan L, Zhao X, Hou C, Ma W, Yu H, Xiao R (2014) Genistein antagonizes inflammatory damage induced by β-amyloid peptide in microglia through TLR4 and NF-κB. Nutrition 30 (1):90–95

Anti-Oxidant Nutrients and Nutraceuticals in Aging

9

Abida Zulfiqar, Sara Ishaq, and Touqeer Ahmed

Abstract

Nutrition plays a strong role in aging. Therefore, the absorption and metabolism of nutrients along with diet are a strong risk factor for aging related diseases. The nutritional requirements for the elderly are different as compared to the younger adults due to factors like drug and nutrient interactions, aging associated conditions, socioeconomic factors, etc. The amount and dosage of minerals, vitamins, and macronutrients are different in the elder population. Assessment of proper nutritional diet is important because in this age group, many people are malnourished that leads to various age-related disorders. Different nutraceuticals like the phytochemicals, carotenoids, vitamin B, D, E, various plant extracts have been reported to have rejuvenating properties. The use of nutraceuticals has beneficial effect on the metabolic and immune systems as they possess anti-oxidant properties.

Keywords

Aging · Nutraceuticals · Reactive oxygen species · Nutrients · Phytochemicals

A. Zulfiqar · S. Ishaq · T. Ahmed (✉)
Neurobiology Laboratory, Department of Healthcare Biotechnology, Atta-ur-Rahman School of Applied Biosciences, National University of Sciences and Technology, Islamabad, Pakistan
e-mail: touqeer.ahmed@asab.nust.edu.pk

S. M. Nabavi et al. (eds.), *Nutrients and Nutraceuticals for Active & Healthy Ageing*,
https://doi.org/10.1007/978-981-15-3552-9_9

9.1 Introduction

9.1.1 Aging and Pathologies Share a Common Mechanism

Aging is considered as the most common cause of decreasing the life span of human beings in many diseases. It was believed, for the longest time, that the physiological condition or mechanism of aging is responsible for most of the diseases. Thus, there is an interlink between aging and the beginning of an ailment, as the basic fundamental procedures behind these two are quite similar (Jaul and Barron 2017).

There are various factors that account for the relation between aging and the beginning of a disease. These factors include the rate at which aging occurs in a cell, cellular signaling patterns, genetic traits of a particular organ, and lifestyle. Consequently, keeping in view the contributing factors, every single activity inside the cell weakens and gradually becomes less effective in fulfilling physiological functions, resulting in clinical pathologies (Franceschi et al. 2018). While both advance in the same way, yet with aging on the one hand and its related diseases on the other, the direction of anticipation of the two is unique and is dependent upon lifestyle of every person and his/her genetic material. There are few considerations that can help in understanding such ideas:

1. Aging has not been chosen during the procedure of development and no gene has been recognized that might directly control aging. DNA repair genes (Sirtuin-6) might have indirect effect by offering better DNA repair.
2. Aging is reliant upon both hereditary and environmental factors which eventually influence the phenotype of the aged person.

While the main function of a gene is to control cell activities, leading to increased life expectancy and improved health of an individual, aging may be perceived as an unpredictable outcome of its function (Franceschi et al. 2018). The environment plays an important role in progressing towards age-related diseases. In response to various environmental stimuli, various epigenetic changes take place throughout an individual's lifetime that affect the major metabolic pathways and may lead to inflammation like conditions that cause a number of diseases associated with aging (Yang et al. 2015).

9.2 Role of Oxidative Stress in Aging

9.2.1 The Free Radical Theory and Pathway of Aging

The free radical theory of aging (FRTA), initially given in 1950s, by Denham Harman, postulated that the formation of oxidative species and development of oxidative stress accelerates the aging process. The harm is caused by the metabolic products of oxygen, called reactive oxygen species (ROS). The FRTA is at present viewed as a prominent concept in the field of aging. Different researches have indicated the role of ROS and how with growing age, the quantity of ROS increases in the body. The investigations additionally showed that a decrease in the accumulation of ROS in the body promotes overall life expectancy of different animal models (Gladyshev 2014).

There has been an increased number of evidences against FRTA, though, that did not influence the hypothesis or its validity. However, the common conflict against the hypothesis is that numerous examinations have proved that the process of aging occurs even under anaerobic conditions, where decreased production of ROS takes place. Studies conducted on yeast gave comparative results where it was seen that yeast produced under anaerobic conditions has a decreased life span as compared to the yeast produced under aerobic conditions (Hekimi et al. 2011).

The investigation of oxidative stress in the field of aging has been crucial as its progression can be visualized using different diagnostic strategies, and can likewise be managed by compounds known for their anti-oxidizing capacities. Thus, along with respiratory and other metabolic procedures, aging also leads to the accumulation of ROS. Cell examination can be used to determine the degree of damage caused by ROS generation (Gemma et al. 2007).

Very few protocols exist to consider the results of different processes and their by-products, which go undetected during ROS generation. Currently, there are very limited findings about the detailed role of proteins, RNAs, their synthesis, and generation regarding ROS metabolites. However, sufficient data can be found on aging and damages caused by different factors, other than oxidative stress and ROS generation. The examination of different stages of aging indicated abnormalities in proteins functions, eventually prompting diseased condition. The idea to work on aging led towards investigation of damage caused by aging to DNA, proteins, and cellular metabolites (Milholland et al. 2017). Currently, various researchers support damage accumulation as a factor for aging; however, the system behind it and the connection between regular clearance and damage accumulation is still unclear. In 1972, the hypothesis of aging was drawn from two postulates which were as follows (Hekimi et al. 2011):

1. The rate at which free radical's reactions (FRRs), start in the mitochondria, increases with age.
2. Life span of an individual is dependent upon the rate at which the free radicals damage the mitochondria.

The impacts of mitochondrial aging have been examined before in 1983. More research in this area helped in determining that better lifestyle aids in a reduction of free radicals reactions. The aging hypothesis proposes approaches to reduce mitochondrial damage which includes:

1. Use of anti-oxidizing agents to lessen the chain length of FRRs,
2. Reducing copper, iron, and other oxidants to decrease the speed of FRRs formation, subsequently decreasing the rate at which aging occurs.

These approaches can reduce the rate of aging even without much efforts on ideal lifestyle. Different studies conducted have also added to these two mentioned parameters (Howes 2006) (Fig. 9.1).

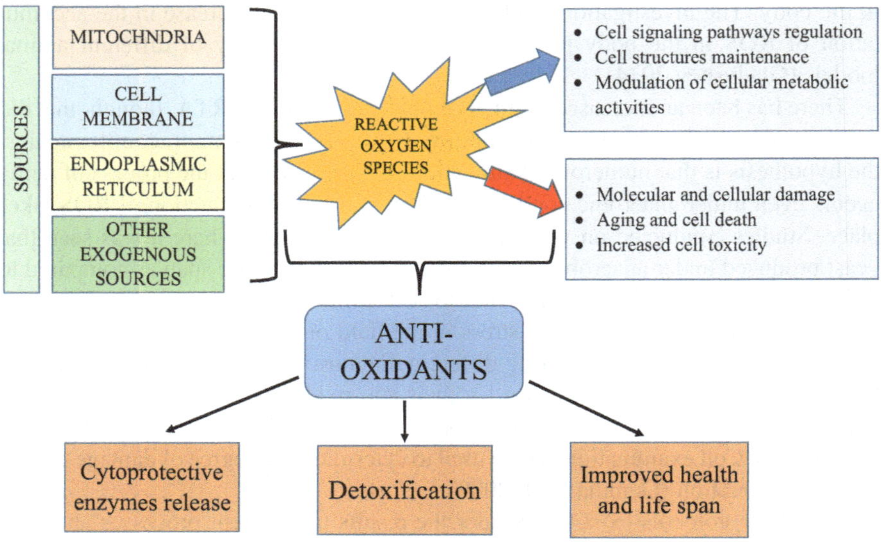

Fig. 9.1 The harmful effects of reactive oxygen species generated from different endogenous and exogenous sources and their counter- effects by anti- oxidants

9.2.2 The Nutrient-Sensing Pathways and Their Role in Longevity

With respect to different physiological changes taking place in organ systems, aging is considered to be an irreversible process. Undoubtedly, the process of aging occurs because of interaction between the genetic and epigenetic factors; however, environmental factors also have a significant role to play in the process. Considering these factors, the life expectancy in people has increased significantly in the past two centuries, while it has been assessed that by 2050, in some nations of Europe, the percentage of life span of people aged 60 years will increase from 20% to 40%, while the ratio of centenarians will be around 3.2 million, everywhere throughout the world (Davinelli et al. 2012).

Regardless of how research has demonstrated the positive role of different variables like hereditary, dietary, and pharmacological aspects, in decreasing the pace of aging and improving life, "Caloric restriction" still appears to be the crucial factor to expand life span in all life forms. The life span of a living being is adjusted by means of nutrient sensors, which react to different physiological as well as environmental conditions. Nutrient sensing mechanisms perform a significant role during aging as different nutrients are directly or indirectly responsible for initiating many nutrient-sensing pathways. Various genes have been distinguished for having a key role as controllers of life expectancy by having distinct roles in nutrient-sensing pathways. These genes are called "nutrient-sensing longevity genes" (Shimokawa et al. 2008).

Dietary considerations are made and the eating pattern is observed so as to decide the health condition and form a link between diet intake and a disease. Disease occurrence due to dietary predisposition is increasing, while, on the other hand, age-related abnormalities are decreasing due to improved human healthcare services. Energy homeostasis is maintained by the capacity of biochemical pathways to detect the nutrient consumption along with accessibility at both cell and individual's levels. However, further investigations are required to completely comprehend the system behind diet and its formation, and how it influences the human life expectancy, particularly concentrating on nutrient-sensing pathways (Marshall 2006).

Furthermore, it has been observed that an improved life expectancy is linked to characteristic genetic variations in nutrient-sensing pathways. During recent years, few of these pathways have been studied in worms or mice; one specific pathway is the IGF-1 pathway. The mechanism of IGF-1 is important in all creatures and research has indicated that an adjusted IGF-1 flagging pathway brings about a decreased life expectancy (Kapahi and Zid 2004). Undoubtedly, changes known to hinder IGF-1 receptor capacity have been demonstrated to be overexpressed in a partner of Ashkenazi Jewish centenarians proposing that centenarians may harbor unique variations in genes encoding segments of the IGF-1 pathway (Suh et al. 2008). Polymorphic variations in genes which are associated with IGF-1 flagging have likewise been associated with life span in a Japanese partner of 122 semi-supercentenarians (105 years of age and more traditional). The FOXO translation factor FOXO3, member of the IGF-1 pathway, is fundamental for caloric restriction impacts and it has been exhibited that polymorphisms in FOXO3 are associated with human life span in a few related groups (Davinelli et al. 2012).

9.3 Nutrients and Nutraceuticals in Aging

9.3.1 Nutrients

Nutrients (additionally known as macronutrients) are common natural compounds that are implied in biochemical processes responsible for life or are constituents of cell biomass. Glucose and related sugars, amino acids and lipids are important nutrients required by cells to perform various vital functions. Some nutrients can be esesential like amino acids, fatty acids, vitamins, and minerals or they can be non-essential like dietary fibers, short-chain fatty acids, etc. Some can be semi-essential as well because essentiality is not really a sign of supplements; for certain amino acids, for example, arginine, cysteine, glutamine, glycine, proline, and tyrosine, vitality is situation dependent. In healthy individuals, the repeated synthesis of these amino acids from different substances meets an organism's needs; however, under specific metabolic requirements, for example, during the growth of infants, they should likewise be acquired from the environment (Efeyan et al. 2015) (Fig. 9.2).

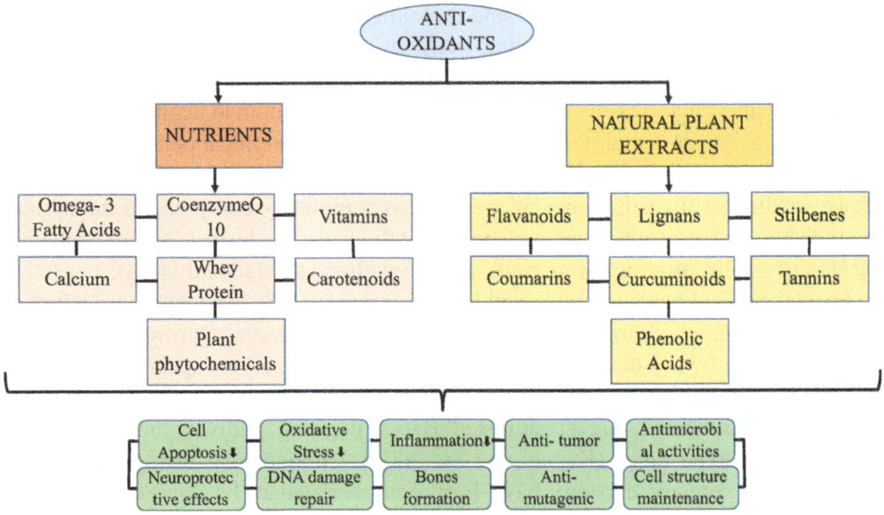

Fig. 9.2 Different anti- oxidant nutrients and natural plant extracts along with their potential beneficial health effects

9.3.2 Caloric Restriction (CR) in Aging

The most well studied and well-established procedure to slow the process of aging is caloric restriction. With a 20–40% decrease in calorie consumption, without resulting in malnutrition, studies have shown an improved life span (Shimokawa et al. 2008). As a consequence of caloric restriction, it is observed that various metabolic pathways are affected, such as the insulin and glucose levels are reduced as well as decrease in thyroid and reproductive hormones. On the other hand, due to its ability to decrease the rate of aging, caloric restriction has been deemed as an imperative area for research to determine the delay in aging in all kinds of organisms (Azzu and Valencak 2017).

9.3.3 Anti-Oxidant Nutrients in Aging

Anti-oxidant nutrients are those which decrease the process of oxidative damage to the cells and tissues of our body. Found in various fruits and vegetable, such as cherries, orange carrots, purple berries, etc., anti-oxidants can fight against cancer and a wide range of other diseases. Anti-oxidants slow the process of aging by reducing the amount of free radicals production which causes damage to our body. Some of the important anti-oxidants and their anti-aging properties are discussed below.

9.3.3.1 Phytochemicals

Phytochemicals are non-essential nutrients found in different plant sources. Regardless of a translational distance between clinical and laboratory research, the present understanding of molecular connection of phytochemicals, safe metabolic pathways, and oxidative stress pathways can help to form strong dietary patterns so as to block the way toward aging and its related illnesses. Studies have shown dietary phytochemicals to cause a condition called hormesis. The condition is caused by low levels of bio-active compounds, for example, phytochemicals that act as mild stress inducers to influence the expression of stress-defensive genes and furthermore increase protection from the procedures that control aging (Puca et al. 2018).

Research carried out at the level of in vivo and in vitro have confirmed that most of the phytochemicals affect the expression of different genes that code for pro-survival proteins, for example, anti-oxidant enzymes, neurotrophic/anti-apoptotic factors (Dillard and German 2000). So as to cause better stress resistance, various cellular adjustments are included which lead to a number of previously mentioned nutrient-sensing pathways for a better life span. The association of hormesis during aging process has been reported by earlier studies conducted by Rattan, where role of caloric restrictions in the process of anti-aging and life span in long-lived species was clarified (Rattan 2008).

A large amount of data can be gathered on the advantages of plant extracts to slow the development of aging and its related diseases (Dillard and German 2000). It has been suggested that signaling of Nrf2 is initiated in the counter aging reactions induced by phytochemicals. Nrf2 is viewed as a key controller of versatile reactions to oxidative stress, but very little research has been done to decide the role of Nrf2 in the regulation of aging pathway and improved life expectancy (Shimokawa and Trindade 2010).

There are also many phytochemicals, like polyphenols, flavonoids, etc., present in fruits, vegetables, and spices as key component and they are crucial in lowering the age-related health deterioration. Interestingly, these phytochemicals ultimately block the production of ROS by activating the Nrf2 pathway, as a result, control oxidative damage and stress in this way (Masoro 2005).

9.3.3.2 Catechins and Plant Polyphenols

Plant polyphenols are thought to be the secondary metabolites and are randomly distributed in higher plants. Some specific properties of polyphenols among several others include anti-oxidizing capabilities and solubility in water. Polyphenols also pose negative effects on human body, that is why nutritionists thought these to be "anti-nutrients." The harmful effects of polyphenols include restricted capabilities of digestive enzymes, decreased energy, and decreased availability of proteins/amino acids along with some other toxic effects. However, they are used as lowering the age-related diseases like CVD, diabetes, neurodegenerative diseases, inflammation, etc. (Forni et al. 2019).

Studies conducted previously have derived a link between increased intake of nutraceutical phenolic compounds and a reduced rate of many types of cancer and coronary diseases. A polyphenol, Epigallocatechin gallate (ECGC), has been observed to protect against DNA damage due to UV radiation along with immune suppression, because it has the ability to decrease the oxidative stress and block NF-kβ (Peng et al. 2014).

9.3.3.3 Carotenoids

Carotenoids, also known as fat-soluble substances, are present in vegetables like carrots, bell peppers, papaya, collard greens, spinach, sweet potatoes, tomatoes, and kale (Forni et al. 2019). There are four types of carotenoids which have been reported to be converted to vitamin A (Retinol):

1. Alpha Carotene.
2. Beta Carotene.
3. Gamma Carotene.
4. Beta Cryptoxanthin.

Some other types of anti-oxidant carotenoids are lutein, zeaxanthin, and lycopene but these are not converted to Retinol. The effect of carotenoids on bone mass density (BMD) is not clear; studies are underway to know how it effects the BMD using serum biomarkers. However, research has shown a relation between high levels of beta carotenes and high BMD in females (Wattanapenpaiboon et al. 2003). The Framingham's study recommended that the incidence of hip fracture was reduced in people with intake of carotenoids. The participants with higher intake of total carotenoid and lycopene were found to have 46% and 34% lowered hip fracture risks, respectively. In this way, increasing the utilization of the carotenoid-rich vegetables may have extra health advantages, despite the fact that the data of how great their ingestion for bones is clear (Sahni et al. 2009).

9.3.3.4 Phytosterols (Plant Sterols)

Many plant-based diets normally contain small amounts of phytosterols, a subclass of terpene. Organic products, nuts, seeds, and vegetables are significant sources of phytosterols. Organic product contains around 15 mg of phytosterols per 100 g wet weight, though, seeds contain around 120 mg of them for every 100 g wet weight and vegetables contain 20 mg of phytosterols per 100 g of wet weight. The dietary ingestion of phytosterols by and large contains 65% sitosterol, 30% campesterol, and 3% stigmasterol. Routine ingestion of plant sterols is around 160–400 mg/day. To bring down the degree of cholesterol in everyday nourishment items, phytosterols are being utilized to enhance the nourishment. Beta-sitosterol and its glycoside are delivered by plants and tend to reduce the overall cholesterol levels which ultimately reduces coronary disease through restraining the retention of cholesterol (Chawla and Goel 2016).

Studies show that anti-neoplastic, anti-inflammatory, anti-pyretic, and immunomodulating activities are manifested by phytosterols. The phytosterols are similar in structure to cholesterol and thus compete with cholesterol in the digestive system for intestinal absorption. In the digestive tract, plant sterols are at first solubilized into a micelle structure. These micelles collaborate with brush-border cells and are moved into enterocytes. Plant sterols are esterified inside the enterocyte, gathered into chylomicrons, and discharged into the lymphatics. They are discharged by means of the biliary framework, thus helping in the removal of cholesterol from the body (Piironen et al. 2003).

They also include another significant subclass of terpenes. Beta-sitosterol and its glycoside are two of the sterol molecules produced by plants. These two substances, in animals, show anti-inflammatory, anti-pyretic, anti-neoplastic, and immunomodulating activity (Chawla and Goel 2016). The concentration of cholesterol in the body is more efficiently decreased by saturated phytosterols than unsaturated substances (Jain and Ramawat 2013).

9.3.3.5 Calcium and Vitamin D

Vitamin D is known to be a vital fat-soluble nutrient required for metabolic activities of calcium. Liver, milk, and fish from salty water are reservoirs of vitamin D. Vitamin D plays an important role in absorption of calcium, balance, bone quality, muscle function, and danger of falling. Moreover, vitamin D is also incorporated into calcium supplements and multivitamin tablets which can be utilized to overcome its deficits, particularly, in the regions where there is an increased shortage of daylight. The prescribed level of vitamin D is 400 IU/day for people under 70 years and 600 IU/day above that by Food and Nutrition Board (FNB) at the Institute of Medicine of the National Academies (USA, Canada). These levels extend between 800 and 1000 IU/day by the National Osteoporosis Foundation for the people more than 50 years. Studies have concluded a clear deficiency of vitamin D in the individuals generally residing in the northern areas due to decreased sun exposure, aged individuals, or individuals with nutrition and absorption issues, or persistent liver/renal issues. Vitamin D insufficiency is more common in dark skinned individuals (Gennari 2001).

Vitamin D along with calcium is extremely important for the well-being of skeletal tissues to decrease fractures in bones. While vitamin D2 (ergocalciferol) is obtained from vegetable sources, vitamin D3 (cholecalciferol) is from animal sources, both have quite the same effectiveness in terms of medical usage over the same dosage of vitamin D (Rosen et al. 2017).

Latest studies also indicate that life expectancy can be increased by using vitamin D with calcium. Calcium and vitamin D combination basically enhances the metabolic activities of bone and thereby preventing the development of osteoporosis. Apart from that, their role is also being studied for their non-skeletal purposes. For instance, the risk of colon cancer and colonic polyps can be decreased effectively with calcium-rich intakes, while diabetes and other cancers can be controlled via intake of vitamin D. Most recent examinations additionally show that life duration can be enhanced by utilizing vitamin D in combination with calcium. The blend of calcium and vitamin D improves the metabolic processes of bone and keeps them protected from osteoporosis. Apart from that, role of calcium and vitamin D is likewise being read for their non-skeletal purposes, such as the danger of colon malignant growth and colonic polyps can be decreased successfully with increased consumption of calcium-rich, while diabetes and different tumors can be controlled by consumption of vitamin D (Del Valle et al. 2011).

9.3.3.6 Vitamin E

These are a class of dietary anti-oxidants including four tocopherols and four tocotrienols. Researches have indicated that provision of vitamin E to rodents resulted in many effects like:

1. Improvement of cognitive abilities.
2. Neuroprotective activity in mice deficient with APOE.
3. Reduced Aβ poisoning in hippocampal cultured neurons.
4. Improved functions of nervous system and mitochondria in aging rodents.

Decreased levels of vitamin E is responsible for impairment of innate and adaptive immunity (Sen et al. 2007). Monocytes release reactive oxygen species (ROS), just as they attach to the endothelium and increase inflammation by the expression of CD11b and late antigen 4. The production of chemokines and pro-inflammatory cytokines are considerably decreased by vitamin E as they effectively hamper the activation of NF-kB (Moriguchi and Muraga 2000). Damages caused by immune system are reduced by vitamin E as it controls the release of IFN-γ and FasL expression by action on T-helper cells. Results from research show that, the immune system of the aging mice was controlled upon treatment with vitamin E as it suppresses prostaglandin E2 generation yielding an improved T cell function and postpones hypersensitivity reactions. The usage of 200 mg/d of vitamin E for 3 months can improve lymphocyte production, IL-2 creation with neutrophil capacities, and natural killer cells activity for individuals. However, literature additionally proposes that interruption of vitamin E consumption reclaims similar capacities to their previous status within half a year. Thus, treatment of vitamin E with aging ought to be taken as regular dosage (Meydani et al. 1990).

9.3.3.7 Vitamin B Group

Vitamin B9 or folates are imperative for the improvement of the central nervous system. Low levels of vitamin B9 is normally seen in aged people. This might be because of a low food intake and furthermore modified/reduced nutritional absorption in the digestive organs. Other medical issues caused due to deficiency of vitamin B9 in the body include increased psychological disabilities, dementia, and increased anxiety in elderly people. Not much supportable information can be found to support the effect of vitamin B9 alone as an addition or in combination with others to reduce or prevent the chances of cognitive decline in neuropsychiatric illnesses in elder patients (Villeponteau et al. 2000).

9.3.3.8 Omega 3 Fatty Acids

Omega 3 fats can be found in different sources, for example, fish, eggs, pecans, and flax seeds. These are polyunsaturated fats having anti-oxidant properties and they may similarly increase calcium absorption in the digestive tract. Research led on 78 healthy, young men provided a positive link between utilization of omega 3 fatty acids and standard BMD (Gogus and Smith 2010). However, a research led on healthy and young children concluded that increased proportion of omega 6 to omega 3 negatively affects BMD and furthermore lowers hip BMD in aged females. The study also shows the potential advantages of omega 3 on skeletal health. The research additionally concluded that the co-administration of omega 3 and calcium may enhance its benefits for skeletal health (Cholewski et al. 2018). Consequently, there is a likelihood that a decreased proportion of omega 6 to omega 3 may be clearly identified associated with skeletal health thus more utilization of omega 3 fatty acids is prescribed, while less utilization of omega 6 is suggested.

9.3.3.9 Whey Protein

Whey is the liquid portion that remains after the milk curdles. It is rich in whey protein which helps in muscle growth, reduces inflammation and blood pressure. A group of elderly men were provided with an experimental diet containing high amount of whey protein (35 g versus 10 g). Reports showed increased amino acid maintenance and an increase in muscle proteins synthesis. Correspondingly, it has been seen that providing elderly men with whey proteins has increasingly positive effect in muscle activity rates when compared with Casein, which is additionally a protein source from real milk. These results demonstrate that utilizing whey protein in diets of elder men may positively impact building of mass which is very extraordinary since Sarcopenia or lost skeletal muscle mass is caused with aging. This, reciprocally, helps battle disability whose one significant explanation is Sarcopenia (Patel 2015).

Amongst a variety of nourishing proteins available, whey proteins are the most acclaimed for their remarkable medical advantages because they contain all the vital amino acids crucial for human body. Long chain amino acids are specifically found in abundance in whey protein (Davoodi et al. 2016). Moreover, it also helps with controlling the loss of muscle and weight. Besides effectively accessible and palatable, these proteins give extra-dietary advantages: for example, alpha lactalbumin (α-lactalbumin), which is another whey protein being promptly accessible, has a high amount of the amino acid tryptophan which is a precursor of the supplement niacin (Vitamin B3) (Gupta and Prakash 2015).

Whey proteins have some important points regarding health, which are as follows:

(a) Abundance of essential amino acids.
(b) Anti-microbial action.
(c) Development of helpful gut microflora: Bifidobacteria.
(d) Immune enhancers.
(e) Check on developmental diseases: malignant growth.
(f) Serum activities.

9.3.3.10 CoenzymeQ10

CoenzymeQ10 is a supplement which has a crucial role in various cellular activities essentially the formation of Adenosine Tri-phosphate (ATP) which is utilized as an essential source of energy by the cell itself. CoenzymeQ10 is an extraordinary cell support molecule which is useful in reducing oxidative stress. Patients experiencing coronary heart disease were given this as supplement and it was discovered that CoenzymeQ10 has a moderating effect by lowering the inflammatory marker IL-6 (Saini 2011). It was found that CoenzymeQ10 supplement enhances endothelial capacity in patients experiencing distinctive heart diseases. CoenzymeQ10 may also be linked with damage indicators as recent reports have indicated low plasma levels of coenzymeQ10 in breast cancer among Chinese women. Additionally, coenzymeQ10 has a role in creatine kinase activity to manage cellular homeostasis and old-aged bipolar disorder (Rodick et al. 2018).

9.3.4 Nutraceuticals

Nutraceuticals are characterized as another class of chemicals which is a cross between medications and nourishment. According to its definition, a nutraceutical is "a supplement or part of diet that benefits health, also improves its nutritional component." Variety of compounds extracted from plants (phytochemicals) or animals in whatever way, when separated and converted into appropriate pharmaceutical form, can make promising tool to help fight and cure many diseased conditions when given in adequate quantity.

One important aspect of nutraceuticals is to treat the metabolic disorder which is characterized by obesity, dyslipidemia, hypertension, and insulin resistance. Diet and lifestyle changes are accepted worldwide to be fundamental in maintaining such disorders. The National Health Systems ensures the best possible and cost-effective restorative method dependent on pharmaceuticals. Misuse of nutraceuticals can lead to many pre-clinical issues and certain health conditions. They can be effectively utilized, by incorporating into daily diet, in the range "beyond the diet, before drugs," as they combine the nutritional and useful properties of diet with the curative properties of the naturally occurring substances (Santini et al. 2017). The common nutrients can act as nutraceuticals when required. Some natural plant extracts which are being used as nutraceuticals are mentioned in the table (Table 9.1).

9.3.5 Important Anti-Aging Plant Nutraceuticals

Plant extracts have been utilized as traditional medicine (TM) throughout the world, and considered to be the fundamental for improving human healthcare and/or disease management. Nearly 10,000 plant species are utilized for this purpose. However, the pharmacological and biomedical testing of natural plant extract mixtures has been done in animal models, yet there is a need to study and evaluate the bioactivity of therapeutic plants. Anti-aging plants are most widely grafted and cultured by various

Table 9.1 Different natural plant extracts used as nutraceuticals

Class	Structural skeleton	Subclass	Source	Health effects	Examples	Refrences
Flavonoids	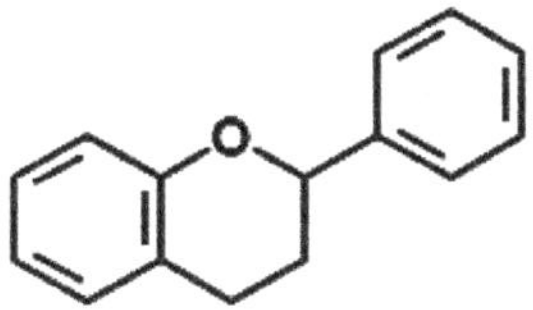	Flavones	Parsley, celery, capsicum pepper	• Anti-microbial • Anti-apoptotic • Anti-inflammatory • Reduce risk of certain cardiovascular diseases • Beneficial against neurological issues • Anti-oxidant • Anti-tumor • Anti-aging • Improve health • Increase life span	Luteolin Apigenin	Pradeep et al. (2014), Ishaq and Jafri (2017), and Rudrapal and Chetia (2017)
		Flavanols	Onion, curly kale, broccoli, leek, cherry, tomato, apple, berries, beans, cocoa, tea, and red wine		Quercetin Kaempferol Myricetin Catechins Gallocatechins	
		Flavanones	Orange, grapefruit and lemon juice		Hesperetin Naringenin Eriodicytol	
		Isoflavone	Soya beans		Genistein Daidzein	
		Anthocyanidins	Berries, blackcurrant, red wine and cherry		Cyanidin DelphiNidin Malvidin Pelargonidin Petunidin	
Phenolic acids	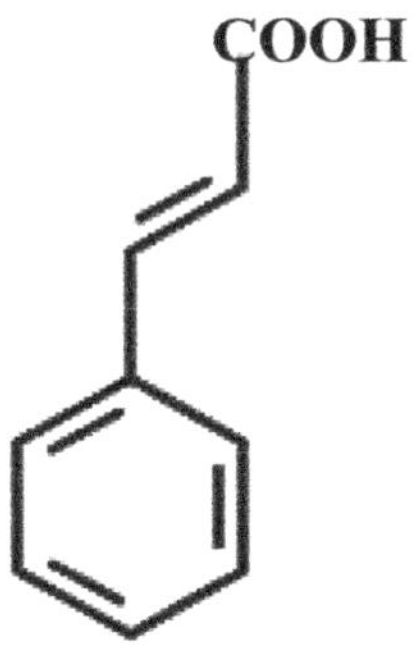	Caffeic acid	Peach, apple, cocoa and apricot	• Anti-oxidant • Anti-inflammatory • Neuroprotective • Signaling molecules		Silva et al. (2000), Robbins (2003), and Andújar et al. (2012)
		Chlorogenic acid				
		Ferulic acid				

(continued)

Table 9.1 (continued)

Class	Structural skeleton	Subclass	Source	Health effects	Examples	Refrences
Stilbenes	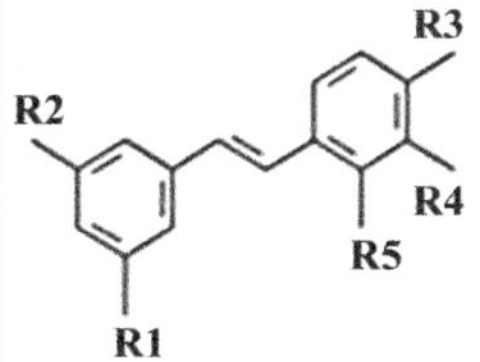	Resveratrol Pterostilbene	Cherry, red wine, blackberry, cocoa, and peanuts	• Anti-microbial • Anti-cancer • Effective against many heart diseases • Improves learning and memory • Improves metabolism • Anti-aging • Increases life span		Sirerol et al. (2016), Chin et al. (2017), and Zakova et al. (2018)
Lignans	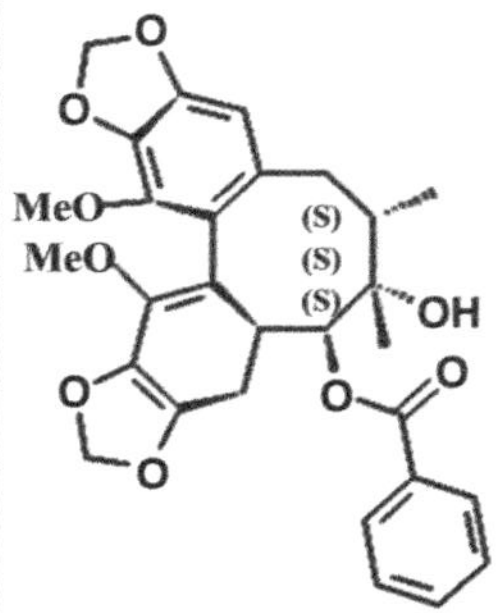	Secoisolariciresinol Lariciresinol Sesamin	Cabbage, carrot, capsicum and peanuts	• Anti-oxidant • Improve immune system • Improve prostate health • Maintain testosterone level • Anti-estrogenic • Reduce the levels of stress hormone (cortisol)	1,4-benzodioxane derivatives	Pilkington and Barker (2015), Fazary et al. (2016), and Rodríguez-García et al. (2019)
Tannins	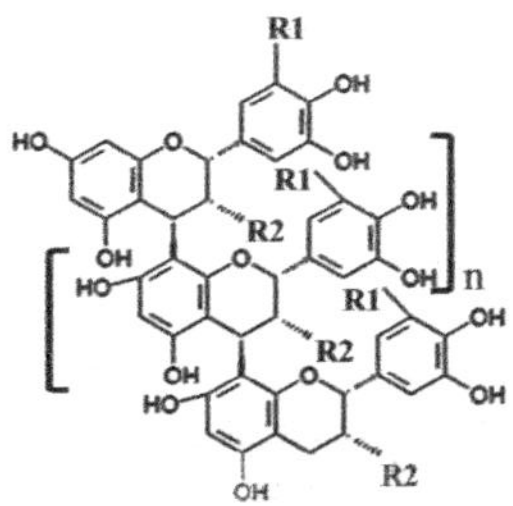	Condensed tannins	Beans, berries, bananas, cocoa, tea, almonds, pomegranate, walnuts, red wine, beer and barley	• Anti-cancer • Anti-mutagenic • Anti-microbial • Anti-oxidant • Reduce nutrient absorption and conversion into novel compounds	Catechin polymers Epicatechin polymers	Yacco et al. (2016), Smeriglio et al. (2017), and Watrelot et al. (2017)
		Hydrolysable tannins			Gallotannins Ellagitannins	

Coumarins	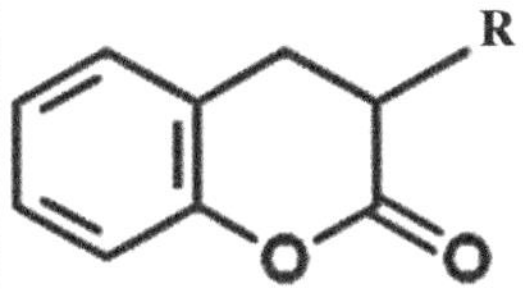	Umbelliferone Aesculetin	Strawberry, cinnamon, tobacco product and lavender	• Anti-coagulant • Anti-inflammatory • Anti-oxidant • Anti-apoptotic • Improve brain health		Fonseca et al. (2018), Srikrishna et al. (2018), and Zhu and Jiang (2018)
Curcuminoids	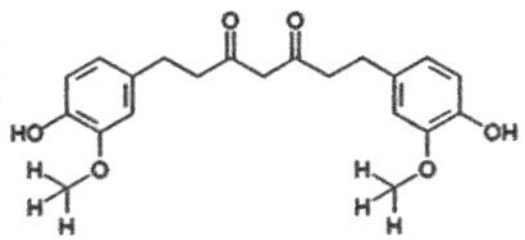	Curcumin	Turmeric	• Anti-inflammatory • Anti-apoptotic • Anti-oxidant • Protect against DNA and mitochondrial damage • Neuroprotective • Help fight against many heart diseases • Decrease lipid peroxidation	Demethoxycurcumin Bisdemethoxycurcumin	Sahebkar et al. (2015), Nabavi et al. (2018), and Ahmed et al. (2019)

markets of dietary and pharmaceutical items, which combine the area of nutraceuticals with nutricosmetics. These are characterized as sustaining nourishment with health and external beauty benefits (German-Baez et al. 2017).

9.3.5.1 Ginkgo Biloba for Neuroprotection

Ginkgo Biloba Extract (EGb-761) is known to be an important source of flavonoids and terpenes. It helps with memory improvement in people with Alzheimer's disease and treats certain other mental illnesses. Apart from this, it was found that a similar concentrate given for a transient period for one and a half month improved different neurocognitive impairments. When rodents were exposed to a certain dosage of EGb-761, it effectively treated mitochondrial damage, reduced action potential, and increased levels of mtDNA damage (Field and Vadnal 1998).

9.3.5.2 Blueberry (*Vaccinium corymbosum*) for Improved Life Expectancy

One of the important cell supplements is *Vaccinium corymbosum*, which is also known as blueberries. These include large variety of polyphenols that provide wide range of health benefits. Cancer preventing agents fight the free radicals responsible for damaging distinctive cell structures such as DNA. Research conducted on specific organisms have shown that blueberry jam enhanced their learning and memory by improving their neuronal capacity (Olas 2018). Studies on Caenorhabditis elegans have, comparatively, indicated that presence of polyphenol in blueberries has increased the life expectancy by decreasing age-related intracellular mass of lipofuscin which acts as a biomarker for age-related damage of cells.

Moreover, these extracts decrease the level of a lipid peroxidation biomarker, called 4-hydroxynenal. Additionally, polyphenols from blueberry enhance the pharyngeal siphoning rates of mature worms and increase resistance to temperature, consequently they improve the worms' health (Gupta and Prakash 2015). It is also known that blueberry compounds have ability to increase the life span of Drosophila by almost 10%. These studies have demonstrated the basic activity of blueberries in invertebrates which can be extended in future to study more of such animals (Hidalgo and Almajano 2017).

9.3.5.3 Cranberry (*Vaccinium macrocarpon*) and Oregano (*Origanum vulgare*) for Increased Life Span

The Cranberry and Oregano plants also provide various health benefits. Cranberry is rich in many phytochemicals. Cranberry juice is reported to be effective in treating urinary tract infections. Many signaling pathways like mitogen-activated protein kinase, Jun kinase (JNK), and NFκB pathways were found to be affected by cranberry, indicating it to have significant anti-oxidant properties. Both cranberry and oregano have shown anti-microbial activities. Oregano is found to possess anti-viral, anti-oxidant, anti-tumor, and anti-apoptotic activities. It was found to be effective against hyperlipidemia and diabetes Zou et al. (2009). Specialists have indicated that the addition of a blend of oregano and cranberry (OC) extracts in the diet results in an increased life expectancy in Mexican common flies (mexfly).

Additionally, enhanced longevity was found to be the result of OC supplementation in middle age. However, even increase in OC did not result in life expectancy during middle age. These results require strong attention towards diet intake and composition to build up a strong developing intervention Zou et al. (2009).

9.3.5.4 Nectarine and Acai

Nectarine and Acai are found to be rich in phytochemicals. Nectarine is used worldwide, while acai is native to Amazon. In high fat diet fed Drosophila, nectarine was seen to be associated with life longevity while acai was linked with improved health. But the exact underlying mechanisms are yet to be identified (Sun et al. 2010; Boyd et al. 2011).

9.3.5.5 Rosa Damascena

Rosa damascena, a hybrid species of rose, is very commonly used to make rose oil and water in health and food industries. Rosa damascena extract contains a number of unstable natural substances such as polyphenols including myricetin, quercetin, gallic acid and kaempferol, and terpenes including heneicosane, citronellol, and disiloxane. Extracts of Rosa damascena have shown to be effective against seizures in rodents, microbial diseases, and amyloid beta accumulation in neurons, thus showing therapeutic potential against Alzheimer's disease. An extract of Rosa damascena was found to increase average life span and longevity in Drosophila. This extract also improves flies' response to stress, protects them from oxidative stress and iron induced stress. Molecular studies have demonstrated that rose extracts reduce heat-initiated activation of a significant heat shock protein HSP70 and a relatively small mitochondrial heat shock protein HSP22. In this way, Rosa damascena was suggested to be helpful in increasing life span of flies by making them resistant against iron induced stress (Dong et al. 2012).

9.3.5.6 Green Tea as a Health Supplement

Epigallochatechin-3-gallate (EGCG) is the basic and dynamic flavonol in green tea, along with this, other important chemicals are epigallochatechin, epicatechin, and epichatechin-3-gallate. Regardless of all the studies concerning the neuroprotective properties of EGCG in people, epidemiological data has shown that higher use of tea/green tea, rich in EGCG, is associated with a reduced risk of neurodegenerative issue and a lower risk of mental illnesses, thus reducing the cases of dementia, anxiety, and Parkinson's disease. Tea catechins have broad range of actions because of their capacity to chelate metal ions, total iron, to start a mitigating reaction, to stimulate cholinergic transmission, and to revive neuronal development (Singh et al. 2010).

9.3.5.7 Cocoa Polyphenols and Enhanced Life Expectancy

Distinctive polyphenols with high disease controlling activities, for example, flavonoids, are obtained from cocoa. According to a study, the cocoa powder is rich in flavonoids which decrease oxidative stress in C. elegans. Genes in C. elegans can be controlled by cocoa polyphenols which reduce oxidative stress, and influence and change the chromatin structure (Andújar et al. 2012).

9.3.5.8 Tannic Acid (TA) and Quercetin as Longevity Supplement

Tannic acid (TA) belongs to a group of tannins; they are optional metabolites of plants with different restorative activities. TA has different pharmacological properties which protect from neurodegeneration, pathogen infection, oxidative damage, and carcinogenesis.

Quercetin is an important cell component and belongs to flavonoid family. It is present in many citrus fruits, tea, apples, parsley, onions, red wine, and sage, and also in different dietary supplements. Quercetin has been found to show beneficial effects on bone mineral density. However, still there is a need of more information to extract flavonoid mixes having beneficial pharmacological effects which can be helpful in reducing age induced aging (David et al. 2016). Transcriptome analysis shows that quercetin impacts the gene regulation in TGF-β members, insulin like family, and p38 MAP kinase pathways. These studies suggest that TGF beta and p38 mitogen associated protein kinase include important and long-lasting roles of TA and quercetin (Gupta and Prakash 2015).

9.3.5.9 Olive Oil as Phenolic Supplement

The utilization of olive oil was found to have numerous beneficial health effects and increases the life span in humans. These effects of olive oil are due to the occurrence of many phenolic compounds. Tyrosol is one of the most common phenol in olive oil; it was found to significantly improve the life expectancy of C. elegans by protecting it from heat and oxidative damage. Tyrosol induced longevity is dependent on HSF-1 and IIS/DAF-16. After treatment with tyrosol, Hsp-12.6, a co-regulated target gene of DAF-16, and HSF-1 are considerably upregulated in grown-up worms. It has been found that the heat shock protein HSP-12.6, improves life expectancy and delay the poly-glutamine protein accumulation in C. elegans. HSF-1 is important for keeping up the homeostasis of various proteins. Together, these examinations recommend that tyrosol as a phenolic compound in olive oil increases life expectancy by increasing resistance to oxidative stress and thermotolerance (Cañuelo et al. 2012; Hsu et al. 2003).

9.3.5.10 Rosmarinic Acid and Caffeic Acid

Rosmarinic acid (RA) and caffeic acid (CA) are commonly found in a group of ordinary herbs and vegetables. RA and CA show anti-carcinogenic, anti-microbial, anti-tumor, anti-apoptotic, and anti-rheumatic properties. RA and CA can improve the life span of C. elegans (Gupta and Prakash 2015). Similarly, some studies also recommend that CA and RA increase life span by controlling the pathways involved in managing stress response (Krajčovičová and Meluš 2013; Magnani et al. 2014).

9.4 Conclusion

It is quite clear from the studies that nutraceuticals advance life expectancy, improve longevity, and provide immunity against developing diseases which can shorten life. More research in this field will help to reveal the underlying mechanisms of action of these naturally occurring substances, so that they can be used as a supplements and/or pharmacological agents for the treatment of old age disorders. Following a balanced diet based on the consumption of these compounds will play a significant impact on an individual's overall health. Promising results of many nutraceuticals have been observed in model organisms, and this data can be further used to decipher their role in aging on molecular levels. The future belongs to nutraceuticals in place of the conventional medicines, because of their maximum potential in curing and preventing age-related diseases with no adverse effects to health.

References

Ahmed T, Nawaz M, Iqbal W (2019) Pharmacological effects of curcuminoids in neurological disorders. In: Curcumin for neurological and psychiatric disorders. Academic Press, London, pp 129–154

Andújar I, Recio M, Giner R, Ríos J (2012) Cocoa polyphenols and their potential benefits for human health. Oxidative Med Cell Longev 2012:1

Azzu V, Valencak TG (2017) Energy metabolism and ageing in the mouse: a mini-review. Gerontology 63(4):327–336

Boyd O et al (2011) Nectarine promotes longevity in Drosophila melanogaster. Free Radic Biol Med 50(11):1669–1678

Cañuelo A et al (2012) Tyrosol, a main phenol present in extra virgin olive oil, increases lifespan and stress resistance in Caenorhabditis elegans. Mech Ageing Dev 133(8):563–574

Chawla R, Goel N (2016) Phytosterol and its esters as novel food ingredients: a review. Asian J Dairy Food Res 35(3):217

Chin YT, Cheng GY, Shih YJ, Lin CY, Lin SJ, Lai HY et al (2017) Therapeutic applications of resveratrol and its derivatives on periodontitis. Ann New York Acad Sci 1403(1):101–108

Cholewski M, Tomczykowa M, Tomczyk M (2018) A comprehensive review of chemistry, sources and bioavailability of omega-3 fatty acids. Nutrients 10(11):1662

David AVA, Arulmoli R, Parasuraman S (2016) Overviews of biological importance of quercetin: a bioactive flavonoid. Pharmacogn Rev 10(20):84

Davinelli S, Willcox DC, Scapagnini G (2012) Extending healthy ageing: nutrient sensitive pathway and centenarian population. Immun Ageing 9(1):9

Davoodi SH, Shahbazi R, Esmaeili S, Sohrabvandi S, Mortazavian A, Jazayeri S, Taslimi A (2016) Health-related aspects of milk proteins. Iran J Pharm Res 15(3):573

Del Valle HB, Yaktine AL, Taylor CL, Ross AC (2011) Dietary reference intakes for calcium and vitamin D. National Academies Press, Washington

Dillard CJ, German JB (2000) Phytochemicals: nutraceuticals and human health. J Sci Food Agric 80(12):1744–1756

Dong Y et al (2012) Nutraceutical interventions for promoting healthy aging in invertebrate models. Oxidative Med Cell Longev 2012:1

Efeyan A, Comb WC, Sabatini DM (2015) Nutrient-sensing mechanisms and pathways. Nature 517 (7534):302

Fazary AE, Alfaifi MY, Saleh KA, Alshehri MA, Elbehairi SEI (2016) Bioactive lignans: a survey report on their chemical structures. Nat Prod Chem Res 4(226):2

Field BH, Vadnal R (1998) Ginkgo biloba and memory: an overview. Nutr Neurosci 1(4):255–267
Fonseca A, Matos MJ, Vilar S, Kachler S, Klotz KN, Uriarte E, Borges F (2018) Coumarins and adenosine receptors: new perceptions in structure–affinity relationships. Chem Biol Drug Des 91(1):245–256
Forni C, Facchiano F, Bartoli M, Pieretti S, Facchiano A, D'Arcangelo D, Norelli S, Valle G, Nisini R, Beninati S (2019) Beneficial role of phytochemicals on oxidative stress and age-related diseases. Biomed Res Int 2019:1
Franceschi C, Garagnani P, Morsiani C, Conte M, Santoro A, Grignolio A, Monti D, Capri M, Salvioli S (2018) The continuum of aging and age-related diseases: common mechanisms but different rates. Front Med 5:61
Gemma C, Vila J, Bachstetter A, Bickford PC (2007) Chap. 15. Oxidative stress and the aging brain: from theory to prevention. In: Brain aging: models, methods, and mechanisms. CRC Press, Boca Raton
Gennari C (2001) Calcium and vitamin D nutrition and bone disease of the elderly. Public Health Nutr 4(2b):547–559
German-Baez LJ et al (2017) Anti-aging and nutraceutical characterization of plant infusions used in traditional medicine. Pak J Nutr 16(4):285–292
Gladyshev VN (2014) The free radical theory of aging is dead. Long live the damage theory! Antioxid Redox Signal 20(4):727–731
Gogus U, Smith C (2010) N-3 omega fatty acids: a review of current knowledge. Int J Food Sci Technol 45(3):417–436
Gupta C, Prakash D (2015) Nutraceuticals for geriatrics. J Tradit Complement Med 5(1):5–14
Hekimi S, Lapointe J, Wen Y (2011) Taking a "good" look at free radicals in the aging process. Trends Cell Biol 21(10):569–576
Hidalgo G-I, Almajano MP (2017) Red fruits: extraction of antioxidants, phenolic content, and radical scavenging determination: a review. Antioxidants 6(1):7
Howes RM (2006) The free radical fantasy: a panoply of paradoxes. Ann N Y Acad Sci 1067 (1):22–26
Hsu A-L, Murphy CT, Kenyon C (2003) Regulation of aging and age-related disease by DAF-16 and heat-shock factor. Science 300(5622):1142–1145
Ishaq S, Jafri L (2017) Biomedical importance of cocoa (Theobroma cacao): significance and potential for the maintenance of human health. Matrix Sci Pharm 1(1):1–5
Jain N, Ramawat KG (2013) Nutraceuticals and antioxidants in prevention of diseases. In: Natural products: phytochemistry, botany and metabolism of alkaloids, phenolics and terpenes. Springer, Heidelberg, pp 2559–2580
Jaul E, Barron J (2017) Age-related diseases and clinical and public health implications for the 85 years old and over population. Front Public Health 5:335
Kapahi P, Zid B (2004) TOR pathway: linking nutrient sensing to life span. Sci Aging Knowl Environ 2004(36):PE34
Krajčovičová Z, Meluš V (2013) Bioactivity and potential health benefits of Rosmarinic acid. Univ Rev 7(2):8–14
Magnani C, Isaac VLB, Correa MA, Salgado HRN (2014) Caffeic acid: a review of its potential use in medications and cosmetics. Anal Methods 6(10):3203–3210
Marshall S (2006) Role of insulin, adipocyte hormones, and nutrient-sensing pathways in regulating fuel metabolism and energy homeostasis: a nutritional perspective of diabetes, obesity, and cancer. Sci. STKE 2006(346):re7
Masoro EJ (2005) Overview of caloric restriction and ageing. Mech Ageing Dev 126(9):913–922
Meydani SN, Barklund MP, Liu S, Meydani M, Miller RA, Cannon JG, Morrow FD, Rocklin R, Blumberg JB (1990) Vitamin E supplementation enhances cell-mediated immunity in healthy elderly subjects. Am J Clin Nutr 52(3):557–563
Milholland B, Suh Y, Vijg J (2017) Mutation and catastrophe in the aging genome. Exp Gerontol 94:34–40

Moriguchi S, Muraga M (2000) Vitamin E and immunity. In: Vitamins and hormones. Academic Press, Ames

Nabavi SF, Sureda A, Dehpour AR, Shirooie S, Silva AS, Devi KP et al (2018) Regulation of autophagy by polyphenols: paving the road for treatment of neurodegeneration. Biotechnol Adv 36(6):1768–1778

Olas B (2018) Berry phenolic antioxidants–implications for human health? Front Pharmacol 9:78

Patel S (2015) Emerging trends in nutraceutical applications of whey protein and its derivatives. J Food Sci Technol 52(11):6847–6858

Peng C, Wang X, Chen J, Jiao R, Wang L, Li YM, Zuo Y, Liu Y, Lei L, Ma KY (2014) Biology of ageing and role of dietary antioxidants. Biomed Res Int 2014:831841

Piironen V, Toivo J, Puupponen-Pimiä R, Lampi AM (2003) Plant sterols in vegetables, fruits and berries. J Sci Food Agric 83(4):330–337

Pilkington LI, Barker D (2015) Synthesis and biology of 1, 4-benzodioxane lignan natural products. Nat Prod Rep 32(10):1369–1388

Pradeep A, Dinesh M, Govindaraj A, Vinothkumar D, Ramesh Babu NG (2014) Phytochemical analysis of some important medicinal plants. Int J Biol Pharm Res 5(1):48–50

Puca AA, Spinelli C, Accardi G, Villa F, Caruso C (2018) Centenarians as a model to discover genetic and epigenetic signatures of healthy ageing. Mech Ageing Dev 174:95–102

Rattan SI (2008) Hormesis in aging. Ageing Res Rev 7(1):63–78

Robbins RJ (2003) Phenolic acids in foods: an overview of analytical methodology. J Agric Food Chem 51(10):2866–2887

Rodick TC, Seibels DR, Babu JR, Huggins KW, Ren G, Mathews ST (2018) Potential role of coenzyme Q 10 in health and disease conditions. Nutr Diet Suppl 10:1

Rodríguez-García C, Sánchez-Quesada C, Toledo E, Delgado-Rodríguez M, Gaforio JJ (2019) Naturally lignan-rich foods: a dietary tool for health promotion? Molecules 24(5):917

Rosen HN, Rosen C, Schmader K, Mulder J (2017) Calcium and vitamin D supplementation in osteoporosis. Up To Date

Rudrapal M, Chetia D (2017) Plant flavonoids as potential source of future antimalarial leads. Syst Rev Pharm 8(1):13

Sahebkar A, Serban MC, Ursoniu S, Banach M (2015) Effect of curcuminoids on oxidative stress: a systematic review and meta-analysis of randomized controlled trials. J Funct Foods 18:898–909

Sahni S, Hannan MT, Blumberg J, Cupples LA, Kiel DP, Tucker KL (2009) Protective effect of total carotenoid and lycopene intake on the risk of hip fracture: a 17-year follow-up from the Framingham Osteoporosis Study. J Bone Miner Res 24(6):1086–1094

Saini R (2011) Coenzyme Q10: the essential nutrient. J Pharm Bioallied Sci 3(3):466–467

Santini A, Tenore GC, Novellino E (2017) Nutraceuticals: a paradigm of proactive medicine. Eur J Pharm Sci 96:53–61

Sen CK, Khanna S, Rink C, Roy S (2007) Tocotrienols: the emerging face of natural vitamin E. Vitam Horm 76:203–261

Shimokawa I, Trindade LS (2010) Dietary restriction and aging in rodents: a current view on its molecular mechanisms. Aging Dis 1(2):89

Shimokawa I, Chiba T, Yamaza H, Komatsu T (2008) Longevity genes: insights from calorie restriction and genetic longevity models. Mol Cells 26(5):427

Silva FA, Borges F, Guimarães C, Lima JL, Matos C, Reis S (2000) Phenolic acids and derivatives: studies on the relationship among structure, radical scavenging activity, and physicochemical parameters. J Agric Food Chem 48(6):2122–2126

Singh R, Akhtar N, Haqqi TM (2010) Green tea polyphenol epigallocatechi3-gallate: inflammation and arthritis. Life Sci 86(25–26):907–918

Sirerol JA, Rodríguez ML, Mena S, Asensi MA, Estrela JM, Ortega AL (2016) Role of natural stilbenes in the prevention of cancer. Oxidative Med Cell Longev 2016:15

Smeriglio A, Barreca D, Bellocco E, Trombetta D (2017) Proanthocyanidins and hydrolysable tannins: occurrence, dietary intake and pharmacological effects. Br J Pharmacol 174 (11):1244–1262

Srikrishna D, Godugu C, Dubey PK (2018) A review on pharmacological properties of coumarins. Mini Rev Med Chem 18(2):113–141

Suh Y, Atzmon G, Cho M-O, Hwang D, Liu B, Leahy DJ, Barzilai N, Cohen P (2008) Functionally significant insulin-like growth factor I receptor mutations in centenarians. Proc Natl Acad Sci 105(9):3438–3442

Sun X et al (2010) Açai palm fruit (Euterpe oleracea Mart.) pulp improves survival of flies on a high fat diet. Exp Gerontol 45(3):243–251

Villeponteau B, Cockrell R, Feng J (2000) Nutraceutical interventions may delay aging and the age-related diseases. Exp Gerontol 35(9–10):1405–1417

Watrelot AA, Schulz DL, Kennedy JA (2017) Wine polysaccharides influence tannin-protein interactions. Food Hydrocoll 63:571–579

Wattanapenpaiboon N, Lukito W, Wahlqvist ML, Strauss BJ (2003) Dietary carotenoid intake as a predictor of bone mineral density. Asia Pac J Clin Nutr 12(4):467

Yacco RS, Watrelot AA, Kennedy JA (2016) Red wine tannin structure–activity relationships during fermentation and maceration. J Agric Food Chem 64(4):860–869

Yang H-J et al (2015) Vision from next generation sequencing: multi-dimensional genome-wide analysis for producing gene regulatory networks underlying retinal development, aging and disease. Prog Retin Eye Res 46:1–30

Zakova T, Rondevaldova J, Bernardos A, Landa P, Kokoska L (2018) The relationship between structure and in vitro antistaphylococcal effect of plant-derived stilbenes. Acta Microbiol Immunol Hung 65(4):467–476

Zhu JJ, Jiang JG (2018) Pharmacological and nutritional effects of natural coumarins and their structure–activity relationships. Mol Nutr Food Res 62(14):1701073

Zou S, Carey JR, Liedo P, Ingram DK, Yu B, Ghaedian R (2009) Prolongevity effects of an oregano and cranberry extract are diet dependent in the Mexican fruit fly (Anastrepha ludens). J Gerontol A Biomed Sci Med Sci 65(1):41–50

Functional Foods and Dietary Patterns for Prevention of Cognitive Decline in Aging

10

Zahra Bayrami, Madiha Khalid, Sedigheh Asgari Dastjerdi, and Motahareh Sadat Masjedi

Abstract

Cognitive decline is known as the prevalent impairments related to aging and recently by the increasing of life expectancy in societies, the number of patients is rising. The impact of several nutrients and whole dietary patterns on cognitive disorders proves that a healthy diet along with other modifiable factors (e.g., physical activity and cognitive activity) and non-modifiable factors (e.g., gender, age, genetic) is an important factor to maintain cognitive function and improve life quality during old age.

Aging and metabolic abnormality are both associated with cognitive decline that have effect on fine motor control, balance, short-term and long-term memories, and executive function. Oxidative stress and inflammation are common features in cognitive decline. Neuro inflammation occurs locally in the brain thus peripheral inflammatory cells and circulating inflammatory mediators (e.g., cytokines) can also infiltrate the brain, and this occurs more frequently as we age. Therefore, strategies like targeting peripheral inflammation can reduce infiltration of inflammatory mediators into the brain and, as a result, reduce the prevalence of a variety of age-related deficits.

Z. Bayrami (✉) · M. Khalid
Toxicology and Diseases Group, Pharmaceutical Sciences Research Center, The Institute of Pharmaceutical Sciences (TIPS), Tehran University of Medical Sciences, Tehran, Iran
e-mail: bayrami@farabi.tums.ac.ir

S. Asgari Dastjerdi
Isfahan Cardiovascular Research Center, Cardiovascular Research Institute, Isfahan University of Medical Sciences, Isfahan, Iran

M. Sadat Masjedi
Hypertension Research Center, Cardiovascular Research Institute, Isfahan University of Medical Sciences, Isfahan, Iran

S. M. Nabavi et al. (eds.), *Nutrients and Nutraceuticals for Active & Healthy Ageing*,
https://doi.org/10.1007/978-981-15-3552-9_10

Functional foods affect the body by enhancing the consumers' health or reducting the chronic diseases risks, and these effects are more than the benefits of usual foods. The average quantity and usual composition of food and beverage used by an individual or a group of people—dietary pattern—are the main source of nutrients for body and it is expected to affect the life quality and health condition during lifetime including age-related cognitive decline.

Nutrients in the food can affect cognitive processes and emotions by changing the chemical composition of our brain and alter our mood. Dietary factors affect multiple brain processes through regulating neurotransmitter pathways, membrane fluidity, synaptic transmission, and signal-transduction pathways. Some plausible mechanisms of action for diets and nutrients shown to be effective on cognitive aging have been schemed in this chapter.

Keywords

Functional food · Dietary pattern · Cognitive decline · Aging

10.1 Introduction

In recent decades, the life expectancy is growing as a result of improving the public health and lower infectious diseases. Based on WHO report, the population over 60 in the world is increasing during recent years and the societies are going old faster in the next decades (Ageing and health 2018).

Aging of population causes the epidemic of chronic diseases associated with elderly people such as diabetes, heart disease, dementia, and Alzheimer's disease (AD). These disorders affect the quality of life in aged population and cause massive financial burden on the health systems, which in developing countries bring more problem because they already engaged in fighting with the infectious diseases. Many efforts have launched to understand the factors of prevention and control of chronic diseases related to aging and enhance the life quality of aged people as well as decrease the treatment costs for the old people, their family, and the governments (Shetty 2012).

Cognitive decline and dementia are known as the prevalent impairments related to aging. These chronic diseases cause disability more than mortality in the patients and impose heavy costs on the families of patients and the societies and the financial burden of these diseases are growing by the increase of aging in the world population. Prevention of the dementia is one of the most important research priorities to reduce the worldwide burden of the disease by 2025. Different risk/protective factors have been evaluated and effect of diet, as one of the modifiable factors on dementia has been defined (Shah et al. 2016). The impact of several nutrients and whole dietary patterns on cognitive disorders has been studied (Richard et al. 2018; Smith and Blumenthal 2016) and the findings explicit the synergetic effects of nutrients and components of diet (Richard et al. 2018; Solfrizzi et al. 2017). According to these information, a healthy diet along with other modifiable factors (e.g., physical activity and cognitive activity) and non-modifiable factors (e.g., gender, age, genetic) is an important factor to maintain cognitive function and improve life quality during old age. In this chapter, authors attempt to explain cognitive impairment related to aging, definition of functional foods and dietary patterns and their classifications, the role of functional foods and dietary patterns on cognitive function, and the possible mechanisms and pathways involved in the effects of dietary patterns on cognitive decline related to aging. For quick access, almost all of neuroprotective nutrients including amino acids, vitamins, flavonoids, fatty acids, minerals, and even gut hormones have been summarized as a table and plausible mechanisms of action for diets and nutrients shown to be effective on cognitive aging have been schemed.

10.2 How Aging Causes Cognitive Impairment?

Aging of the world population is an undeniable current actuality (Dominguez and Barbagallo 2018) so today aging-related impairments are becoming more prevalent and costly. Among these diseases, decline in cognitive function is a specific concern (Tucker 2016) because cognitive skills play a crucial role in the daily functioning of older people (Klimova et al. 2017) and dementia is one of the greatest causes of disability (Dominguez and Barbagallo 2018). Cognitive decline occurs in mental domains such as processing speed, reasoning, memory, and executive functions (Deary et al. 2009).

On the other hand, aging increases risks of cognitive impairment and dementia (Abbatecola et al. 2018). Worldwide, 47 million persons are currently affected by dementia with about eight million newly diagnosed cases annually (Dominguez and Barbagallo 2018). Prevalence of dementia has been increased to the point that more than 13% of adults aged 72 years and older have dementia, and by 85 years of age and older, 32% may have AD (Tucker 2016). Most of them, 60–75%, have AD followed by vascular and Lewy bodies dementia. Because cognitive decline occurs generally in old age and the causal brain pathology develops years before, thus preventive strategies or delaying the onset of dementia would have intense effects on public health (Dominguez and Barbagallo 2018).

Cognitive decline occurs both in normal aging and in pathological conditions, such as neurodegenerative diseases. When considering about aging and cognitive decline one certain interest is the hippocampus. With age-related cognitive decline, neurobiological alterations occurred in hippocampus including increased oxidative stress and neuro inflammation, altered intracellular signaling and gene expression as well as reduced neurogenesis and synaptic plasticity (Bettio et al. 2017).

Adult neurogenesis is an integral component of neural plasticity, brain homeostasis, maintenance, and tissue remodeling in the central nervous system (CNS). In mammalians, adult neural stem cells or precursor cells in the hippocampus can differentiate and proliferate into new neurons. The process begins with the proliferation of progenitor cells, is followed by commitment to a neuronal phenotype and morphological and physiological maturation with the development of functional neuronal characteristics, and ends with the existence of a newly functioning integrated neuron.

The process is a highly complex and multi-step that determined by both intrinsic and extrinsic factors, including neurotrophins, antidepressants, opioids, seizures, physical activity, glucocorticoids, sex hormones, growth factors, excitatory neurotransmission, learning, physical exercise, stress, and diet.

There is impaired adult neurogenesis in patients with neurological diseases, including AD, Parkinson disease (PD), Huntington disease (HD), epilepsy, ischemia, autism spectrum disorders (ASD), and prion diseases, leading to continuous loss of neurons and subsequent cognitive and motor disabilities. However, some strategies for adult neurogenesis have vast therapeutic potential in combating the exponential rise in the occurrence of neurodegenerative diseases (poulose et al. 2017).

There are several risk factors for cognitive decline that divided into modifiable and non-modifiable risk factors. The non-modifiable risk factors include age, race and ethnicity, gender, and genetics; and the modifiable risk factors involve diabetes, head injuries, lifestyle, socio-economic, and environmental parameters including nutrition and education (Dominguez and Barbagallo 2018; Klimova et al. 2017). Prevention strategies for treatment of age-related cognitive decline are very important (Dominguez and Barbagallo 2018). Neurogenesis plays a critical role in neural plasticity, brain homeostasis, and maintenance in the central nervous system. Adult neurogenesis is known to be affected by many intrinsic and extrinsic factors. Intrinsic factors such as aging, neuro inflammation, oxidative stress, and brain injury, as well as lifestyle factors such as high-fat and high-sugar diets and alcohol and opioid addiction, negatively affect adult neurogenesis (poulose et al. 2017).

Aging and metabolic abnormality are both associated with cognitive decline that have an effect on fine motor control, balance, short-term and long-term memory, or executive function. Oxidative stress and inflammation are common features in cognitive decline. Neuro inflammation occurs locally in the brain thus peripheral inflammatory cells and circulating inflammatory mediators (e.g., cytokines) can also infiltrate the brain, and this occurs more frequently as we age. Therefore, strategies like targeting peripheral inflammation can reduce infiltration of inflammatory mediators into the brain and, as a result, reduce the prevalence of a variety of age-related deficits (Spencer et al. 2017).

10.3 Functional Foods and Dietary Patterns

Functional food or physiologically functional food is defined as a food with some functions and physiological effects in the body by enhancement of consumers' health or reduction of the chronic diseases risks, and these effects are more than the benefits of usual foods. Functional food is known as a food enriched in one or more constituents in order to promote the health of the body (Institute of Medicine 1994) by affecting the immune, nervous, digestive, endocrine, and circulatory system. Arai S. introduced functional food for the first time in 1984, as the food with the tertiary function in the body known as physiological function besides to primary and secondary functions related to nutrients and preferences (Arai 1996). Based on this explanation, functional food enhances the health of consumers further than usual and traditional foodstuff (Ohama et al. 2006). Beneficial effects of

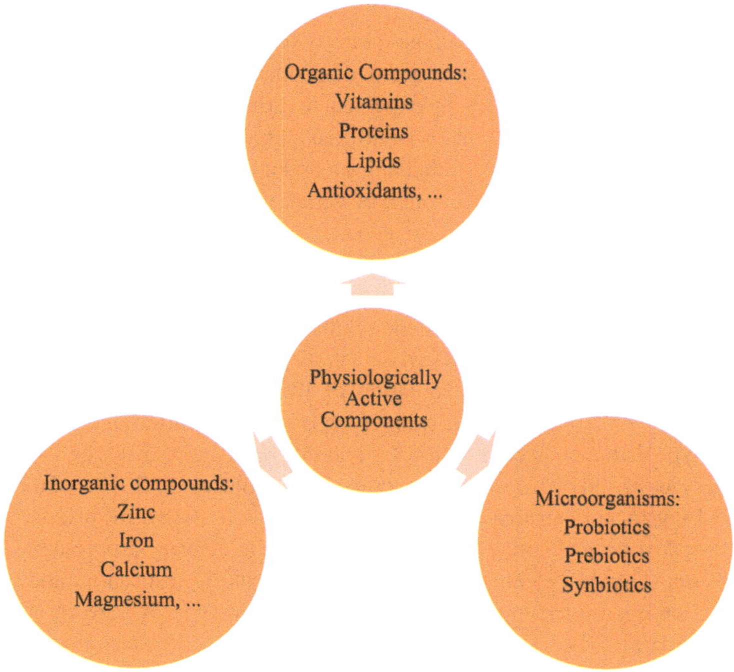

Fig. 10.1 Classification of functional foods based on physiologically active components

functional food have been shown in scientific literatures (Nicoletti 2012; Szakály et al. 2012; Roberfroid 1999) and its consumption is raising tremendously (Khan et al. 2013), thus various publications attempted to explain it (Diplock et al. 1999; Poulsen 1999; Aronson 2017). Besides all proposed explanations, there is not a distinct internationally accepted definition for the term "functional food" yet (Aronson 2017).

Functional foods are classified based on different features such as:

- Their origin (e.g., plants: oats, tomatoes, broccoli and other cruciferous vegetables, garlic, tea, cranberry wine and grapes; and animal: fish, dairy products, and beef) (Hasler 1998).
- Diseases that are supposed to be affected by them (e.g., diabetes, colon cancer, osteoporosis).
- Physiological effects (e.g., antitumor activity, immunology, digestibility).
- Processing methods (e.g., chromatography, encapsulation, freezing), properties (e.g., color, solubility, texture, physical properties).
- Physiologically active components (e.g., inorganic compounds, organic compounds, microorganisms; as illustrated in Fig. 10.1) (Juvan et al. 2005).

Dietary patterns defined as the average quantity and usual composition of food and beverage used by an individual or a group of people. Since dietary food is the main source of nutrients for body, it is expected to affect the life quality and health condition during lifetime and causes health risks especially chronic diseases.

Evaluation of dietary pattern is challenging and there are some limitations such as measurement of exact food consumption or variations in dietary pattern thus various strategies applied to overcome these defects. Categorization of similar foods in broader groups repeated measurement of food intake during the study, and assessment of the results from the systematic reviews of long-term studies has been used as some of applied approaches to study the relation between dietary pattern and major chronic diseases (Schulze et al. 2018).

According to the statistical evaluation of the results from several systematic reviews, dietary pattern has a significant effect on the prevention of chronic diseases including type 2 diabetes, cardiovascular and coronary heart diseases, obesity, and cancer (Ley et al. 2014; Mozaffarian 2016; Bechthold et al. 2019; World Cancer Research Fund/American Institute for Cancer Research 2018). In different studies, the beneficial effects of some dietary patterns such as Dietary Approaches to Stop Hypertension (DASH), Healthy Diet Index (HDI), caloric restriction, and Mediterranean diet on cognitive function and dementia have shown (Wengreen et al. 2013; Corrêa Leite et al. 2001; Le Bourg 2012; Cao et al. 2016; Petersson and Philippou 2016). In addition, following Mediterranean and Mediterranean-DASH Intervention for Neurodegenerative Delay (MIND) diets have known to be useful to prevent age-related cognitive decline (Calil et al. 2018; Feart et al. 2010). Physiological properties of dietary pattern interfere in the metabolic pathways and affect the health condition according to the composition of foods. For instance, antioxidant and anti-inflammatory properties of Mediterranean diet (Sureda 2016; Estruch et al. 2016) cause the beneficial effects on the health and reduction of health risks such as its neurodegenerative diseases (Gardener and Caunca 2018).

Functional food consumption as nutrient supplement and healthy dietary pattern are both useful for health condition but the dietary pattern is known to be more effective in promotion of health and reduction of health risks (Tucker 2010).

10.4 Role of Functional Foods and Dietary Patterns on Cognitive Function?

Undoubtedly direct connection is between nutrition, brain function, and behavior (Gómez-Pinilla 2008). Nutrients in the food can affect cognitive processes and emotions by changing the chemical composition of our brain and alter our mood (Banjari et al. 2014). Dietary factors affect multiple brain processes through regulating neurotransmitter pathways, membrane fluidity, synaptic transmission, and signal-transduction pathways (Gómez-Pinilla 2008). Foods are made up of more than one nutrient that their interactions can have impact on cognitive possibilities through alertness and the production or release of neurotransmitters (Banjari et al. 2014).

Modifiable lifestyle factors have positive or negative effects on the susceptibility of cognitive abilities to pathological aging conditions. Good factors including dietary interventions, physical exercise, and environmental enrichment have positive effects on oxidative stress, levels of phosphorylated tau and Aβ peptide, neurotrophic secretion, synaptic transmission, and hippocampal neurogenesis. Whereas bad factors include metabolic dysregulation, impairments in hormone signaling patterns, elevated stress, increased caloric intake, sedentary lifestyles, and poor cognitive reserve (Bettio et al. 2017). Modifiable lifestyle factors, including physical activity, cognitive engagement, and diet, are a key strategy for maintaining brain health during aging (Phillips 2017). Thus, changing lifestyle may reduce the risk of developing aging-associated neuropathologies and neurodegenerative diseases (Bettio et al. 2017).

Recently researches on combined effects, such as multi-nutrient approaches, or a healthy dietary pattern are growing. The complexity of the diet and possible interaction and synergy between nutrients are important (van de Rest et al. 2015) because combinations of foods and nutrients may act synergistically to provide stronger benefits than those conferred by individual dietary components (Dominguez and Barbagallo 2018). Studies showed that multi-nutrient approaches using the Mediterranean diet, DASH, and MIND could low-off risk of cognitive impairment, mild cognitive impairment, and AD in older persons. This multi-nutrient style seems to hold better effects than single nutrient intervention (Abbatecola et al. 2018).

At present, healthy diets, antioxidant supplements, and the prevention of nutritional deficiencies in patients with dementia could be considered the first line of protection against the development and progression of cognitive decline (Solfrizzi et al. 2003). Collectively, significant slowing and reduction of cognitive decline may be reached by following a healthy dietary pattern contain limits intake of added sugars, while maximizing intakes of fish, fruits, vegetables, nuts, complex carbohydrates, fibers, cereals, red wine, non-animal fat, seeds (Tucker 2016; Solfrizzi et al. 2003), and functional foods (Sivamaruthi et al. 2018).

Brain is highly susceptible to alterations in body chemistry subsequent from nutrient intake and deficiency. Epigenetic studies approve that some nutrients alter our brain development and vulnerability to diseases (Banjari et al. 2014). The mild brain chemical balance is controlled by the blood–brain barrier that can be realized through brain's capability of receiving, storage, and integrating sensory information while initiating and controlling motor responses. These functions are related to mental events and form the basis of behavior (Banjari et al. 2014). Neural impulses are mainly resulting from sodium-potassium exchange, but factors such as complex carbohydrates, amino acids (tryptophan and tyrosine), fatty acids, predominantly omega-3 fatty acids, affect the permeability of cell membrane, neurotransmitter metabolism, and glial cells (Banjari et al. 2014).

Food intake is basically motivated behavior with the potential to modulate brain structure and function. Whereas the brain covers 2% of total body weight, it consumes 20% of the total energy derivative from nutrients. The extreme request for energy is because of requisite needs of neurons to maintain ionic gradients across their membranes to facilitate neurotransmission via oxidative metabolism.

Consequently, neurons are extremely sensitive to mitochondrial dysfunction and oxidative stress. The importance of feeding behavior for survivability makes it seem possible that improved food intake lead to positive function of brain that occurred through ability of dietary factors to modulate synaptic plasticity by altering neurogenesis, inflammation, antioxidant defense mechanisms, neurotrophin levels, and energy metabolism (Phillips 2017).

Many of these nutrients can improve learning and memory by affecting the hippocampal brain region. These compounds can induce adaptive stress–response molecules and modify the specific microenvironments in adult progenitor cells area and protect brain after more severe stress in the occurrence of larger insults by promoting cell repair and survival, via inducing and activating trophic factors, antioxidant and DNA-repair enzymes, and proteins involved in mitochondrial biogenesis (poulose et al. 2017).

Finally, the brain is the most metabolically active organ in the human body that consumes 20% of the total energy derivative from nutrients (Phillips 2017). Intake foods may have good or bad properties on cognitive function that it depends on specific nutrient and synergism of nutrients between together. Nutrients affect on cognition function by some pathways including sodium-potassium exchange, affect permeability of cell membrane, neurotransmitter metabolism, and glial cells, altering neurogenesis, inflammation, antioxidant defense mechanisms, neurotrophin levels, energy metabolism, modify the specific microenvironments in adult progenitor cells area, inducing and activating trophic factors, antioxidant and DNA-repair enzymes, and proteins involved in mitochondrial biogenesis, and maintain ionic gradients across their membranes to facilitate neurotransmission via oxidative metabolism (Poulose et al. 2017; Banjari et al. 2014; Phillips 2017). For enhancing cognition, there are numerous dietary patterns and lifestyles that may more or less affect neurodegenerative disease in particular cognitive decline. In order to protect against a variety of age-related conditions that can impair your memory and the general functioning of your brain, a good first step is to concentrate on incorporating three nutrients into your diet: omega-3 fatty acids, flavonoids, and vitamin E. Other good nutrients include Phosphatidylserine, Taurine, Creatine, *Melissa officinalis*, *Salvia officinalis*, Alpha-lipoic acid (ALA), Acetyl-L-carnitine (ALCAR), Coenzyme Q10 (ubiquinone), and Docosahexaenoic acid (DHA) (Scholey et al. 2011). On the other hand, high intakes of nutrients that contribute to hypertension, atherosclerosis, and poor glycemic control may have negative effects on cognition. Therefore the 7 worst foods for your brain that you must avoid them to prevent cognition decline include sugary drinks, refined carbs,[1] foods high in trans fats, highly processed foods, aspartame, alcohol, and fish high in mercury (Scholey et al. 2011; Parrott and Greenwood 2007).

Table 10.1 demonstrates almost all of nutrients that have an effect on cognition function include amino acids, vitamins, flavonoids, fatty acids, minerals, and even gut hormones.

[1]Refined carbohydrates include sugars and highly processed grains, such as white flour.

Table 10.1 Nutrients, their food sources, and the pathways of their effects on brain and cognitive functions are summarized with their references

Nutrient	Food sources	Pathway	References
Omega-3 and omega-6 fatty acids, linolenic acid (LA, 18:2 omega-6) found in most plants, coconut and palm, and α-linolenic acid (ALA, 18:3 omega-3) in green leafy vegetables, flax, and walnuts	Fish (salmon, trout, sardines, mackerel, halibut, and herring), oils (flaxseed oil, chia seed oil, cod liver oil, and krill oil), leafy greens (brussels sprouts, spinach, arugula, mint, kale, and watercress), kiwi fruit, butternuts, walnuts	Effects on endocannabinoid and inflammatory pathways in specific brain regions Improvement of cognitive decline in the elderly Promote the efficient electrical signaling between nerve cells	Spencer et al. (2017) and Gómez-Pinilla (2008)
Saturated fat	Butter, ghee, suet, lard, coconut oil, cottonseed oil, palm kernel oil, dairy products (cream, cheese), meat	Exacerbation of cognitive decline in aging humans	Gómez-Pinilla (2008)
Curcumin	Turmeric (curry spice)	Antioxidant and anti-inflammatory in people with Alzheimer's may benefit memory to help clear the amyloid plaques and eases depression by boosting serotonin and dopamine and helps new brain cells grow by boosting brain-derived neurotrophic factor	Gómez-Pinilla (2008)
Caffeine	Coffee, green tea	Caffeine blocks adenosine, a chemical messenger that makes you sleepy, boosts some of neurotransmitters, such as serotonin, sharpened concentration and also reduced risk of neurological diseases, such as Parkinson's and Alzheimer's	
Flavonoids/polyphenol	Coffee, cocoa, green tea, ginkgo tree, citrus fruits, wine (higher in red wine), dark chocolate berries: Blueberries, strawberries, black berries leafy greens: Spinach, kale, water cress	Improvement of cognitive function in the elderly Increasing the number of connections between neurons to disrupt the development of amyloid plaques Reducing inflammation	Spencer et al. (2017), Gómez-Pinilla (2008), and Miquel et al. (2018)

(continued)

Table 10.1 (continued)

Nutrient	Food sources	Pathway	References
	Colorful fruit or vegetable-bearing plants Other colorful produce: Butternut squash, avocados, plums, and red grapes	and oxidative stress in the brain Anti-inflammatory bioactives, suppressing the activation of microglia through the down-regulation of cytokine expression and the modulation of signaling pathways involved in the resolution of inflammation, and indirectly synthesis of bioactive mediators with pro-resolutive activities such as resolvins	
B vitamins, folate, B12, and B6	Egg, vitamin B12 is not available from plant products	Supplementation with vitamin B6, vitamin B12, or folate has positive effects on memory performance in women of various ages	Gómez-Pinilla (2008) and Calvaresi and Bryan (2001)
Vitamin D	Fish liver, fatty fish, mushrooms, fortified products, milk, soy milk, cereal grains	Important for preserving cognition in the elderly	Gómez-Pinilla (2008)
Vitamin E	Asparagus, avocado, nuts, peanuts, olives, red palm oil, seeds, spinach, vegetable oils, wheat germ	Reduces cognitive decay in the elderly	Gómez-Pinilla (2008)
Combination of vitamins (C, E, carotene)	Vitamin C: Citrus fruits, several plants, and vegetables: Bell peppers, guava, kiwi, tomatoes and strawberries, calf and beef liver. Vitamin E: See above	Antioxidant vitamin intake delays cognitive decline in the elderly	Gómez-Pinilla (2008)
Choline	Egg yolks, soy beef, chicken, veal, Turkey liver, lettuce	Uses to create acetylcholine, a neurotransmitter that helps regulate mood and memory Evidence is for a causal relationship between dietary choline and cognition in humans	Gómez-Pinilla (2008)

(continued)

Table 10.1 (continued)

Nutrient	Food sources	Pathway	References
Calcium, selenium	Calcium: Milk, coral; selenium: Nuts, cereals, meat, fish, eggs	High serum calcium is associated with faster cognitive decline in the elderly; lifelong low selenium level is associated with lower cognitive function in humans	Gómez-Pinilla (2008)
Zinc	Oysters, a small amount in beans, nuts, almonds, whole grains (oatmeal, whole-grain breads, and brown rice), sunflower seeds, pumpkin seeds	It is crucial for nerve signaling Reduction of zinc in diet helps to reduce cognitive decay in the elderly	
Magnesium	Pumpkin seeds	Essential for learning and memory	
Copper	Oysters, beef/lamb liver, Brazil nuts, blackstrap molasses, cocoa, black pepper, pumpkin seeds	Help to control nerve signals. Cognitive decline in patients with Alzheimer's disease correlates with low plasma concentrations of copper	Gómez-Pinilla (2008)
Iron	Red meat, fish, poultry, lentils, beans, pumpkin seeds	Iron treatment normalizes cognitive function in young women	Gómez-Pinilla (2008)
Phosphatidylserine	Bovine and vegetarian, soy bean	Anti-inflammatory and antioxidant properties Enhances both cholinergic function and neuroplasticity	Scholey et al. (2011)
Tryptophan	Nearly all protein-containing foods	Influences serotonin synthesis and releases in the brain	Miquel et al. (2018)
Tyrosine	Protein-rich foods such as meat, fish, dairy products, also nuts, seeds, beans, and be synthesized by gut microbiota	Precursor of the catecholamines (e.g., dopamine and noradrenaline)	Miquel et al. (2018)
L-theanine amino acid	Green tea	Increase the activity of the neurotransmitter GABA Increases the frequency of alpha waves in the brain	
Taurine	Semi-essential amino acid, seafood, and meat, available "energy" drinks		

(continued)

Table 10.1 (continued)

Nutrient	Food sources	Pathway	References
Vitamin K, lutein, folate, and beta-carotene	Kale, spinach, collards, broccoli green, leafy vegetables	Anti-inflammatory and antioxidant effects	
Zingiber ginkgo		Antioxidants and anti-amyloid effects	Goel and Maurya (2019)
Gut hormones and some metabolites produced by microorganisms	Fermented foods (tofu, tempeh, or tahoe and other genistein-containing foods, fermented milk, probiotic milks)	Several gut hormones can enter the brain, or are produced in the brain itself, influence cognitive ability	Gómez-Pinilla (2008) and Sivamaruthi et al. (2018)

Accordingly, it is increasingly thought that bioactive substances in food represent a novel target for lifestyle interventions that may support healthy brain aging and preserve cognitive function, especially in aging adults at risk for nutritional deficits. Given that dietary alterations are reflected by many to be safer and more simpler integrated into lifestyle changes than conventional pharmacotherapeutics (Phillips 2017).

Nutritional phytochemicals, which have many neurogenic properties, play a beneficial role in brain aging and neurodegenerative disease. Today investigation on nutrient compound especially herbal drugs with enhanced cognitive function properties is growing. Some herbal drugs for cognitive enhancement included *Ginkgo biloba* (Ginkgoaceae), *Bacopa monnieri* (Plantaginaceae), *Centella asiatica* (Apiaceae), *Acorus calamus* (Acoraceae), *Evolvulus alsinoides* L. (Convolvulaceae), *Caesalpinia crista* Linn. (Caesalpiniaceae), *Tinospora Cordifolia* (Menispermaceae), *Zingiber Officinale* (Zingiberaceae), *Ilex Paraguariensis* (Aquifoliaceae), *Huperzia Sqururus* (Lycopodiaceae), *Commiphora Wightii* (Burseraceae), *Emblica Officinalis* (Euphorbiaceae), *Salvia Lavandulaefolia* (Lamiaceae), *Foeniculum Vulgare* (Umbelliferae), *Magnolia Officinalis* (Magnoliaceae), Lepidium Meyenii (Brassicaceae), *Rosa Alba* (Rosaceae), *Thespesia Populnea* (Malvaceae), *Sesamum Indicum* (Pedaliacae), *Salvia officinalis* (Lamiaceae), *Rosmarinus officinalis* (Lamiaceae), *Apium graveolens* (Apiaceae), *Rosa damascena* (Rosaceae), and *Citrus* species (Rutaceae) (Goel and Maurya 2019; Ebrahim Esfandiary et al. 2018).

Compounds such as curcumin, resveratrol, blueberry polyphenols, sulforaphanes, salvionic acids, PUFAs (e.g., omega-3 and DHA), the LMN diet (a patented diet by the company La Morella Nuts enriched with polyphenols and PUFAs), and flours rich in soluble fibers have been shown to induce neurogenesis in the adult brain. Although the molecular mechanisms by which these compounds effect neurogenesis have yet been established, these compounds reduce oxidative stress and neuro inflammation, enhance cell signaling, activate autophagy, and affect growth factors. In the following part mechanisms and pathways involved in the effects of dietary patterns on cognitive decline in aging are described concisely by details.

10.5 Mechanisms and Pathways Involved in the Effects of Dietary Patterns on Cognitive Decline in Aging

Different dietary pattern found to affect cognitive function, attention, memory, and learning via different mechanisms. However, most of the diet components are simply known to affect cognition, while the exact mechanism is still unknown. Following are a few examples with the plausible mechanism of action on cognitive aging.

10.5.1 High-Fat Diet

The diet or metabolic syndrome induces cognitive dysfunction via increased reactive oxygen species (ROS) and oxidative stress. Topiramate is a mitochondrial carbonic anhydrase (mCA) inhibitor decreases ROS and oxidative stress and thereby protects blood–brain barrier (BBB) from developing central nervous system dysfunction. High-fat diet (HFD) increases permeability of ^{14}C-sucrose in mice hypothalamus, hippocampus (HPC), and ^{99}mTc-albumin in the whole brain. However, topiramate attenuates absorption of both help decreasing oxidative stress by increasing expression of tight junction protein ZO-1 and claudin-12 (Salameh et al. 2019). A clinical trial reported similar result. HFD demonstrated decline in glucose transporter-1 (GLUT1) while increase of vascular endothelial growth factor (VEGF) that may be responsible for negatively affecting cognitive function and memory (Schuler et al. 2018).

10.5.2 Ketogenic Diet

One mechanism for cognitive decline is inability of hippocampus (HPC) and prefrontal cortex (PFC) to metabolize glucose for energy production. However, neuronal glycolysis may be bypass by the use of ketogenic diet (KD). Rat fed on KD showed improved brain function by increasing monocarboxylate transporters, i.e. MCT1 and MCT4 while decreasing GLUT1. The number of vesicular GABA transporter (VGAT) also increased in PFC and HPC. MCT1 and MCT4 help transport ketone bodies, while GLUT1 is responsible for transporting glucose across BBB (Hernandez et al. 2018).

10.5.3 Bread and Rice in Breakfast

Taki et al. reported nutritional importance for brain maturation during childhood and adolescence. Individuals with bread in breakfast showed larger regional gray and white matter volumes. Perceptual organization index (POI) found to be higher among individuals on rice suggesting that breakfast staple type affects cognitive function via affecting brain gray and white matter volumes (Taki et al. 2010). Matrix metalloproteinases (MMPs) are extracellular matrix proteins recently reported as not a major factor of white matter damage in individuals of vascular cognitive impairment dementia. However, MMP-9 can interfere with learning by a mechanism, i.e. yet unknown warranting future research (Raz et al. 2018).

10.5.4 Magnesium

Functional plasticity declines in aged brain. Short-term synaptic changes are dependent on transmitter release and neuronal excitability, while long-term synaptic changes are dependent on activated NMDA receptors. Magnesium (Mg^{2+}) is known to modulate neuronal excitability, NMDA receptor activation, and release of transmitter. Mg^{2+} is thus capable of affecting long and short-term changes in synaptic strength, neuronal plasticity, and memory. Aged animals fed on Mg^{2+} rich diet demonstrated improved learning and memory by correcting and maintaining brain Mg^{2+} homeostasis (Billard 2011).

10.5.5 Berberine

Berberine is an isoquinoline alkaloid helps in glucose metabolism. Berberine administration in aged rats found to improve cognitive and muscular function by decreasing ROS level, increasing protein expression of adenosine monophosphate-activated protein kinase (p-AMPK), sirtuin type 1 (SIRT1), and proliferator-activated receptor γ coactivator 1α (PGC-1α) and ATP production (Yu et al. 2018).

10.5.6 Blueberry

Blueberry supplementation to mice fed on HFD demonstrated fewer microglia and increased neuroplasticity in comparison to those fed on HFD alone. This suggests blueberry protective effect against memory deficits. However, further researches are warranted to explore the exact neuroprotective mechanism.

10.5.7 Caloric and Dietary Restriction

Caloric restriction (CR) helps in normalizing CNS function and memory. Hippocampal DNA of male mice fed on CR diet demonstrated age-related changes in CG and CH hippocampal methylation suggesting neuroprotection via epigenetic mechanism of CR diet (Hadad et al. 2018). CR triggers the circulating factors that help to upregulate the vascular endothelial growth factor (VEGF) signaling, increase the proliferation of endothelial cells, and stimulate the angiogenic processes, thereby helps promoting cardiac function and prevent vascular cognitive impairment (Csiszar et al. 2013). Another study on CR-fed transgenic mice reported downregulation of miRNAs including −30e, −34a, and −181a with increased B-cell lymphoma-2 (Bcl-2), while decreased Bcl-2 associated X protein (Bax) expression and Caspase cleaving. This chain of events suggesting a mechanism that contribute to neuronal survival in CR-fed mice (Khanna et al. 2011). CR also reported to prevent age-related reduced expression level of subunit of NMDA receptor in rats suggesting that life-long CR may ameliorate cognitive deficits associated with aging (Eckles-Smith et al. 2000). Insulin like growth factor-1 (IGF1), phosphatidylinositol 3-kinase (PI3K) pathway, and protein kinase B (Akt/PKB) signaling cascade are vital for cellular and tissue function in aging process. Mice on dietary restriction (DR) demonstrated reduced level of PI3K, pAkt, IGF1 and its receptor. Aged mice showed down-regulated IGF1 leading to compensatory increase in Forkhead box protein O (FoxO) activity thus suggesting neuroprotective effect of DR (Hadem and Sharma 2017). In another animal study, ghrelin agonist used to induce hunger in the absence of CR demonstrated prevention of AD pathology and cognitive decline by reducing amyloid beta (Aβ) level and microglial activation (Dhurandhar et al. 2013). Another animal study reported beneficial effects of dietary energy restriction (DER) and exercise on behavior and brain structural plasticity via brain-derived neurotrophic factor (BDNF) through warranting further research to explore the exact mechanism (Rothman et al. 2012). Long-term intermittent fasting diet (L-IFD) reported increasing NR2B subunits of NMDA receptor in mice brain that might explain the molecular mechanism for synaptic plasticity and improved cognition (Fontan-Lozano et al. 2007).

10.5.8 Decalepis Hamiltonii

Antioxidants are rich in the root extract of *Decalepis hamiltonii* (Dh). Aged Drosophila flies and their offspring cognitive ability impair upon declining antioxidant enzymes. 0.1% Dh diet demonstrated increased superoxide dismutase (SOD) and catalase activity that protected the aged flies and their offspring from cognitive aging and memory impairment (Haddadi et al. 2013).

10.5.9 Ethanol

Both young and aged rats on ethanol (EtOH) diet demonstrated withdrawal and anxiety-like behavior. This suggests that EtOH may lead to cognitive impairment among both age groups (Novier et al. 2016).

10.5.10 Polyunsaturated Fatty Acids

The aged individual on nutrition rich in polyunsaturated fatty acids (PUFAs) and vitamin E demonstrated improvement in white matter integrity and its microstructure suggesting its beneficial role on cognition (Gu et al. 2016).

10.5.11 Saturated Fatty Acids

Hsu et al. explained the possible mechanism of HPC dysfunction induced by the western diet (WD). Metabolism of WD components, i.e. carbohydrates and saturated fatty acids in small intestine increases circulating Aβ that contribute to BBB damage via declining gene expression of occludin and claudin 5 (Hsu and Kanoski 2014).

10.5.12 Vitamin A

Retinoic acid (RA) is the active metabolite of vitamin A that protects the brain from defective spatial memory and HPC neurogenesis. Vitamin A regulates free plasma and HPC corticosterone (CORT) levels via modulating corticosteroid binding globulin (CBG) capacity and 11β-hydroxysteroid dehydrogenase type 1 (11βHSD1) activity. While vitamin A deficiency increases CORT level which in turn impairs spatial memory, HPC neurogenesis and induces anxiety-like behavior in rats (Bonhomme et al. 2014).

10.5.13 Folic Acid

Deficiency of folic acid or vitamin B12 leads to megaloblastic anemia, depression, dementia, and cognitive impairment. Vitamin B12 and folic acid have a vital role during all stages of CNS development and function. They are also important for the synthesis of DNA, nucleotides, thymidine, and purine; for genomic and nongenomic methylation; and for tissue growth, differentiation, and repair (Reynolds 2014).

10.5.14 Vitamin K

Lifelong low vitamin K supplementation caused cognitive impairment in aged rats. There observed a higher concentration of ceramides in HPC ($P < 0.05$), and lower gangliosides in midbrain and pons-medulla ($P < 0.05$). This suggests the contribution of vitamin k in cognitive aging. Though the mechanism of action remains to be elucidated requiring further research (Carrie et al. 2011).

10.5.15 Vitamin D

Vitamin D regulates proline dehydrogenase (PRODH) expression. Chromosome 22q11 linked to PRODH and also for genetic risk of schizophrenia. Vitamin D deficiency leads to schizophrenia, i.e. treated with vitamin D supplementation. Clelland JD et al. found that vitamin D deficiency leads to schizophrenia by increasing proline level, due to reduced PRODH expression and dysregulated neurotransmission (Clelland et al. 2014).

10.5.16 Vitamin E

Peroxiredoxin II (PrxII) prevents age-related mitochondrial decay in HPC CA1 pyramidal neurons by scavenging intracellular ROS. Dietary vitamin E prevents PrxII deficiency, thereby prevent mitochondrial decay, oxidative damage, and cognitive decline (Kim et al. 2011).

10.5.17 Dietary Inflammatory Index

Inflammatory potential of diet can be determined by using the dietary inflammatory index (DII). Cognitive performance of French adults of middle age has been studied for 13 years. There observed an inverse relationship between DII and overall cognitive function. The study reported that the proinflammatory diet during middle age may lower cognitive functioning later in life, while diet with anti-inflammatory properties maintains cognitive health with increasing age (Kesse-Guyot et al. 2017).

10.5.18 Tryptophan Diet

Tryptophan (TrP) diet improves memory with aging by inducing hyperserotonemia. Aged rat fed on a diet rich in TrP for a month showed increased expression of BDNF and 5-HT level in frontal cortex (FC) and HPC. This suggests a protective effect of TrP on the serotonergic system thereby improving memory with aging (Musumeci et al. 2015).

10.5.19 Others

Flavanols of cocoa extracts know to promote healthy brain aging (Dubner et al. 2015). Neuroprotective effect of anthocyanins has been reported in pigs that were fed with 2% whole freeze-dried, powdered blueberry in the diet for 8 weeks. Neuroprotection may be due to modulation of signal transduction, gene expression rather than direct antioxidant radical quenching. However, requiring further research to explore the exact mechanism (Milbury and Kalt 2010). Anthocyanin-rich mulberry demonstrated antioxidant and improved cognition in senescence-accelerated mice (SAMP). Brain and liver reported higher antioxidant enzyme activity and lower lipid oxidation level supporting memory improvement in aging animals (Shih et al. 2010). Apple juice being rich in antioxidants maintains adequate acetylcholine levels in frontal cortex and hippocampus of mice and can prevent cognitive aging (Chan et al. 2006). Another study reported alleviation of a compensatory increase in glutathione synthase transcription activity and its antioxidant potential to prevent neurodegeneration (Tchantchou et al. 2004). *Ginkgo biloba* leaves extract reported to act as cognitive enhancer among aged mice. Synaptic plasticity and improved cognition, spatial learning and memory were observed (Wang et al. 2006). Another study reported synaptic plasticity and excitability in HPC slices of aged mice that

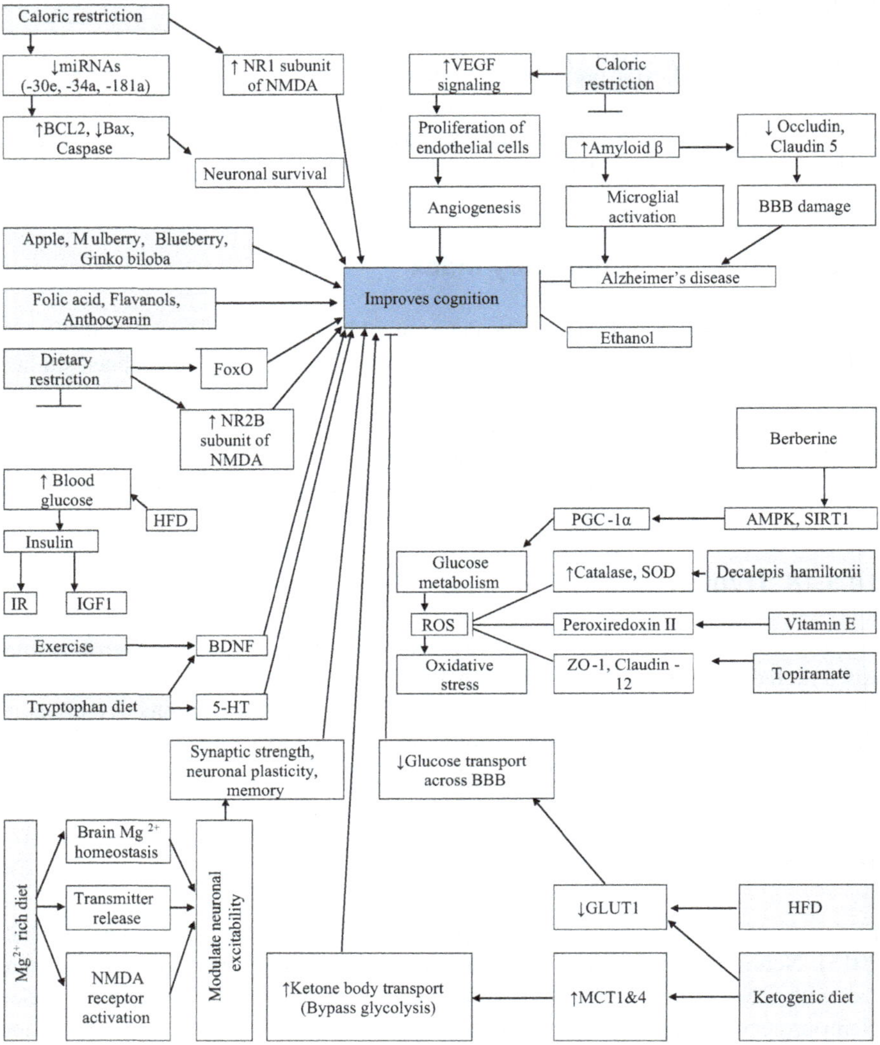

Fig. 10.2 Plausible scheme of diet affecting cognition

may be due to direct interaction with the glutamatergic system, suggesting its effect on the cognitive system (Williams et al. 2004). Iodine deficiency and hypothyroidism decrease c-fos and c-jun protein expression in rat HPC. Cognitive function, learning, attention, and memory need healthy thyroid hormone secretion, while iodine deficiency impairs cognition (Dong et al. 2005). Some of diet, foods, and nutrients affecting cognition and their possible pathways are illustrated in Fig. 10.2, as it shows that many different mechanisms are involved and each diet or nutrient can affect cognition by more than one mechanism.

References

Abbatecola AM et al (2018) Dietary patterns and cognition in older persons. Curr Opin Clin Nutr Metab Care 21(1):10–13

Ageing and health (2018). https://www.who.int/news-room/fact-sheets/detail/ageing-and-health

Arai S (1996) Studies on functional foods in Japan--state of the art. Biosci Biotechnol Biochem 60 (1):9–15

Aronson JK (2017) Defining 'nutraceuticals': neither nutritious nor pharmaceutical. Br J Clin Pharmacol 83(1):8–19

Banjari I, Vukoje I, Mandić ML (2014) Brain food: how nutrition alters our mood and behaviour. Hrana u zdravlju i bolesti: znanstveno-stručni časopis za nutricionizam i dijetetiku 3(1):13–21

Bechthold A et al (2019) Food groups and risk of coronary heart disease, stroke and heart failure: a systematic review and dose-response meta-analysis of prospective studies. Crit Rev Food Sci Nutr 59(7):1071–1090

Bettio LE et al (2017) The effects of aging in the hippocampus and cognitive decline. Neurosci Biobehav Rev 79:66–86

Billard JM (2011) Brain free magnesium homeostasis as a target for reducing cognitive aging. In: Vink R, Nechifor M (eds) Magnesium in the central nervous system. University of Adelaide Press, Adelaide

Bonhomme D et al (2014) Vitamin A status regulates glucocorticoid availability in Wistar rats: consequences on cognitive functions and hippocampal neurogenesis? Front Behav Neurosci 8:20

Calil SRB et al (2018) Adherence to the Mediterranean and MIND diets is associated with better cognition in healthy seniors but not in MCI or AD. Clin Nutr ESPEN 28:201–207

Calvaresi E, Bryan J (2001) B vitamins, cognition, and aging: a review. J Gerontol B Psychol Sci Soc Sci 56(6):P327–P339

Cao L et al (2016) Dietary patterns and risk of dementia: a systematic review and meta-analysis of cohort studies. Mol Neurobiol 53(9):6144–6154

Carrie I et al (2011) Lifelong low-phylloquinone intake is associated with cognitive impairments in old rats. J Nutr 141(8):1495–1501

Chan A, Graves V, Shea TB (2006) Apple juice concentrate maintains acetylcholine levels following dietary compromise. J Alzheimers Dis 9(3):287–291

Clelland JD et al (2014) Vitamin D insufficiency and schizophrenia risk: evaluation of hyperprolinemia as a mediator of association. Schizophr Res 156(1):15–22

Corrêa Leite ML et al (2001) Nutrition and cognitive deficit in the elderly: a population study. Eur J Clin Nutr 55:1053

Csiszar A et al (2013) Circulating factors induced by caloric restriction in the nonhuman primate Macaca mulatta activate angiogenic processes in endothelial cells. J Gerontol A Biol Sci Med Sci 68(3):235–249

Deary IJ et al (2009) Age-associated cognitive decline. Br Med Bull 92:135–152

Dhurandhar EJ et al (2013) Hunger in the absence of caloric restriction improves cognition and attenuates Alzheimer's disease pathology in a mouse model. PLoS One 8(4):e60437

Diplock A et al (1999) Scientific concepts of functional foods in Europe. Consensus document. Br J Nutr 81(Suppl 1):S1–S27

Dominguez LJ, Barbagallo M (2018) Nutritional prevention of cognitive decline and dementia. Acta Biomed Atenei Parmensis 89(2):276–290

Dong J et al (2005) Congenital iodine deficiency and hypothyroidism impair LTP and decrease C-fos and C-jun expression in rat hippocampus. Neurotoxicology 26(3):417–426

Dubner L et al (2015) Recommendations for development of new standardized forms of cocoa breeds and cocoa extract processing for the prevention of Alzheimer's disease: role of cocoa in promotion of cognitive resilience and healthy brain aging. J Alzheimers Dis 48(4):879–889

Ebrahim Esfandiary ZA et al (2018) Novel effects of Rosa damascena extract on patients with neurocognitive disorder and depression: a clinical trial study. Int J Prev Med 9:57

Eckles-Smith K et al (2000) Caloric restriction prevents age-related deficits in LTP and in NMDA receptor expression. Brain Res Mol Brain Res 78(1-2):154–162

Estruch R et al (2016) Long-term immunomodulatory effects of a mediterranean diet in adults at high risk of cardiovascular disease in the PREvención con DIeta MEDiterránea (PREDIMED) randomized controlled trial. J Nutr 146(9):1684–1693

Feart C, Samieri C, Barberger-Gateau P (2010) Mediterranean diet and cognitive function in older adults. Curr Opin Clin Nutr Metab Care 13(1):14–18

Fontan-Lozano A et al (2007) Caloric restriction increases learning consolidation and facilitates synaptic plasticity through mechanisms dependent on NR2B subunits of the NMDA receptor. J Neurosci 27(38):10185–10195

Gardener H, Caunca MR (2018) Mediterranean diet in preventing neurodegenerative diseases. Curr Nutr Rep 7(1):10–20

Goel B, Maurya NK (2019) Memory booster herb (natural cognitive enhancers): an overview

Gómez-Pinilla F (2008) Brain foods: the effects of nutrients on brain function. Nat Rev Neurosci 9 (7):568

Gu Y et al (2016) White matter integrity as a mediator in the relationship between dietary nutrients and cognition in the elderly. Ann Neurol 79(6):1014–1025

Hadad N et al (2018) Caloric restriction mitigates age-associated hippocampal differential CG and non-CG methylation. Neurobiol Aging 67:53–66

Haddadi M et al (2013) Decalepis hamiltonii root extract attenuates the age-related decline in the cognitive function in Drosophila melanogaster. Behav Brain Res 249:8–14

Hadem IKH, Sharma R (2017) Differential regulation of hippocampal IGF-1-associated signaling proteins by dietary restriction in aging mouse. Cell Mol Neurobiol 37(6):985–993

Hasler CM (1998) Functional foods: their role in disease prevention and health promotion. Food Technol 52:63–147

Hernandez AR et al (2018) A ketogenic diet improves cognition and has biochemical effects in prefrontal cortex that are dissociable from hippocampus. Front Aging Neurosci 10:391

Hsu TM, Kanoski SE (2014) Blood-brain barrier disruption: mechanistic links between Western diet consumption and dementia. Front Aging Neurosci 6:88

Institute of Medicine (1994) In: Thomas PR, Earl R (eds) Opportunities in the nutrition and food sciences: research challenges and the next generation of investigators. The National Academies Press, Washington, DC, p 328

Juvan S, Bartol T, Boh B (2005) Data structuring and classification in newly-emerging scientific fields. Online Inf Rev 29(5):483–498

Kesse-Guyot E et al (2017) Long-term association between the dietary inflammatory index and cognitive functioning: findings from the SU. VI. MAX study. Eur J Nutr 56(4):1647–1655

Khan RS et al (2013) Functional food product development – opportunities and challenges for food manufacturers. Trends Food Sci Technol 30(1):27–37

Khanna A et al (2011) Gain of survival signaling by down-regulation of three key miRNAs in brain of calorie-restricted mice. Aging (Albany NY) 3(3):223–236

Kim SU et al (2011) Peroxiredoxin II preserves cognitive function against age-linked hippocampal oxidative damage. Neurobiol Aging 32(6):1054–1068

Klimova B, Valis M, Kuca K (2017) Cognitive decline in normal aging and its prevention: a review on non-pharmacological lifestyle strategies. Clin Interv Aging 12:903

Le Bourg É (2012) Dietary restriction studies in humans: focusing on obesity, forgetting longevity. Gerontology 58(2):126–128

Ley SH et al (2014) Prevention and management of type 2 diabetes: dietary components and nutritional strategies. Lancet 383(9933):1999–2007

Milbury PE, Kalt W (2010) Xenobiotic metabolism and berry flavonoid transport across the blood-brain barrier. J Agric Food Chem 58(7):3950–3956

Miquel S et al (2018) Poor cognitive ageing: vulnerabilities, mechanisms and the impact of nutritional interventions. Ageing Res Rev 42:40–55

Mozaffarian D (2016) Dietary and policy priorities for cardiovascular disease, diabetes, and obesity: a comprehensive review. Circulation 133(2):187–225
Musumeci G et al (2015) Changes in serotonin (5-HT) and brain-derived neurotrophic factor (BDFN) expression in frontal cortex and hippocampus of aged rat treated with high tryptophan diet. Brain Res Bull 119(Pt A):12–18
Nicoletti M (2012) Nutraceuticals and botanicals: overview and perspectives. Int J Food Sci Nutr 63 (sup1):2–6
Novier A et al (2016) Differences in behavioral responding in adult and aged rats following chronic ethanol exposure. Alcohol Clin Exp Res 40(7):1462–1472
Ohama H, Ikeda H, Moriyama H (2006) Health foods and foods with health claims in Japan. Toxicology 221(1):95–111
Parrott MD, Greenwood CE (2007) Dietary influences on cognitive function with aging: from high-fat diets to healthful eating. Ann N Y Acad Sci 1114(1):389–397
Petersson SD, Philippou E (2016) Mediterranean diet, cognitive function, and dementia: a systematic review of the evidence. Adv Nutr 7(5):889–904
Phillips CJ (2017) Lifestyle modulators of neuroplasticity: how physical activity, mental engagement, and diet promote cognitive health during aging. Neural Plast 2017:1
Poulose SM et al (2017) Nutritional factors affecting adult neurogenesis and cognitive function. Adv Nutr 8:804–811
Poulsen J (1999) Danish consumers' attitudes towards functional foods (MAPP Working Paper 62). The Aarhus School of Business, Aarhus, Denmark
Raz L et al (2018) MMP-9 inhibitors impair learning in spontaneously hypertensive rats. PLoS One 13(12):e0208357
Reynolds EH (2014) The neurology of folic acid deficiency. Handb Clin Neurol 120:927–943
Richard EL et al (2018) Dietary patterns and cognitive function among older community-dwelling adults. Nutrients 10(8):1088
Roberfroid MB (1999) What is beneficial for health? The concept of functional food. Food Chem Toxicol 37(9):1039–1041
Rothman SM et al (2012) Brain-derived neurotrophic factor as a regulator of systemic and brain energy metabolism and cardiovascular health. Ann N Y Acad Sci 1264:49–63
Salameh TS et al (2019) Disruption of the hippocampal and hypothalamic blood-brain barrier in a diet-induced obese model of type II diabetes: prevention and treatment by the mitochondrial carbonic anhydrase inhibitor, topiramate. Fluids Barriers CNS 16(1):1
Scholey A et al (2011) Functional foods and cognition. In: Functional foods. Elsevier, Cambridge, pp 277–308
Schuler R et al (2018) VEGF and GLUT1 are highly heritable, inversely correlated and affected by dietary fat intake: consequences for cognitive function in humans. Mol Metab 11:129–136
Schulze MB et al (2018) Food based dietary patterns and chronic disease prevention. BMJ 361: k2396
Shah H et al (2016) Research priorities to reduce the global burden of dementia by 2025. Lancet Neurol 15(12):1285–1294
Shetty P (2012) Grey matter: ageing in developing countries. Lancet 379(9823):1285–1287
Shih PH et al (2010) Antioxidant and cognitive promotion effects of anthocyanin-rich mulberry (Morus atropurpurea L.) on senescence-accelerated mice and prevention of Alzheimer's disease. J Nutr Biochem 21(7):598–605
Sivamaruthi B, Kesika P, Chaiyasut C (2018) Impact of fermented foods on human cognitive function—a review of outcome of clinical trials. Sci Pharm 86(2):22
Smith PJ, Blumenthal JA (2016) Dietary factors and cognitive decline. J Prev Alzheimers Dis 3 (1):53–64
Solfrizzi V, Panza F, Capurso A (2003) The role of diet in cognitive decline. J Neural Transm 110 (1):95–110

Solfrizzi V et al (2017) Relationships of dietary patterns, foods, and micro- and macronutrients with Alzheimer's disease and late-life cognitive disorders: a systematic review. J Alzheimers Dis 59 (3):815–849

Spencer SJ et al (2017) Food for thought: how nutrition impacts cognition and emotion. npj Sci Food 1(1):7

Sureda A et al (2016) Mediterranean diets supplemented with virgin olive oil and nuts enhance plasmatic antioxidant capabilities and decrease xanthine oxidase activity in people with metabolic syndrome: the PREDIMED study. Mol Nutr Food Res 60(12):2654–2664

Szakály Z et al (2012) The influence of lifestyle on health behavior and preference for functional foods. Appetite 58(1):406–413

Taki Y et al (2010) Breakfast staple types affect brain gray matter volume and cognitive function in healthy children. PLoS One 5(12):e15213

Tchantchou F et al (2004) Dietary supplementation with apple juice concentrate alleviates the compensatory increase in glutathione synthase transcription and activity that accompanies dietary- and genetically-induced oxidative stress. J Nutr Health Aging 8(6):492–496

Tucker KL (2010) Dietary patterns, approaches, and multicultural perspective. Appl Physiol Nutr Metab 35(2):211–218

Tucker KL (2016) Nutrient intake, nutritional status, and cognitive function with aging. Ann N Y Acad Sci 1367(1):38–49

van de Rest O et al (2015) Dietary patterns, cognitive decline, and dementia: a systematic review. Adv Nutr 6(2):154–168

Wang Y et al (2006) The in vivo synaptic plasticity mechanism of EGb 761-induced enhancement of spatial learning and memory in aged rats. Br J Pharmacol 148(2):147–153

Wengreen H et al (2013) Prospective study of Dietary Approaches to Stop Hypertension– and Mediterranean-style dietary patterns and age-related cognitive change: the Cache County Study on Memory, Health and Aging. Am J Clin Nutr 98(5):1263–1271

Williams B et al (2004) Age-related effects of Ginkgo biloba extract on synaptic plasticity and excitability. Neurobiol Aging 25(7):955–962

World Cancer Research Fund/American Institute for Cancer Research (2018) Diet, nutrition, physical activity and cancer: a global perspective. Continuous update project expert report

Yu Y et al (2018) Berberine improves cognitive deficiency and muscular dysfunction via activation of the AMPK/SIRT1/PGC-1a pathway in skeletal muscle from naturally aging rats. J Nutr Health Aging 22(6):710–717

Mediterranean Diet for Active and Healthy Aging

11

Nida Noreen, Muhammad Ajmal Shah, Fazlullah Khan, Kamal Niaz, Faqir Muhammad, Ismail Shah, and Mohammad Abdollahi

Abstract

The significant rise in burden of age-related chronic degenerative disorders is increasing the need for products that support active and healthy aging. Today's sedentary lifestyle enhances the propensity to aging related diseases and premature death. Accumulating data establishes a beneficial relationship between food and health. The modern consumer has become aware of the valuable impact of Mediterranean diet (MD) on healthy aging. MD is largely plant based and rich in sources of unsaturated fatty acids like nuts and extra virgin olive oil, legumes, whole grains, fish, and fresh vegetables and fruits and it discourages the use of red and processed meat, added sugars as well as refined grains, has been associated

N. Noreen · F. Muhammad
Institute of Pharmacy, Physiology and Pharmacology, University of Agriculture, Faisalabad, Pakistan

M. A. Shah
Department of Pharmacognosy, Faculty of Pharmaceutical Sciences, Government College University, Faisalabad, Pakistan

F. Khan (✉)
Department of Toxicology and Pharmacology, The Institute of Pharmaceutical Sciences, Faculty of Pharmacy, Tehran University of Medical Sciences (TUMS), Tehran, Iran

K. Niaz
Department of Pharmacology and Toxicology, Cholistan University of Veterinary and Animal Sciences (CUVAS), Bahawalpur, Pakistan

I. Shah
Department of Pharmacy, Abdul Wali Khan University, Garden Campus, Mardan, Pakistan

M. Abdollahi
The Institute of Pharmaceutical Sciences, Faculty of Pharmacy, Tehran University of Medical Sciences (IC-TUMS), Tehran, Iran

S. M. Nabavi et al. (eds.), *Nutrients and Nutraceuticals for Active & Healthy Ageing*,
https://doi.org/10.1007/978-981-15-3552-9_11

with decreased risk of developing various chronic degenerative age-related disorders and increased life expectancy. With the advent of modern medicine and technological advancements the life expectancy has increased in the past few decades but the gap between the healthy life years and the extra years added to the life still remains there. Aging is a process that increases the vulnerability of an organism to challenges. During this process the oxidative stress leads towards various degenerative cascades that result in functional decline in aging population that is mainly associated with under nutrition in older people. Cognitive function declines, reduced mobility and sensory alterations are seen, oral and GI functions and health become compromised, and chronic diseases and age-related illness like osteoarthritis, diabetes type II, cardiovascular diseases, and certain types of cancer are also exhibited. The active and healthy aging is a prerequisite in order to enhance the quality of life as people age. The MD has proven to be the best tool to counteract the degenerative processes and promote an active healthy aging. Major effectors of MD are reduced caloric intake, decreased consumption of saturated fatty acids, microbiota derived metabolites, less amino acid utilization, and an increase in phytochemical consumption. MD protects against oxidative damage, injury, and inflammation and platelet aggregation, lowers lipid levels, modifies the hormones as well as growth factors that are involved in cancer pathogenesis, and inhibits the nutrient sensing pathway via restriction of specific amino acid as well as produces certain metabolites by gut microbiota and it influences the metabolic health. Thus, the molecular and metabolic health is chiefly associated with what we eat. Restriction of the calories can enhance the life span as well as the health span only if it is coupled with sufficient intake of all the essential nutrients and the micronutrients.

Keywords

Mediterranean diet · Nutraceuticals · Healthy aging · Nutrition

11.1 Introduction

The valuable impact of a healthy diet on human physiology remained greatly underappreciated for a long time but now the consumers have become more aware of the beneficial relationship between the food and health. Nearly every degenerative disorder can be influenced by dietary interventions. With respect to aging, the accumulating evidence suggests that quality of life can be improved by modulating the extrinsic factors that affect many aging processes. Nutrition is one of these modifiable factors and it appears to be one of the strongest components that influence the rate of aging as well as the prevalence of age associated disorders like atherosclerosis and neurodegenerative diseases (Battino and Ferreiro 2004).

Consumption of Mediterranean diet (MD) that is rich in minimally processed plant food has been strongly associated with decreased risk of developing various chronic diseases and enhance healthy life expectancy. Data from considerable randomized clinical trials have exhibited beneficial effects in primary and secondary avoidance of type II diabetes, cardiovascular disease, breast cancer, and atrial fibrillation. The mechanism through which adherence to the traditional MD exerts its beneficial effect is not fully known. However, growing evidence illustrates that the MD induces the most important adaptations like protecting against oxidative stress and inflammation, lowering lipid levels, modifying hormones and growth factors involved in cancer pathogenesis, gut microbiota mediated production of metabolites that influence the metabolic health, and restriction of specific amino acids to inhibit the nutrient sending pathways (Tosti et al. 2017). More research studies now support that a single modification of nutrients of the typical MD interacts with energy intake and expenditure as well as the microbiome in order to modulate the key mechanisms that improve cellular, tissue, and organ health during the process of aging. Thus, nutrition is the basic element for the promotion of health and avoidance of the most common age-related chronic diseases. Both the quality and quantity of whatever we eat are crucial to improve metabolic and molecular health. Restriction of the calories can enhance the life span as well as the health span only if it is coupled with sufficient intake of all the essential nutrients and the micronutrients (Longo and Fontana 2010).

11.2 The MD—Pyramid and Description

Originally the conventional dietary pattern in the Mediterranean region was termed as Mediterranean diet. Historically a great abundance and diversification of minimally processed whole grain cereals, non-starchy vegetables, nuts, legumes, and seeds were the staple foods in majority of the countries bordering Mediterranean Sea. Fish, milk, cheese, meat, eggs, and milk were luxurious foods. In Southern Italy till 1950s a very little meat was consumed and milk was only used in coffee or for infants. White potatoes and sugar were only eaten in very small quantities and cream or butter was never used (Vasto et al. 2014a). The principal source of fat was the cold pressed extra virgin olive oil. Since 1950s the original composition of MD has changed completely and the quantity and quality of the food that people in Italy, Greece, and Spain consume nowadays have little to do with conventional Mediterranean diet. The incidence of certain cancers and coronary heart disease has elevated substantially in these countries. Possibly, other lifestyle factors like excessive calorie intake, pollution, sedentary lifestyle, and psychological stress might have added to the enhanced occurrence of chronic degenerative diseases. However, compiling data from consolidated epidemiological, animal studies, human clinical trials, and molecular research indicates that diet is a key factor in the avoidance of obesity, cardiovascular diseases, some common cancers, and type II diabetes (Tosti et al. 2017; Longo and Fontana 2010; Vasto et al. 2014a).

Therefore, during 1960s Ancel Keys first time defined the MD being low in saturated fats and high in the vegetable oils. This definition was based on his observation of dietary patterns of Greece and Southern Italy in the previous decades. This dietary pattern, in the seven-country study, was correlated with decreased risk

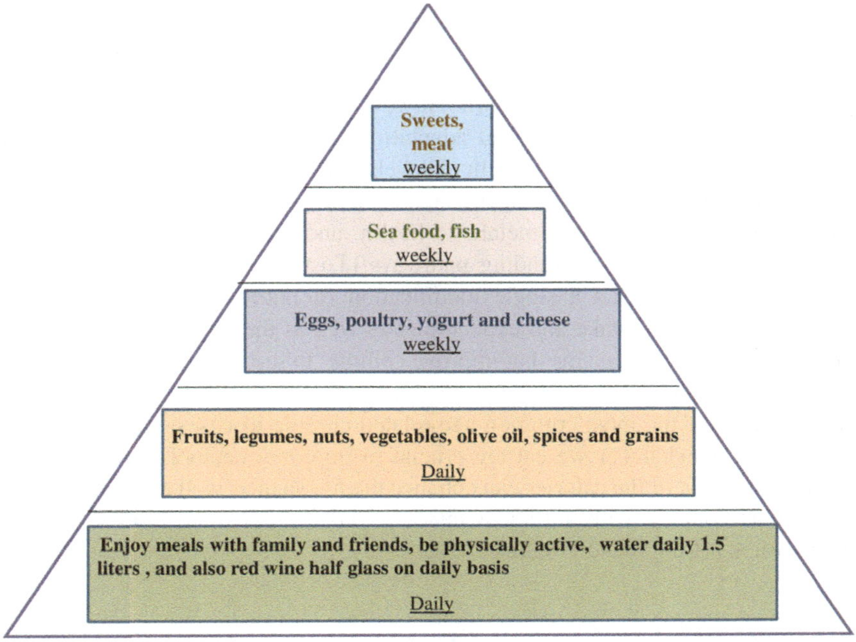

Fig. 11.1 Pyramid representation of Mediterranean diet adapted from Vasto et al. (2014a)

of degenerative disorders and coronary heart disease as compared to the USA and European countries after a follow-up of 25 years (Davis et al. 2015). Over the past various decades the study of MD has progressed so much that the original definition given by Keys has grown and altered in several aspects. There are several ways to define this dietary pattern that includes a priori scoring system, food and nutrient content, dietary pyramids, general descriptions, and a posteriori dietary pattern formation. Out of all these systems a priori scoring system has gained most popularity in the past decade as it simplifies the analysis of adherence to diet with respect to the primary outcomes. Several food groups are made containing separated dietary intakes related to health outcomes and points are given for more intake of health promoting foods and lower intake of health harming food and a single adherence score is calculated. However, various priori MD scores with different scoring criteria are available these days (Vasto et al. 2012a) (Fig. 11.1).

Everyday routine main meals must contain three essential elements: vegetables, fruits, and whole grain cereals. A daily intake of 1.5 L of water should be insured. Half glass of red wine daily adds further nutrition to the daily diet. Dairy products must be favored in the form of cheese, low fat yogurt, and other fermented ones. Olive oil is preferred to be the principal source of dietary lipids and it is placed in the middle of the pyramid. A reasonable utilization of seeds, nuts, and olives should be selected for healthier snacks. Moreover, a variety of animal and plant origin foods like legumes, white meat, fish, and eggs must be consumed every week. The red meat should be utilized in less quantity and in smaller frequency. The foods that should be consumed in small amounts and occasionally are represented at the top of the pyramid. These include candies, pastries, sugar, and beverages such as soft drinks and sweet fruit juices. However, the portion size of the diet must base on moderation in order to embrace a healthier life style. So, the bottom of the pyramid represents the foods that must sustain the routine diet and the top and upper levels represent the foods that must be consumed in moderate quantity (Vasto et al. 2012b, 2014b).

11.3 Longevity and the Gap Between Life Expectancy and Healthy Life Years

Though it seems a good thing to live longer, and definitely the modern technological advances, healthcare and medicines have enabled humans to do so, yet what seems not so good about this is the gap between the life expectancy and the healthy life years. In 2009 the mean life expectancy in Europe at birth was 82.6 years for women and 76.7 years for men. The healthy life years were 62.5 years and 61.6 years, respectively. This implies that the people born in 2009 are expected to live fine in their 70s and 80s but they will probably be ill and/or having some disability during the last 15–20 years of their life (Rechel et al. 2013).

Moreover, if the life expectancy continues to increase as predicted but there is no rise in the healthy life years at the same rate, then the duration of life that will be spent in ill health will be even more in the near future. This will not only affect an individual's quality of life and his/her family but it will also put a lot of pressure on healthcare services and public health, consequently there will be substantial economic and social impact on the society. Thus, it is imperative to make appropriate strategies in order to help the population in healthy aging. United multilevel efforts are needed at the community and national level. From the individual's viewpoint the absolute beneficial investment that one can make is in one's own health. Appropriate life style including physical activities and proper nutrition is the primary step in avoidance of chronic degenerative diseases and disability in the old age (Mak and Caldeira 2014).

11.4 Aging—Increases the Vulnerability of an Organism to Challenges

Aging is defined as decreased performance, lack of fitness with advancing age, and finding difficulty in adapting to the new environmental conditions. Aging is not actually a programmed pathway but it is a stochastic procedure that results from accretion of somatic damage along with a gradual loss of molecular fidelity. Thus, aging is natural and multi-factorial procedure that takes place at cellular, organ as well as organism level and it affects almost all the living things. It was believed that aging is programmed in living beings as some kind of biological clock but that perspective is almost changed now. According to the most common and accepted definition aging is a process that enhances the vulnerability of any organism to challenges during the lifespan and it grows the potential for death (Battino and Ferreiro 2004; Tosti et al. 2017; Longo and Fontana 2010; Vasto et al. 2007, 2012a, b, 2014a, b; Davis et al. 2015; Rechel et al. 2013; Mak and Caldeira 2014). As the scientists are of the opinion that aging is a result of increasing cell and tissue damage in the bodies of living organisms and some of these microscopic flaws always impair normal functioning and can lead to diseases. If scientists can figure out cell aging and find out ways of decreasing the growing cell damage or they are able to improve the effectiveness of natural repair systems of the body, then it is possible to delay the onset of diseases and to improve the quality of old age. The rate of aging of all the organisms is influenced by both extrinsic as well as intrinsic factors like genetic makeup, their lifestyles, and environment (Kirkwood 2008). Growing evidence indicates that the quality of life can be improved by modulating the extrinsic factors which influence several aging processes. The intrinsic factors, on the other hand, are pre-determined and various efforts of genetic manipulation have not yet influenced the aging rate. Thus, environmental factors and lifestyle changes can interfere in aging process and might alter the sensitivity of organisms to age associated degenerative diseases. From environmental factors nutrition is one of the most important factors that can influence the rate of aging and can decrease the possibility of age associated diseases like neurodegenerative diseases and atherosclerosis. During aging process the damage is a fact of life because of relative lack of repair maintenance systems like oxidative stress and inflammation control (Battino and Ferreiro 2004; Tosti et al. 2017; Longo and Fontana 2010; Vasto et al. 2007, 2012a, b, 2014a, b; Davis et al. 2015; Rechel et al. 2013; Mak and Caldeira 2014; Kirkwood 2008; López-Otín et al. 2013).

11.4.1 Oxidative Stress Theory of Aging

The oxidative stress theory of aging was outlined by Harman (1956). According to which various biological oxidants that arise from external environment or are produced endogenously are one of the major factors that are responsible for accumulated damage that leads to aging. Most of the cellular damage happens as a byproduct of normal living. The oxygen that we breathe produces highly reactive molecules termed as free radicals that can cause damage to cell structures (Harraan 1956; Battino et al. 2002a). Normal reaction of reduction and oxidation continuously produces free radicals in trace amounts. These are highly unstable compounds containing one or more unpaired electrons in the outer shell and can react with any susceptible compound nearby like lipids, carbohydrates, proteins, or DNA. Though these free radicals last only for milliseconds yet they can begin a chain reaction that results in oxidation of hundred thousands of particles over an enormous distance as the transfer of unpaired electron from one molecule to another takes place in a game of "hot potato" (Harraan 1956; Battino et al. 2002a; Gaziano 2000). Antioxidants against the free radicals are an excellent natural defense mechanism of the body.

They minimize the un-intended production of free radicals. Antioxidants can act at various stages and at different cell sites during the damage cascade caused due to the free radicals. They can inhibit the production of free radicals, can neutralize their effect, and can also repair the damage caused by them (Niki 1996). There is a balance in oxidant attack and antioxidant action but the defense mechanism is inevitably less than perfect and the damage produced by free radicals accumulates gradually. It is suggested that this oxidant-antioxidant imbalance is the main cause of aging in itself. Thus, aging process is simply termed as the sum of random alteration caused by free radical reactions (Harman 1992).

11.4.2 Degenerative Processes

All the cellular structures are considered to be the target of the actions of free radicals. Nonetheless, all the cell membranes like plasma membranes and all other organelle membranes bear the most amount of oxidative insult. This occurs principally in the hydrophobic core of membrane that is composed chiefly of fatty acids. Therefore, the fatty acid portion is most susceptible to oxidative damage. The attack by free radicals causes serious alterations in the membranes that affect both their molecular structure and function. The activities of mitochondria mainly depend on the complete integrity of its inner membrane and they are exceptionally susceptible to any nutritional, environmental, physiological, or pharmacological stimulus. Brain mitochondria are presumably the best system to study degenerative process causing aging. Accumulated evidence in the last decade indicates that aging affects the brain mitochondria to various extents depending upon the region of brain and the cellular district (Battino et al. 1991, 1995, 1996, 2000, 2002b). This phenomenon is especially visible at synaptic level and it makes possible to study some specific population of aged mitochondria. Their normal activities are extremely impaired like Kreb's cycle alteration (decrease in matrix activities) takes place, slow working of electron transport chain, and the trans-membrane systems are also affected as their ability to regulate the traffic of various substances across the inner membrane gets affected. Consequently, the continuous leakage of electrons along with the formation of free radicals and partial uncoupling of organelles that cannot produce enough ATP occurs in a vicious, continuous, and lethal circle. The membrane structure loses most of its antioxidants like co-enzyme Q and Vitamin-E. Finally, the typical shape of membranes is also altered. Thus, the degeneration process greatly alters the classified aspects of the cell membranes (Yehuda et al. 1997, 1998, 1999, 2000). Therefore, the treatments based on fatty acid administration can help in improving biochemical and cognitive functions. Solfrizzi et al. evaluated the correlation of intake of dietary macronutrients and age-related alterations in cognition in the elderly population of Italy who was consuming typical Mediterranean diet. They established that enhanced intake of mono unsaturated fatty acids appeared to be more protective in case of age-related cognitive function decline (Solfrizzi et al. 1999).

11.4.3 Functional Deterioration in Aging Associated with Under Nutrition

In older people under nutrition is a common problem that results due to the decreased intake of nutrients and impaired metabolism. It is correlated with a variety of age-related complications, disorders, and mortality even in developed countries. Functional alterations due to aging can greatly influence food consumption and the extent of under nutrition in older population (Bousquet et al. 2016). Gradual drop of physical capability like mobility and movement, alterations in physiological functioning in the aging body, chronic disease, and some other age-related disorders can cause change in choice of food, eating habits, and dietary consumption that can result in the increased risk of malnutrition. The comorbidities and functional decline being a major cause of under nutrition result in chronic nutrient deficiency that can further aggravate the age-related conditions (Bousquet et al. 2016; Schilp et al. 2011).

11.4.4 Cognitive Function Decline

Aging causes dementia, a disease characterized by progressive decline in intellect that affects memory, learning, language, judgment, thinking, comprehension as well as behavior changes and decreased capability to perform routine activities like cooking, working, and shopping. The risk of dementia development doubles every 5 years after age of 65 years (Batsch and Mittelman 2012; Prince and Jackson 2009). The most common type of dementia is Alzheimer's disease that affects one in four people aged more than 85 years (Batsch and Mittelman 2012). Moreover, cerebrovascular disorders and diet-related diseases like hypertension, obesity, elevated lipid levels, smoking, and diabetes have been exhibited to enhance the risk of Alzheimer's disease (Cardoso et al. 2013). It is suggested that people having dementia or Alzheimer's disease may forget to drink or eat or they can forget about their meals they have already taken (Shatenstein and Ferland 2000). These behaviors are different at different stages of dementia like at middle stages the patients show excessive hunger, while in later stages, where they have limited activity or are totally dependent, difficulty in eating and drinking is seen. So, due to the aging the cognitive functions are greatly declined and there is most likeliness to become undernourished (Hsiao et al. 2013).

11.4.5 Reduced Mobility and Sensory Alterations

Decreased mobility is a universal concern in aged people. Mobility is one's ability to move independently, carefully, and safely from one place to another place. It is seriously influenced by advancement of some chronic conditions like physical injuries (e.g., fractures from falling) and obesity as well as physiological alterations over time chiefly in the neurological, cardio respiratory, and musculoskeletal systems which lead to limited movement. Sarcopenia that is characterized by loss of muscle mass and strength is a common multi-factorial condition in aged people and it has a main role in pathogenesis of functional impairment and weakness (Rantakokko et al. 2013; Morley et al. 2001). Skeletal muscle mass reaches its peak in early mid-adulthood and after that it starts declining at a rate of 0.5–1% every year. It influences around 30% of adults of over age 60 years and more than 50% of over age 80 years. Other such disorders like osteoporosis, osteoarthritis, and rheumatoid arthritis are known to be the major reason for functional limitations and pain in joints and bones in older people, more commonly in women (Paddon-Jones et al. 2008). Limited movement and decline mobility are more likely to seriously affect the accessibility to food like meal preparations, food shopping and resultantly the consumption of food is reduced (Woolf and Pfleger 2003). Similarly the senses of taste and smell decline as people age. Due to the gradual decrease in number of olfactory fibers and receptors in olfactory bulb and also the increased occurrence of receptor cell death the olfactory functions deteriorate (Boyce and Shone 2006). The taste cell membrane physiology also changes over time and the functions of ion channels and receptors become compromised. Medication use can also affect sensitivity of taste as it is an established fact that deficiency of zinc can impact acuity of taste (Roberts and Rosenberg 2006). When the senses are curbed it becomes hard to recognize and detect certain flavors and tastes like sugar and salt. So, the eating experience turns into a less enjoyable action and sense of thirst becomes less acute. The interest in food and the motivation to eat also decline (Stewart-Knox et al. 2008). Finally, the sensory alterations in older ones affect dietary habits like adding extra sugar and salt to beverages and foods in order to enhance the flavor as well it alters the variety, quality, and quantity of foods they consume (Ahmed and Haboubi 2010). Thus, it increases the risk of under nutrition in older people that leads to the development of chronic conditions like hypertension and diabetes type II. Prevalence of eye disorders like diabetic retinopathy, age-related macular degeneration, glaucoma, and cataract are higher in older people. In the UK the age-related macular degeneration is a cause of blindness in around 42% of people of age 65–74 years. Poor vision in older adults affects their overall life quality because they are more dependent on taking care of their daily needs like food shopping and meal preparations (Rasmussen and Johnson 2013; Coleman et al. 2008).

11.4.6 Oral and Gastro-Intestinal (GI) Functions and Health

Oral health is the most important determinant of consumption of food. The physiological changes in the oral cavity take place with aging and older people commonly experience difficulties like dry mouth due to decreased salivary secretion and flow, declined elasticity of connective tissue and muscles of oral cavity, and also the loss of teeth. These issues cause difficulty in swallowing and chewing (Sura et al. 2012; Petersen and Yamamoto 2005). Resultantly, older adults avoid such foods that are difficult to chew like vegetables and fruits and they also change the ways in which they cook their food such as vegetables might be overcooked to compensate for difficulty in swallowing and chewing. Thus, they compromise their consumption of essential nutrients like vitamins and dietary fibers (Walls et al. 2000). Similarly, the physiological changes take place in gastro-intestinal tract with aging and it has direct influence on functions of gut like appetite and satiety regulation. The decreased food intake and reduced appetite on older people are termed as anorexia of aging. They eat more slowly, often consume smaller meals and less snacks as compared to younger people (Moss et al. 2012). Appetite is mainly regulated by the hormones in the guts. These hormones are released in response to some nutritional stimuli. Evidence supports that with increasing age the release of these hormones is changed that leads to suppression of hunger as well as feeling of satiety. Moreover, physiological changes in the stomach are also observed during the aging process. Delayed gastric emptying emphasizes the feeling of fullness (satiation) as well as early satiety in older adults (Britton and McLaughlin 2013). Alteration in colon also influences the wish to eat like the decrease in colon neurons causes reduction of neural transmitters that affect the propulsive and peristaltic activities of colon and increase bowel transit time (Winge et al. 2003). This results in constipation that is commonly observed complaint in older people and this abdominal discomfort changes their appetite (Franceschi et al. 2009). The physiological changes along with lifestyle factors like diet, smoking, and medication lead to pathological consequences in upper gastrointestinal tract. These include peptic ulcer, gastric cancer, and gastroesophageal reflux disease in old people. Such disorders influence food consumption and result in weight loss, anorexia, vomiting, anemia, and dysphagia that increase in severity with increasing age (Woodmansey 2007).

11.4.7 Chronic Diseases and Age-Related Illness

Old people having chronic conditions like infections, diseases, and other disabilities that need long-term medical care are at a risk of developing under nutrition. In addition to this the side effects of medications like loss of appetite, vomiting, anorexia, malabsorption of nutrients, nausea, diarrhea and delayed gastric emptying, and also the hospitalization are some other major causes that cause undernutrition in older people (Hickson 2006).

11.5 Active and Healthy Aging: Enhancing the Quality of Life as People Age

Healthy aging is all about enabling old people to enjoy good quality of life. Regular physical activity and healthy diet makes longevity possible with reduced levels of morbidity and less years of disability with a high quality of life (Meslé and Vallin 2011; Masi et al. 2011). Thus, healthy aging is defined as the optimization of equal opportunities for health in order to enable old people to partake actively and independently in the society (Schöllgen et al. 2010). The earlier the adoption of good dietary habits that affect the health outcomes, the greater the benefit in the old age. The promotion of health has an important role in assuring the healthy aging. Good nutrition along with physical activity is the basic component of aging well. Nutritional habits must be modified after 50 years of age because the caloric needs of the body reduce (depends on activity level) with aging, while nutritional needs of body increase as there are a lot of physiological changes taking place (Pruchno et al. 2010; Nosikov and Gudex 2003). One example is osteoporosis that is much more pronounced in old women than in the men. Further, evidence advocates that the osteoporosis onset can be halted or even averted for women if they employ a variety of preventive actions during the course of life such as consumption of balanced diet along with some exercise and not smoking (Baert et al. 2015). It has been said that frequent physical activity is the best preventive medicine for older age. All the adults need minimum 30 minutes regular physical activity on most days of week in order to stay healthy. It benefits both mental and physical health and reduces the risk of disorders related to inactivity like type II diabetes and heart diseases (Verschuuren et al. 2013; Van Buuren and Tennant 2004). Hippocrates said, "Let food be your medicine and medicine be your food." It proves that nutrition and diet have been used to keep people healthy ever since. His statement reflects that nutraceuticals are the substances that are a food or a part of a food providing medical and health benefits as well as the prevention and treatment of various age-related diseases. In a recent heart study it has been reported that levels of serum cholesterol, glucose intolerance, sex, educations status, systolic blood pressure, and dietary habits are the most important and predictive factors for survival and morbidity free survival up to the age of 85 years (Statistical Office of the European Communities 2003; Bogers et al. 2005).

11.6 Role of Med Diet in Active and Healthy Aging

The MD that is rich in unsaturated fatty acids like nuts and olive oil emphasizes the use of whole grains, fish, fresh fruits, vegetables, legumes and limits the use of red and processes meat, added sugar and refined grains which is the healthiest eating pattern. Nutrition science has proven that this eating pattern is best for the active and healthy aging (Tosti et al. 2017; Longo and Fontana 2010). The water, macronutrients like proteins, carbohydrates, and lipids, and micronutrients like vitamins enable the body to counter degenerative processes associated with the aging.

11.6.1 Energy Requirements During Aging

Due to decreasing total energy expenditure (TEE) from the age of 50 years the energy requirements start declining gradually. TEE is measured by two major factors: first one is the basal metabolic rate and the second is physical activity level (United Nations University, World Health Organization 2004). Basal metabolic rate is the internal energy requirement of body for normal physiological functioning when a person is at rest and the physical activity level refers to the extra energy requirement of body when doing extra things like exercising. Basal metabolic rate reduces with age because of alterations in body composition like decreased lean body mass chiefly in muscles and increased body fat mass (Phillips 2003). As muscles are metabolically more active than that of fats, the old people require fewer calories at rest as compared to their young counterparts. Various studies have proposed that with every decade of life the basal metabolic rate of an individual having normal body mass index reduces by 2% in women and 2.9% in men (Milne et al. 2009). In addition physical activity level also declines with aging. This is because the old people are physically less active than young people for reasons like reduced mobility and frailty. It is essential that old people get enough energy from the routine diet to avoid under nutrition and other conditions such as impaired immune response, delayed wound healing, greater length of hospital stay, impaired respiratory and muscle functions, longer rehabilitation, and increased mortality (EFSA Panel on Dietetic Products Nutrition and Allergies (NDA) 2013). Calorie restriction is termed as the decrease in energy consumption without compromising intake of other key nutrients or malnutrition. It has been suggested to enhance healthy life years in a wide range of organisms like rats, yeast, worms, monkeys, and flies. It also protected them against age-related diseases and functional decline. There is evidence that calorie restriction might reduce risk factors for cardiovascular diseases, diabetes, and cancer (Fontana et al. 2010).

11.6.2 Daily Water Intake

Adequate consumption of water is the key factor to avoid certain chronic conditions as well as to fight infections in old people. Water is fundamental to perform normal body functions like absorption of nutrients, distribution, thermoregulation, and excretion of waste. Dehydration is the most common cause of mortality and morbidity in older people (Manz and Wentz 2005; Ferry 2005). Research studies have exhibited that even moderate dehydration might cause adverse effects on recurrence of urinary stone and bronchial, pulmonary, and renal diseases. Dehydration is very common in older people but it is often overlooked. As old people become less sensitive to thirst and they have declined food consumption, as vegetables and fruits intake contributes a great proportion of daily water consumption, it is recommended for them to drink 1.5 L of water per day (Bouby and Fernandes 2003; Chidester and Spangler 1997). They should be offered small amount of liquid and water at equal intervals throughout the day and they must not take large amounts of liquids/fluids at once (Schols et al. 2009).

11.6.3 Proteins: Building Blocks of Muscle Fibers and Bones

In order to maintain a lean body mass and to slow down or avoid musculoskeletal conditions like osteoporosis and sarcopenia the sufficient utilization of proteins is very important (Gaffney-Stomberg et al. 2009). Proteins that are made up of amino acids are one of the major building blocks of bones and muscle fibers. Studies have proposed that eating moderate to large amount of amino acids can develop muscle proteins synthesis in young as well as old healthy subjects. However, the response of skeletal muscles to proteins becomes less efficient with increasing age, particularly when there is a very low level of ingested essential amino acids. Older people require more intakes of essential amino acids to have the same extent of stimulation for muscle proteins synthesis as in their young counterparts (Paddon-Jones and Rasmussen 2009). Various research studies have evaluated the effect of supplementation of leucine (an essential amino acid that plays a significant role in stimulation of insulin secretion and synthesis of muscle protein) on the prevention of sarcopenia in old adults. As it is suggested that protein metabolism becomes less efficient with increase in age that is why the old people are recommended to increase their protein consumption to prevent muscle loss and avoid malnutrition. A study estimated the effect of protein and energy supplementation in older people who were at risk of malnutrition. They did it in 62 clinical trials and the results exhibited that supplementation produced a small weight gain (2.2%) in old people in 42 trials. In a small sample of old people who were already malnourished the rate of mortality was decreased when they were supplemented with protein and energy. Still, there is a continuous debate about protein consumption and recommended amounts for old people. A study in 2013 reviewed the dietary protein requirements in old people and recommended average daily consumption of 1–2 g/kg/day for people of 65 years and above (EFSA Panel on Dietetic Products Nutrition and Allergies (NDA) 2012; Bauer et al. 2013).

11.6.4 Fats: Energy Dense Nutrients

A major component of the MD and it is most energy dense nutrient as it consists of most calories per gram (Maclean et al. 2003). It is the most important major energy source as it facilitates the absorption process of vitamin A, E, K, and D and it also has vital regulatory and structural functions in the body. However, overconsumption of fats can result in excessive total energy utilization because its energy density is very high. This can promote obesity and overweight. In addition, the intake of trans fatty acids is being related to adverse effects on cardio vascular health (Swinburn et al. 2004). On the contrary, the poly and mono unsaturated fatty acids are proposed to exert beneficial effect on human metabolic health like enhancing insulin sensitivity and improving cardiovascular conditions. Though, the present evidence is more in the favor of poly unsaturated fatty acids than mono unsaturated fatty acids (Brouwer et al. 2013). Lately, the long chain omega 3 fatty acids have been suggested to have protective effect on brain health via decreasing oxidative stress and inflammation. Thus, they have implications on brain functioning in aging population. Still the data comes chiefly from longitudinal and cross-sectional observations that describe some promising effects of fatty acids on cognitive functioning in healthy old people (Brouwer et al. 2013; Mori 2014). On review found that out of 26 studies of different study designs 19 studies recognized a positive correlation between consumption of fish or intake of n-3 fatty acid and cognitive status, while other studies established either little or no such beneficial effects. Long chain n-3 fatty acids have also been suggested to have other health promoting characteristics in normal aging process such as immune function as well as bone and muscle health (Brouwer et al. 2013; Mori 2014; Schwingshackl and Hoffmann 2012). Various clinical trials have established that even a low dose of n-3 fatty acid supplement can enhance immune response in old people (Wallin et al. 2012). A systemic review of 23 clinical studies proved a modest but fairly persistent benefit of fish oil that contains n-3 poly unsaturated fatty acid on inflamed joints and pain in case of rheumatoid arthritis. Further, it exhibited a decreased duration of stiffness in morning as well as improvements in other disease indicators were observed (Nettleton and Katz 2005; van de Rest et al. 2012). According to a very few studies protective effect of n-3 fatty acids on musculoskeletal health has been observed. Currently, no specific recommendations are available for dietary fat intake in older people (Ubeda et al. 2012; Dangour et al. 2012). For European adults, EFSA has suggested a reference consumption range of 20–35% of daily total energy from fat source. For saturated and trans fatty acids the consumption should be as low as possible within the context of a nutritionally adequate diet (Sydenham et al. 2012; Miles and Calder 2012; EFSA Panel on Dietetic Products Nutrition and Allergies (NDA) 2010).

11.6.5 Carbohydrates

Very little is known about the role of carbohydrates in active and healthy aging. Though, glucose is a significant energy source for brain yet a Cochrane review in 2011 concluded that the evidence is insufficient about the carbohydrate use for improving the cognitive performance in older people with normal or having mild cognitive impairment, because of limited study in this area (Ooi et al. 2011). On the contrary, evidence from some observational studies links diabetes with cognitive function decline consisting executive functions and memory, especially in old (65 years and above) diabetic patients who exhibited cognitive decrements and lower score in cognitive tests as compared to the younger patients (Exalto et al. 2012; McCrimmon et al. 2012; Pasquier et al. 2006). However, some studies suggest that blood pressure control may help in preventing cognitive function decline (Messier 2005).

11.6.6 Micronutrients

Old people are especially susceptible to micronutrient deficiencies because of decreased consumption of foods rich in vitamins and minerals. The pervasiveness of under nutrition and deficiency is highest among the women, very old and those people who are in care institutions. Micronutrients are a key to maintain a normal physiological and cognitive functioning in the aging body and insufficient consumption in return will lead to deterioration of health and will promote development of certain disorders (Arvanitakis et al. 2009, 2013). There has been an increasing interest in supplementing vitamin B6, B12, and folic acid in a large number of age-related vascular diseases because these vitamins have an important role in metabolism of an amino acid (homocysteine) (Feng et al. 2006). At high levels the homocysteine is thought to be an independent risk factor for various vascular diseases. Previous studies on status of vitamin B and cognition function suggested that old people having higher levels of homocysteine tend to have low status of vitamin B as well as low scores on cognitive tests. They also exhibited high risk of certain vascular diseases like Alzheimer's disease and dementia as compared to those people who had normal status of vitamin B (Seshadri et al. 2002; Miller 2003; Ravaglia et al. 2005; Kado et al. 2005). This observation provoked the theory that sufficient consumption of these vitamins can decrease the levels of homocysteine that can result in prevention of the diseases. A review on the estimation of effects of vitamin B6 supplementation on cognition found only two relevant trials in healthy old people. One study stated there is no significant effect of supplementation on cognition or mood in old women, while the other study said that there was a modest but significant effect on long-term memory in old men by vitamin B6 (Bryan et al. 2002; Deijen et al. 1992). Another review investigated the effect of vitamin B6 supplementation in preventing cardiovascular disease recurrence in some clinical trials. The overall results failed to exhibit positive influence in spite of the relative consistent relation between low status of vitamin B6 and incidence of cardio vascular disease in epidemiological studies (Lotto et al. 2011). Previous studies on vitamin B12 supplementation did not show any improvements in cognitive functions in old people having dementia. Long-term use of folic acid supplementation exhibited to improve cognitive functions in healthy old people having higher levels of homocysteine (Malouf and Areosa Sastre 2003; McCracken 2010). A small intervention trial found that multi-vitamin B supplementation slows the rate of accelerated brain atrophy and it also slows cognitive decline in patients with increased homocysteine levels (Malouf and Evans 2008). In a larger trial on old people with depressive symptoms, a combination of vitamin B12 and folic acid for two consecutive years improved cognitive functions in some aspects. Furthermore, the combination of nutrients received from food may have synergistic and interactive effects on health, while these beneficial effects may not be seen in supplementation trials (Morris et al. 2007; Balk et al. 2007).

Calcium and vitamin D has an important role in bone health. Calcium is the basic structural component of bones and teeth. Vitamin D plays vital role in absorption of calcium and maintenance of serum calcium and phosphorous homeostasis. When level of vitamin D is low it disturbs the absorption of calcium and stimulates release of parathyroid hormones that are responsible for bone resorption and fasten the bone loss (Lips et al. 2010). Normally, vitamin D is formed in the skin by UVB light from sun but in old people the synthesis of vitamin D from sunlight exposure is limited because of the decrease in the precursor of vitamin D in aging skin and also due to the less time spent outdoor in old age (Mosekilde 2005). Thus, in old people prone to deficiency an increase intake of vitamin D is essential for bone health. Vitamin D alone has little effect on risk of fracture. Studies have shown that supplementation of vitamin D and calcium reduces hip fracture by 16%. Cholecalciferol, a specific form of vitamin D, appears to reduce mortality in predominantly old women (Avenell et al. 2009; Bjelakovic et al. 2014). Further, vitamin D and calcium supplements are used in postmenopausal women in order to prevent osteoporosis. Antioxidants are needed in order to protect the cells from oxidative damage. Antioxidants neutralize the free radicals (Bolland et al. 2010). They come from two major sources, one from internal production of body and second from consumption of antioxidant nutrients like vitamin A, C and E and also a number of other poly-phenolic compounds. Consumption of a diet rich in antioxidants may decrease the risk of cardiovascular diseases, cognitive decline, dementia, and age-related eye diseases (Bratic and Larsson 2013). Other antioxidants like zinc and selenium have shown some positive effects on healthy aging. Selenium is a trace element that has anti-inflammatory as well as antioxidant properties. Major dietary sources of selenium include fish, meat, eggs, cereals, offal, grains, and certain vegetables and fruits (Fusco et al. 2007). Low selenium levels may be linked with higher risk of mortality, poor immune functioning, and decline in cognition (Kieliszek and Błażejak 2013; Rees et al. 2013). Zinc is another micronutrient that can show antioxidant properties in the body. Zinc deficiency has been associated with neuronal damage seen in Alzheimer's disease because zinc may be essential in decreasing the copper level in brain. Excess of copper can function as pro-oxidant and may enhance the probability of Alzheimer's disease (Brewer 2012). Zinc has an important role in immune system as well (Vishwanathan et al. 2013). It is a well-established fact that immune system alters with advancing age as the ability to respond to infection and development of immunity after vaccination decreases with increasing age and it leads to enhanced risk of mortality due to the infections in old people (Aw et al. 2007).

11.6.7 Phytochemical: Virgin Olive Oil

Dietary olive oil is a tool for counteracting the degenerative process mediated by the free radical production and effect on various cells, tissues, and organs. Dietary manipulation has been proved to be a great tool for partially modifying the structure and features of biological membranes (Landvik et al. 1996). There is a possibility to support the cell membranes by some specific membrane components that can counteract activity of free radicals. The virgin olive oil is enriched with antioxidants that prevent the attack of free radicals. It has very high content of mono unsaturated fatty acids and it greatly enriches the biological membranes and it partially substitutes other fatty acids like poly unsaturated fatty acids. Resulting in the formation of such membranes that are less susceptible to oxidative damage due to the increased level of oleic acid, which is important to maintain optimal fluidity of membrane for proper functioning. In addition, it slows down the lipid per oxidation (Landvik et al. 1996; Grignaffini et al. 1994).

Virgin olive oil also contains large amount of antioxidant molecules such as tocopherol, some phenolic compounds, and coenzyme-Q. These are known as free radical scavengers that counteract the toxic radicals and in some cases even prevent their formation. Thus, they protect the biological structures from oxidative injury (Grignaffini et al. 1994; Visioli et al. 1995). The efficacy of dietary interventions based on virgin olive oil has been widely demonstrated because olive oil is proven to be a unique source of fat for modifying the fatty acid pattern of biological membranes and it makes them less prone to suffer oxidative modifications (Petroni et al. 1995).

11.6.8 Microbiota: Probiotics and Prebiotics

The microbiota or gut flora is all the microorganisms that live in the human gut and also the metabolites generated by these microorganisms. Microbiota is essential for gut health maintenance. Alterations in the composition of microbiota have been associated with inflammation and metabolic disorders like irritable bowel syndrome, colorectal cancer, inflammatory bowel disease, diabetes, CVD, and frailty in older adults. Healthy people have a stable microbiota composition in gut, while the older people have somewhat altered compositions of microbiota. Thus, improving the microbial balance and composition in old people may help in reducing the risk of metabolic diseases and inflammation. Probiotics and probiotics both have been suggested to promote the health of gut in older people (Duncan and Flint 2013). Probiotics are living microorganisms that on administration can cause health benefits in host. Common types include *Lactobacilli* and *Bifidobacteria* and they are found in yogurt and fermented milk products and in some fortified juices. Prebiotics are non-digestible substrate molecules for probiotics and selectively stimulate the growth and activity of these bacteria in gut. Probiotics are naturally present in vegetables as well as synthetic forms of non-digestible carbohydrates (Hamilton-Miller 2004).

11.7 A Tool for Counteracting Degenerative Process in Aging—The Mediterranean Diet

The major effectors of MD are reduced caloric intake, decreased consumption of saturated fatty acids, microbiota derived metabolites, less amino acid utilization, and an increase in phytochemical consumption. Various interdependent and protruding factors have been thought to play a significant role in healthy and active aging

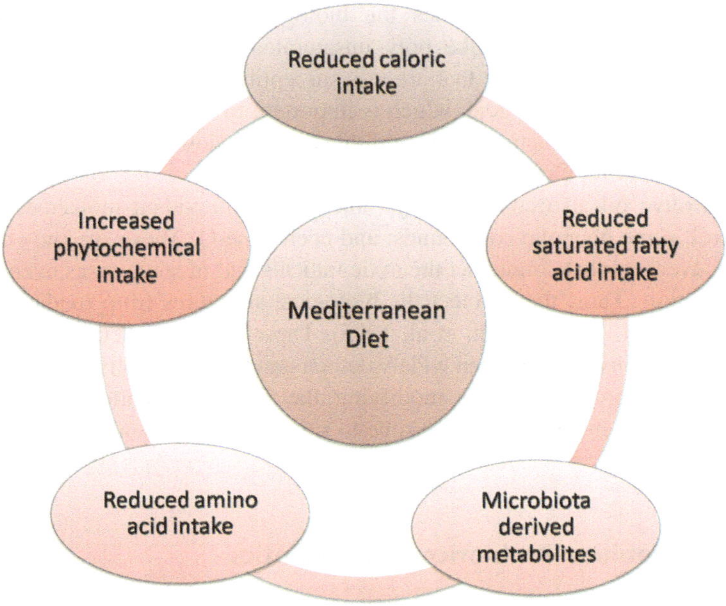

Fig. 11.2 Major effectors of Mediterranean diet adapted from Tosti et al. (2017)

through consumption of Mediterranean diet. The first and foremost effect is the protection against oxidative damage, injury, and inflammation and platelet aggregation (Fig. 11.2).

The second way to mediate pro-health and longevity effect of MD is through lowering lipid levels. It also modifies the hormones as well as growth factors that are involved in cancer pathogenesis. This inhibits the nutrient sensing pathway via restriction of specific amino acid. MD has proven to be the best way to produce certain metabolites by gut microbiota and it influences the metabolic health (Sköldstam et al. 2003; Goldstein and Brown 2015; Beauchamp et al. 2005; Harris et al. 2012; Lamming et al. 2015; Levine et al. 2014; David et al. 2014). A recent study utilized the preferential full, randomized, and controlled study design called PREDIMED (PREvención con DIeta MEDiterránea), it involved total 7447 individuals including both men (aged 55–80) and women(aged 60–80) who were at a high risk of CVD but did not have the disease at the time of their enrollment in the study. Subjects were assigned randomly to one of the three groups. The first group was on controlled diet, the second one received MD plus four tablespoon of extra virgin olive oil daily, and the third group received MD plus one ounce of nuts daily. By the end of the study both the MD groups experienced a significant reduction (30%) in the combined risk of stroke, acute heart attack, and cardiovascular death over the five-year period as compared to the control group. What is compelling about these findings is that the people in the MD groups had a very few dietary restrictions in comparison with the control group and no group restricted their caloric intake or boosted their exercise at all. Most importantly this validated the MD as a true health intervention that can significantly decrease the risk of CVD. This is a radical advance over many previous studies, which could only show correlation between diet and heart disease risk (Estruch et al. 2013). Polyphenols reduce the blood pressure. The researchers also examined the impact of the two MD (first with extra virgin olive oil and second one with nuts) on blood pressure at the start of the study and also after one year of being on diet. The goal was to evaluate if there was a connection between polyphenol levels and blood pressure. It was found that those subjects on either of the two MD experienced significant reductions in both diastolic and systolic blood pressure compared with those on the control. These blood pressure reductions were associated with significant increases in total urinary polyphenols as well as plasma nitric oxide levels. Endothelial nitric oxide is the body's natural blood-vessel relaxant, and is a biomarker of good cardiovascular function and healthy aging. This study gives the first clue that polyphenols in the MD promote healthy aging and protect against CVD by directly improving vascular function and decreasing blood pressure. An additional follow-up research study provides insight into another key mechanism of action (Medina-Remon et al. 2015) (https://www.newscientist.com/article/2116971-brain-shrinks-less-in-older-people-who-eat-mediterranean-diet/. Accessed January 5, 2017). Polyphenols, a larger part of MD, reduce inflammation as well. The researchers followed a sub-group of 1139 people in the PREDIMED study to evaluate if the levels of polyphenols were connected with inflammatory markers. Chronic inflammation is found to be the fundamental contributor to CVD, rheumatoid arthritis, and osteoarthritis and the polyphenols are known as anti-inflammatory compounds, so it was a sensible area of investigation. After one year, the subjects who had greatest increase in urinary polyphenols, they showed significantly low levels of five important inflammatory markers that correlate with CVD and other diseases. They include intracellular adhesion molecule- I, tumor necrosis factor- alpha, vascular cell adhesion molecule-I, interleukin-6, and monocyte chemotactic protein-I. In addition, the subjects who had highest increase in polyphenol levels exhibited significantly less systolic and diastolic blood pressure and also had significant increase in beneficial HD cholesterol levels. It is evident from these studies that the polyphenols in the MD account for a large proportion of its value in preventing catastrophic heart disease, stroke, and cardiovascular deaths (Medina-Remon et al. 2017).

11.8 Conclusion

Accumulating data suggests that the nutrition is an essential factor for promoting the health and preventing the most common age-related chronic conditions. The molecular and metabolic health is chiefly associated with what we eat. The caloric restriction increases the healthy life span only when it is combined with sufficient intake of all other essential macro- and micronutrients. It can be concluded that certain dietary interventions can be a newer, interesting, and promising technique to prevent and treat various age-related diseases, as an effective adjuvant to the pharmacological therapy. As MD provides vast amounts of stable and non-oxidizable fatty acids and remarkable amount of powerful antioxidants, it can slow down the degenerative processes during aging and can promote an active and healthy aging (Fontana and Partridge 2015).

Unlike the typical American and European diet, the conventional MD incorporates a wide variety of minimally processed and fiber rich plant food, full of vitamins, phytochemicals, and minerals. Less intake of eggs, fish, meat, and cheese serves other essential components like vitamin B12 that cannot be obtained from exclusively plant-based food. In past decades the great amount of energy was obtained by intake of energy dense foods like extra virgin olive oil and dried fruits. The moderate restriction of energy obtained by more intakes of fiber rich and energy poor plant foods and also the restriction of specifically sulfur and other amino acids and saturated fatty acids play a vital role in the mediation of health and longevity benefits of Mediterranean diet. To sum up, the close adherence to MD can cause higher rate of longevity in people. It implies a daily intake of macro- and micronutrients as well as phytochemicals protects from age-related diseases and promotes active and healthy aging (Bertozzi et al. 2017).

Conflict of Interest The authors declare no conflict of interest.

References

Ahmed T, Haboubi N (2010) Assessment and management of nutrition in older people and its importance to health. Clin Interv Aging 5:207–216

Arvanitakis M, Coppens P, Doughan L, Van Gossum A (2009) Nutrition in care homes and home care: recommendations–a summary based on the report approved by the Council of Europe. Clin Nutr 28(5):492–496

Arvanitakis M, Vandewoude M, Perkisas S, Van Gossum A (2013) Undernutrition in community dwelling elderly. e-SPEN J 8(5):e213–e2e5

Avenell A, Gillespie WJ, Gillespie LD, O'Connell D (2009) Vitamin D and vitamin D analogues for preventing fractures associated with involutional and post-menopausal osteoporosis. Cochrane Database Syst Rev (2):CD000227

Aw D, Silva AB, Palmer DB (2007) Immunosenescence: emerging challenges for an ageing population. Immunology 120(4):435–446

Baert V, Gorus E, Mets T, Bautmans I (2015) Motivators and barriers for physical activity in older adults with osteoporosis. J Geriatr Phys Ther 38(3):105–114

Balk EM, Raman G, Tatsioni A, Chung M, Lau J, Rosenberg IH (2007) Vitamin B6, B12, and folic acid supplementation and cognitive function: a systematic review of randomized trials. Arch Intern Med 167(1):21–30

Batsch NL, Mittelman MS (2012) World Alzheimer report 2012. Overcoming the stigma of dementia. [consultado 2012 ene 21 Disponible en:]. http://www.alz.co.uk/research/world-report-2012.2012

Battino M, Ferreiro MS (2004) Ageing and the Mediterranean diet: a review of the role of dietary fats. Public Health Nutr 7(7):953–958

Battino M, Bertoli E, Formiggini G, Sassi S, Gorini A, Villa RF et al (1991) Structural and functional aspects of the respiratory chain of synaptic and nonsynaptic mitochondria derived from selected brain regions. J Bioenerg Biomembr 23:345–363

Battino M, Gorini A, Villa RF, Genova ML, Bovina C, Sassi S et al (1995) Coenzyme Q content in synaptic and nonsynaptic mitochondria from different brain regions in the ageing rat. Mech Ageing Dev 78:173–187

Battino M, Littarru GP, Gorini A, Villa RF (1996) Coenzyme Q, peroxidation and cytochrome oxidase features after Parkinson's- like disease by MPTP toxicity in intra-synaptic and non-synaptic mitochondria from Macaca fascicularis cerebral cortex and hippocampus: action of dihydroergocriptine. Neurochem Res 21:1505–1514

Battino M, Quiles JL, Huertas JR, Mataix JF, Villa RF, Gorini A (2000) Cerebral cortex synaptic heavy mitochondria may represent the oldest synaptic mitochondrial population: biochemical heterogeneity and effects of L-acetylcarnitine. J Bioenerg Biomembr 32:163–173

Battino M, Ferreiro MS, Gallardo I, Newman HN, Bullon P (2002a) The antioxidant capacity of saliva. J Clin Periodontol 29:189–194

Battino M, Ferreiro MS, Littarru GP, Quiles JL, Ramirez- Tortosa MC, Huertas JR et al (2002b) Structural damage induced by peroxidation may account for functional impairment of heavy synaptic mitochondria. Free Radic Res 36:479–484

Bauer J, Biolo G, Cederholm T, Cesari M, Cruz-Jentoft AJ, Morley JE, Phillips S, Sieber C, Stehle P, Teta D, Visvanathan R (2013) Evidence-based recommendations for optimal dietary protein intake in older people: a position paper from the PROT-AGE Study Group. J Am Med Dir Assoc 14(8):542–559

Beauchamp GK, Keast RS, Morel D et al (2005) Phytochemistry: ibuprofen- like activity in extra-virgin olive oil. Nature 437:45–46. https://doi.org/10.1038/437045a

Bertozzi B, Tosti V, Fontana L (2017) Beyond calories: an integrated approach to promote health, longevity, and well-being. Gerontology 63:13–19

Bjelakovic G, Gluud LL, Nikolova D, Whitfield K, Wetterslev J, Simonetti RG, Bjelakovic M, Gluud C (2014) Vitamin D supplementation for prevention of mortality in adults. Cochrane Database Syst Rev (1):CD007470

Bogers R, Tijhuis M, van Gelder B, Kromhout D (2005) Final report of the HALE Project, Healthy Ageing: a Longitudinal study in Europe. Report 260853003. RIVM, Bilthoven, Netherlands

Bolland MJ, Avenell A, Baron JA, Grey A, MacLennan GS, Gamble GD, Reid IR (2010) Effect of calcium supplements on risk of myocardial infarction and cardiovascular events: meta-analysis. BMJ 341:c3691

Bouby N, Fernandes S (2003) Mild dehydration, vasopressin and the kidney: animal and human studies. Eur J Clin Nutr 57(Suppl 2):S39–S46

Bousquet J, Farrell J, Crooks G, Hellings P, Bel EH, Bewick M, Chavannes NH, Sousa JC, Cruz AA, Haahtela T, Joos G (2016) Scaling up strategies of the chronic respiratory disease programme of the European Innovation Partnership on Active and Healthy Ageing (Action Plan B3: Area 5). Clin Transl Allergy 6(1):29

Boyce JM, Shone GR (2006) Effects of ageing on smell and taste. Postgrad Med J 82(966):239–241

Bratic A, Larsson NG (2013) The role of mitochondria in aging. J Clin Invest 123(3):951–957

Brewer GJ (2012) Copper excess, zinc deficiency, and cognition loss in Alzheimer's disease. Biofactors 38(2):107–113

Britton E, McLaughlin JT (2013) Ageing and the gut. Proc Nutr Soc 72(01):173–177

Brouwer IA, Wanders AJ, Katan MB (2013) Trans fatty acids and cardiovascular health: research completed. Eur J Clin Nutr Rev 67(5):541–547

Bryan J, Calvaresi E, Hughes D (2002) Short-term folate, vitamin B-12 or vitamin B-6 supplementation slightly affects memory performance but not mood in women of various ages. J Nutr 132 (6):1345–1356

Cardoso BR, Cominetti C, Cozzolino SM (2013) Importance and management of micronutrient deficiencies in patients with Alzheimer's disease. Clin Interv Aging 8:531–542

Chidester JC, Spangler AA (1997) Fluid intake in the institutionalized elderly. J Am Diet Assoc 97 (1):23–28;quiz 9-30

Coleman HR, Chan C-C, Ferris FL III, Chew EY (2008) Age-related macular degeneration. Lancet 372(9652):1835–1845

Dangour AD, Andreeva VA, Sydenham E, Uauy R (2012) Omega 3 fatty acids and cognitive health in older people. Br J Nutr 107(Suppl 2):S152–S158

David LA, Maurice CF, Carmody RN et al (2014) Diet rapidly and reproducibly alters the human gut microbiome. Nature 505:559–563

Davis C, Bryan J, Hodgson J, Murphy K (2015) Definition of the Mediterranean diet; a literature review. Nutrients 7(11):9139–9153

Deijen JB, van der Beek EJ, Orlebeke JF, van den Berg H (1992) Vitamin B-6 supplementation in elderly men: effects on mood, memory, performance and mental effort. Psychopharmacology 109(4):489–496

Duncan SH, Flint HJ (2013) Probiotics and prebiotics and health in ageing populations. Maturitas 75(1):44–50

EFSA Panel on Dietetic Products Nutrition and Allergies (NDA) (2010) Scientific opinion on dietary reference values for fats, including saturated fatty acids, polyunsaturated fatty acids, monounsaturated fatty acids, trans fatty acids, and cholesterol. EFSA J 8(3):1461

EFSA Panel on Dietetic Products Nutrition and Allergies (NDA) (2012) Scientific opinion on dietary reference values for protein. EFSA J 10(2):2557

EFSA Panel on Dietetic Products Nutrition and Allergies (NDA) (2013) Scientific opinion on dietary reference values for energy. EFSA J 11(1):3005

Estruch R, Ros E, Salas-Salvado J et al (2013) Primary prevention of cardiovascular disease with a Mediterranean diet. N Engl J Med 368(14):1279–1290

Exalto LG, Whitmer RA, Kappele LJ, Biessels GJ (2012) An update on type 2 diabetes, vascular dementia and Alzheimer's disease. Exp Gerontol 47(11):858–864

Feng L, Ng TP, Chuah L, Niti M, Kua EH (2006) Homocysteine, folate, and vitamin B-12 and cognitive performance in older Chinese adults: findings from the Singapore Longitudinal Ageing Study. Am J Clin Nutr 84(6):1506–1512

Ferry M (2005) Strategies for ensuring good hydration in the elderly. Nutr Rev 63:S22–S29

Fontana L, Partridge L (2015) Promoting health and longevity through diet: from model organisms to humans. Cell 161:106–118

Fontana L, Partridge L, Longo VD (2010) Extending healthy life span–from yeast to humans. Science 328(5976):321–326

Franceschi M, Di Mario F, Leandro G, Maggi S, Pilotto A (2009) Acid-related disorders in the elderly. Best Practice Res Clin Gastroenterol 23(6):839–848

Fusco D, Colloca G, Lo Monaco MR, Cesari M (2007) Effects of antioxidant supplementation on the aging process. Clin Interv Aging 2(3):377–387

Gaffney-Stomberg E, Insogna KL, Rodriguez NR, Kerstetter JE (2009) Increasing dietary protein requirements in elderly people for optimal muscle and bone health. J Am Geriatr Soc 57 (6):1073–1079

Gaziano JM (2000) Dietary antioxidants and cardiovascular disease. Vitam Horm 58:299–320

Goldstein JL, Brown MS (2015) A century of cholesterol and coronaries: from plaques to genes to statins. Cell 161:161–172. https://doi.org/10.1016/j.cell.2015.01.036

Grignaffini P, Roma P, Galli C, Catapano AL (1994) Protection of low density lipoprotein from oxidation by 3,4-dihydroxyphenylethanol. Lancet 343:1296–1297

Hamilton-Miller JM (2004) Probiotics and prebiotics in the elderly. Postgrad Med J 80 (946):447–451

Harman D (1992) Role of free radicals in aging and disease. Ann N Y Acad Sci 673:126–141

Harraan D (1956) Aging: a theory based on free radical and radiation chemistry. J Gerontol 11:298–300

Harris RE, Beebe J, Alshafie GA (2012) Reduction in cancer risk by selective and nonselective cyclooxygenase-2 (COX-2) inhibitors. J Exp Pharmacol 4:91–96. https://doi.org/10.2147/JEP.S23826

Hickson M (2006) Malnutrition and ageing. Posgrad Med J 82(963):2–8

Hsiao HC, Chao HC, Wang JJ (2013) Features of problematic eating behaviors among community-dwelling older adults with dementia: family caregivers' experience. Geriatr Nurs 34(5):361–365

Kado DM, Karlamangla AS, Huang MH, Troen A, Rowe JW, Selhub J, Seeman TE (2005) Homocysteine versus the vitamins folate, B6, and B12 as predictors of cognitive function and decline in older high-functioning adults: MacArthur Studies of Successful Aging. Am J Med 118(2):161–167

Kieliszek M, Błażejak S (2013) Selenium: significance, and outlook for supplementation. Nutrition 29(5):713–718

Kirkwood TB (2008) Understanding ageing from an evolutionary perspective. J Intern Med 263:117–127

Lamming DW, Cummings NE, Rastelli AL et al (2015) Restriction of dietary protein decreases mTORC1 in tumors and somatic tissues of a tumor-bearing mouse xenograft model. Oncotarget 6:31233–31240

Landvik SV, Diplock AT, Packer L (1996) Efficacy of vitamin E inhuman health and disease. In: Cadenas E, Packer L (eds) Handbook of antioxidants. Marcel Dekker, New York, pp 63–87

Levine ME, Suarez JA, Brandhorst S et al (2014) Low protein intake is associated with a major reduction in IGF-1, cancer, and overall mortality in the 65 and younger but not older population. Cell Metab 19:407–417

Lips P, Bouillon R, Van Schoor NM, Vanderschueren D, Verschueren S, Kuchuk N, Milisen K, Boonen S (2010) Reducing fracture risk with calcium and vitamin D. Clin Endocrinol 73 (3):277–285

Longo VD, Fontana L (2010) Calorie restriction and cancer prevention: metabolic and molecular mechanisms. Trends Pharmacol Sci 31:89–98. https://doi.org/10.1016/j.tips.2009.11.004

López-Otín C, Blasco MA, Partridge L, Serrano M, Kroemer G (2013) The hallmarks of aging. Cell 153:1194–1217

Lotto V, Choi SW, Friso S (2011) Vitamin B6: a challenging link between nutrition and inflammation in CVD. Br J Nutr 106(2):183–195

Maclean W, Harnly J, Chen J, Chevassus-Agnes S, Gilani G, Livesey G, Warwick P (2003) Food energy–methods of analysis and conversion factors. In Food and Agriculture Organization of the United Nations Technical Workshop Report 77

Mak TN, Caldeira S (2014) The role of nutrition in active and healthy ageing: for prevention and treatment of age-related diseases: evidence so far. Publications Office of the European Union, Brussels

Malouf R, Areosa Sastre A (2003) Vitamin B12 for cognition. Cochrane Database Syst Rev (3): CD004326

Malouf R, Evans JG (2008) Folic acid with or without vitamin B12 for the prevention and treatment of healthy elderly and demented people. Cochrane Database Syst Rev (4):CD004514

Manz F, Wentz A (2005) The importance of good hydration for the prevention of chronic diseases. Nutr Rev 63:S2–S5

Masi CM, Chen HY, Hawkley LC, Cacioppo JT (2011) A meta-analysis of interventions to reduce loneliness. Personal Soc Psychol Rev 15(3):219–266

McCracken C (2010) Challenges of long-term nutrition intervention studies on cognition: discordance between observational and intervention studies of vitamin B12 and cognition. Nutr Rev 68(S1):S11–S15

McCrimmon RJ, Ryan CM, Frier BM (2012) Diabetes and cognitive dysfunction. Lancet 379 (9833):2291–2299

Medina-Remon A, Tresserra-Rimbau A, Pons A et al (2015) Effects of total dietary polyphenols on plasma nitric oxide and blood pressure in a high cardiovascular risk cohort. The PREDIMED randomized trial. Nutr Metab Cardiovasc Dis 25(1):60–67

Medina-Remon A, Casas R, Tressserra-Rimbau A et al (2017) Polyphenol intake from a Mediterranean diet decreases inflammatory biomarkers related to atherosclerosis: a substudy of the PREDIMED trial. Br J Clin Pharmacol 83(1):114–128

Meslé F, Vallin J (2011) Historical trends in mortality. In: International handbook of adult mortality. Springer, Dordrecht, pp 9–47

Messier C (2005) Impact of impaired glucose tolerance and type 2 diabetes on cognitive aging. Neurobiol Aging 26(S1):26–30
Miles EA, Calder PC (2012) Influence of marinen-3 polyunsaturated fatty acids on immune function and a systematic review of their effects on clinical outcomes in rheumatoid arthritis. Br J Nutr 107(Suppl 2):S171–S184
Miller AL (2003) The methionine-homocysteine cycle and its effects on cognitive diseases. Altern Med Rev 8(1):7–19
Milne AC, Potter J, Vivanti A, Avenell A (2009) Protein and energy supplementation in elderly people at risk from malnutrition. Cochrane Database Syst Rev 2009(2):CD003288
Mori TA (2014) Dietary n-3 PUFA and CVD: a review of the evidence. Proc Nutr Soc 73(1):57–64
Morley JE, Baumgartner RN, Roubenoff R, Mayer J, Nair KS (2001) Sarcopenia. J Lab Clin Med 137(4):231–243
Morris MC, Tangney CC, Dugra J, Van Boxtel MP, Schouten EG, Kok FJ, Jolles J, Katan MB, Verhoef P (2007) Effect of 3-year folic acid supplementation on cognitive function in older adults in the FACIT trial: a randomised, double blind, controlled trial. Lancet 369 (9557):208–216
Mosekilde L (2005) Vitamin D and the elderly. Clin Endocrinol 62(3):265–281
Moss C, Dhillo WS, Frost G, Hickson M (2012) Gastrointestinal hormones: the regulation of appetite and the anorexia of ageing. J Hum Nutr Diet 25(1):3–15
Nettleton JA, Katz R (2005) N-3 long-chain polyunsaturated fatty acids in type 2 diabetes: a review. J Am Diet Assoc 105(3):428–440
Niki E (1996) a-Tocopherol. In: Cadenas E, Packer L (eds) Handbook of antioxidants. Marcel Dekker, New York, pp 3–25
Nosikov A, Gudex C (2003) Development of a common instrument for quality of life. In: EUROHIS: developing common instruments for health surveys, vol 57. IOS Press, Amsterdam, p 145
Ooi CP, Loke SC, Yassin Z, Hamid TA (2011) Carbohydrates for improving the cognitive performance of independent-living older adults with normal cognition or mild cognitive impairment. Cochrane Database Syst Rev (4):CD007220
Paddon-Jones D, Rasmussen BB (2009) Dietary protein recommendations and the prevention of sarcopenia. Curr Opin Clin Nutr Metab Care 12(1):86–90
Paddon-Jones D, Short KR, Campbell WW, Volpi E, Wolfe RR (2008) Role of dietary protein in the sarcopenia of aging. Am J Clin Nutr 87(5):1562S–1566S
Pasquier F, Boulogne A, Leys D, Fontaine P (2006) Diabetes mellitus and dementia. Diabetes Metab 32(5 Pt 1):403–414
Petersen PE, Yamamoto T (2005) Improving the oral health of older people: the approach of the WHO Global Oral Health Programme. Community Dent Oral Epidemiol 33(2):81–92
Petroni A, Blasevich M, Salami M, Papini N, Montedoro GF, Galli C (1995) Inhibition of platelet aggregation and eicosanoid production by phenolic components of olive oil. Thromb Res 78:151–160
Phillips F (2003) Nutrition for healthy ageing. Nutr Bull 28(3):253–263
Prince M, Jackson J (2009) World Alzheimer's report. Alzheimer's Disease International, London
Pruchno RA, Wilson-Genderson M, Rose M, Cartwright F (2010) Successful aging: early influences and contemporary characteristics. Gerontologist 50(6):821–833
Rantakokko M, Mänty M, Rantanen T (2013) Mobility decline in old age. Exerc Sport Sci Rev 41 (1):19–25
Rasmussen HM, Johnson EJ (2013) Nutrients for the aging eye. Clin Interv Aging 8:741–748
Ravaglia G, Forti P, Maioli F, Martelli M, Servadei L, Brunetti N, Porcellini E, Licastro F (2005) Homocysteine and folate as risk factors for dementia and Alzheimer disease. Am J Clin Nutr 82 (3):636–643
Rechel B, Grundy E, Robine J-M, Cylus J, Mackenbach JP, Knai C et al (2013) Ageing in the European Union. Lancet 381(9874):1312–1322

Rees K, Hartley L, Day C, Flowers N, Clarke A, Stranges S (2013) Selenium supplementation for the primary prevention of cardiovascular disease. Cochrane Database Syst Rev (1):CD009671

Roberts SB, Rosenberg I (2006) Nutrition and aging: changes in the regulation of energy metabolism with aging. Physiol Rev 86(2):651–667

Schilp J, Wijnhoven HA, Deeg DJ, Visser M (2011) Early determinants for the development of undernutrition in an older general population: Longitudinal Aging Study Amsterdam. Br J Nutr 106(5):708–717

Schöllgen I, Huxhold O, Tesch-Römer C (2010) Socioeconomic status and health in the second half of life: findings from the German Ageing Survey. Eur J Ageing 7(1):17–28

Schols JMGA, Groot CPGM, Cammen TJM, Olde Rikkert MGM (2009) Preventing and treating dehydration in the elderly during periods of illness and warm weather. J Nutr Health Aging 13 (2):150–157

Schwingshackl L, Hoffmann G (2012) Monounsaturated fatty acids and risk of cardiovascular disease: synopsis of the evidence available from systematic reviews and meta-analyses. Nutrients 4(12):1989–2007

Seshadri S, Beiser A, Selhub J, Jacques PF, Rosenberg IH, D'agostino RB, Wilson PW, Wolf PA (2002) Plasma homocysteine as a risk factor for dementia and Alzheimer's disease. N Engl J Med 346(7):476–483

Shatenstein B, Ferland G (2000) Absence of nutritional or clinical consequences of decentralized bulk food portioning in elderly nursing home residents with dementia in Montreal. J Am Diet Assoc 100(11):1 354–1 360

Sköldstam L, Hagfors L, Johansson G (2003) An experimental study of a MD intervention for patients with rheumatoid arthritis. Ann Rheum Dis 62:208–214

Solfrizzi V, Panza F, Torres F, Mastroianni F, Del Parigi A, Venezia A (1999) High monounsaturated fatty acids intake protects against age-related cognitive decline. Neurology 52:1563–1569

Statistical Office of the European Communities (2003) Health statistics: key data on health 2002: data 1970–2001. European Commission

Stewart-Knox BJ, Simpson EE, Parr H, Rae G, Polito A, Intorre F et al (2008) Taste acuity in response to zinc supplementation in older Europeans. Br J Nutr 99(1):129–136

Sura L, Madhavan A, Carnaby G, Crary MA (2012) Dysphagia in the elderly: management and nutritional considerations. Clin Interv Aging 7:287–298

Swinburn BA, Caterson I, Seidell JC, James WP (2004) Diet, nutrition and the prevention of excess weight gain and obesity. Public Health Nutr 7(1A):123–146

Sydenham E, Dangour AD, Lim WS (2012) Omega 3 fatty acid for the prevention of cognitive decline and dementia. Cochrane Database Syst Rev (6):CD005379

Tosti V, Bertozzi B, Fontana L (2017) Health benefits of the Mediterranean diet: metabolic and molecular mechanisms. J Gerontol A 73(3):318–326

Ubeda N, Achon M, Varela-Moreiras G (2012) Omega 3 fatty acids in the elderly. Br J Nutr 107 (Suppl 2):S137–S151

United Nations University, World Health Organization (2004) Human energy requirements: report of a joint FAO/WHO/UNU expert consultation: Rome, 17–24 October 2001. Food and Agriculture Organization

Van Buuren S, Tennant A (2004) Response conversion for the health monitoring program. TNO-report 2004.145. TNO Preventie en Gezondheid, Leiden

van de Rest O, van Hooijdonk LW, Doets E, Schiepers OJ, Eilander A, de Groot LC (2012) B vitamins and n-3 fatty acids for brain development and function: review of human studies. Ann Nutr Metab 60(4):272–292

Vasto S, Candore G, Balistreri CR, Caruso M, Colonna-Romano G, Grimaldi MP, Listi F, Nuzzo D, Lio D, Caruso C (2007) Inflammatory networks in ageing, age-related diseases and longevity. Mech Ageing Dev 128:83–91

Vasto S, Scapagnini G, Rizzo C, Monastero R, Marchese A, Caruso C (2012a) Mediterranean diet and longevity in Sicily: survey in a Sicani Mountains population. Rejuvenation Res 15:184–188

Vasto S, Rizzo C, Caruso C (2012b) Centenarians and diet: what they eat in the Western part of Sicily. Immun Ageing 9(1):10

Vasto S, Buscemi S, Barera A, Di Carlo M, Accardi G, Caruso C (2014a) Mediterranean diet and healthy ageing: a Sicilian perspective. Gerontology 60(6):508–518

Vasto S, Barera A, Rizzo C, Di Carlo M, Caruso C, Panotopoulos G (2014b) Mediterranean diet and longevity: an example of nutraceuticals? Curr Vasc Pharmacol 12(5):735–738

Verschuuren M, Gissler M, Kilpeläinen K, Tuomi-Nikula A, Sihvonen AP, Thelen J, Gaidelyte R, Ghirini S, Kirsch N, Prochorskas R, Scafato E (2013) Public health indicators for the EU: the joint action for ECHIM (European Community Health Indicators & Monitoring). Arch Public Health 71(1):12

Vishwanathan R, Chung M, Johnson EJ (2013) A systematic review on zinc for the prevention and treatment of age-related macular degeneration. Invest Ophthalmol Vis Sci 54(6):3985–3998

Visioli F, Bellom G, Montedoro GF, Galli C (1995) Low density lipoprotein oxidation is inhibited in vitro by olive oil constituents. Atherosclerosis 117:25–32

Wallin A, Di Giuseppe D, Orsini N, Patel PS, Forouhi NG, Wolk A (2012) Fish consumption, dietary long-chain n-3 fatty acids, and risk of type 2 diabetes: systematic review and meta-analysis of prospective studies. Diabetes Care 35(4):918–929

Walls AWG, Steele JG, Sheiham A, Marcenes W, Moynihan PJ (2000) Oral health and nutrition in older people. J Public Health Dent 60(4):304–307

Winge K, Rasmussen D, Werdelin LM (2003) Constipation in neurological diseases. J Neurol Neurosurg Psychiatry 74(1):13–19

Woodmansey EJ (2007) Intestinal bacteria and ageing. J Appl Microbiol 102(5):1 178–1 186

Woolf AD, Pfleger B (2003) Burden of major musculoskeletal conditions. Bull World Health Organ 81:646–656

Yehuda S, Rabinovitz S, Mostofsky DI, Huberman M, Sredni B (1997) Essential fatty acid preparation improves biochemical and cognitive functions in experimental allergic encephalomyelitis rat. Eur J Pharmacol 328:23–29

Yehuda S, Rabinovitz S, Mostofsky DI (1998) Modulation of learning and neuronal membrane composition in the rat by essential fatty acid preparation: time-course analysis. Neurochem Res 23:627–634

Yehuda S, Rabinovitz S, Mostofsky DI (1999) Essential fatty acids are mediators of brain biochemistry and cognitive functions. J Neurosci Res 56:565–570

Yehuda S, Rabinovitz S, Carasso RL, Mostofsky DI (2000) Fatty acid mixture counters stress changes in cortisol, cholesterol, and impair learning. Int J Neurosci 101:73–87

Technological Advances in Improving Bioavailability of Phytochemicals for the Treatment of Alzheimer's Disease

12

Mehtap Ozkur, Necla Benlier, Ilker Saygili, and Eda Ogut

Abstract

Alzheimer's disease (AD) is globally the most prevalent age-related neurodegenerative disease. There are drugs available for symptomatic treatment of the disease to delay the progression of symptoms of neurocognitive and physical decline. Studies have demonstrated many phytochemicals as candidates for the treatment of AD with anti-cholinesterase, anti-amyloidogenic, antioxidant, and anti-inflammatory effects. However, their properties differ from those of drugs. Many factors may play a critical role in limiting the bioavailability of these plant-originated chemicals. Therefore, studies have focused on developing new drug delivery systems like phytosomes, nanotechnology-based technologies, etc., to enable effective and reliable delivery of phytochemicals to the central nervous system (CNS). However, we still do not have any data on long-term treatment with phytochemicals delivered by these systems in AD.

Keywords

Alzheimer's disease · Phytochemicals · Bioavailability · Drug delivery

M. Ozkur (✉) · N. Benlier
Department of Medical Pharmacology, Faculty of Medicine, University of Sanko, Gaziantep, Turkey
e-mail: mozkur@sanko.edu.tr

I. Saygili · E. Ogut
Department of Medical Biochemistry, Faculty of Medicine, University of Sanko, Gaziantep, Turkey

S. M. Nabavi et al. (eds.), *Nutrients and Nutraceuticals for Active & Healthy Ageing*,
https://doi.org/10.1007/978-981-15-3552-9_12

Alzheimer's disease (AD) is globally the most prevalent age-related neurodegenerative disease associated with progressive memory impairment. AD is typical with overproduction of β-amyloid (Aβ) peptides and malfunction of the cholinergic system followed by cognitive decline leading to brain atrophy and death (Aminoff et al. 2015). It is the most prevalent cause of dementia. While there is no radical solution for AD, several drugs are available for symptomatic treatment including inhibitors of cholinesterase enzyme and *N*-methyl-D-aspartate receptor (NMDAR), to delay the progression of symptoms of neurocognitive and physical decline (Goodman 1996). Although many of the approved AD drugs are plant-derived compounds (galantamine, physostigmine, huperzine A), their limited effects and adverse side effects as well as lack of extensive knowledge on the neurobiological mechanisms underlying AD prompted search into new natural anti-AD pharmaceuticals (Huegel 2015). Chemical studies have demonstrated plenty of phytochemicals such as alkaloids, terpenes, and polyphenols that have multiple pharmacological activities including anti-cholinesterase, anti-amyloidogenic, anti-oxidant, and anti-inflammatory effects (Uddin et al. 2019). These multi-targeting features make these chemicals attractive candidates for the treatment of AD (Akram and Nawaz 2017; Kimura 2006). Moreover, regular intake of these phytochemicals may slow down the onset and the progression of AD and improve health status (Uddin et al. 2019). However, their properties differ from those of drugs. Many factors may play a critical role in limiting the bioavailability of these plant-derived chemicals such as solubility of the compound, metabolism by bowel microbiota, stability under gastric and intestinal pH conditions, absorption from the intestines, active efflux pumps, and presystemic degradation (Yang et al. 2008). Additionally, problems in passage of the compound through the blood–brain barrier and its subsequent delivery to the brain may also contribute to poor bioavailability (Van der Schyf et al. 2006; Shoji and Nakashima 2004). To overcome these limitations, studies have focused on new technologies to enable safe, suitable, and targeted delivery of phytochemicals to the CNS with the aim to find an optimal therapy for AD (Table 12.1) (Fig. 12.1) (Harvey et al. 2010). Currently, there is an ongoing research to devise new herbal drug delivery systems like sustained-controlled release formulations (e.g., drug-loaded gels, various nanocarriers), phytosomes (nano or not), and nanoparticles (Harvey et al. 2010; Devi et al. 2010).

12.1 Drug Delivery Systems for Alzheimer's Disease

12.1.1 Phytosome

Phytosome is a lipid-based drug delivery system or drug carrier that is formed through complexation of phytochemicals with a phospholipid. There are many phytosomal formulations in the market including *Ginkgo biloba* (gingko flavonoids), olive fruits or leaves extracts (polyphenols), curcumin and silybin (flavonoids). Phytosomal formulations confer high lipophilicity to the compound and protection of the phytochemical(s) from destruction by digestive enzymes and gut microflora. Thus, phytosomal products enhance the bioavailability of active

Table 12.1 Novel drug delivery systems for the most prevalent phytochemicals/plants/plant extracts in the treatment of AD

Phytochemical	Technology	Study model	References
Curcumin	Polymeric nanoparticles	LAG cell line	Mathew et al. (2012)
		Red blood cells	
		HT22 cells	Fan et al. (2018)
	Solid lipid nanoparticles	Rat	Sadegh Malvajerd et al. (2019)
	Metal nanoparticles	Murine Neuro-2a (N2a) cells	Kochi et al. (2015)
		Mouse	Bicer et al. (2018)
		In vitro model	Picciano and Vaden (2013)
	Micelles	In vitro model, U87MG cell line, rat	Desai and Patravale (2018)
	Liposomes	Post-mortem brain samples of AD patients	Mourtas et al. (2014)
	Nanoemulsion	SK-N-SH cells, sheep nasal mucosa	Sood et al. (2014)
Resveratrol	Solid lipid nanocarriers	Human brain capillary-like endothelial cells	Loureiro et al. (2017)
	Hydrogels	In vitro model, rats	Rajput et al. (2018)
	Transfersome/ nanoemulsion	Rat	Salem et al. (2019)
	Polymeric nanoparticles	PC12 cells	Lu et al. (2009)
Huperzine A	Polymeric nanoparticles	16HBE cell line, mouse	Meng et al. (2018)
	Solid lipid nanocarriers	Mouse	Patel et al. (2013)
Piperine	Polymeric nanoparticles	Rat	Elnaggar et al. (2015a)
	Cubosome	Rat	Elnaggar et al. (2015b)
	Microemulsion	Rat	Etman et al. (2018)
	Solid lipid nanocarriers	Rat	Yusuf et al. (2013)
Quercetin	Solid lipid nanocarriers	Zebrafish	Rishitha and Muthuraman (2018)
	Polymeric nanoparticles	Mouse	Sun et al. (2016)
	Metal nanoparticles	SH-SY5Y cells	Liu et al. (2019)
	Liposomes	SK-N-MC cells	Kuo and Tsao (2017)
Rivastigmine	Solid lipid nanocarriers	An in vitro model	Malekpour-Galogahi et al. (2018)
	Polymeric nanoparticles	An in vitro model	Wilson et al. (2011)
	Liposomes	Mouse	Mutlu et al. (2011)
	Microemulsion	Ex vivo goat nasal mucosa	Shah et al. (2015)

(continued)

Table 12.1 (continued)

Phytochemical	Technology	Study model	References
Galantamine	Solid lipid nanocarriers	Rat	Misra et al. (2016)
	Polymeric nanoparticles	Rat	Wahba et al. (2016)
	Liposomes	PC12 neuronal cells	Mufamadi et al. (2013)
	Gel type drug reservoir/patch	An in vitro model	Woo et al. (2015)
Nicotine	Polymeric nanoparticles	An in vitro model	Singh et al. (2006)
Ginsenoside Rg3	Polymeric nanoparticles	An in vitro model	Aalinkeel et al. (2018)
Ginkgo biloba extracts	Polymeric nanoparticles	An in vitro model	Wang et al. (2018)

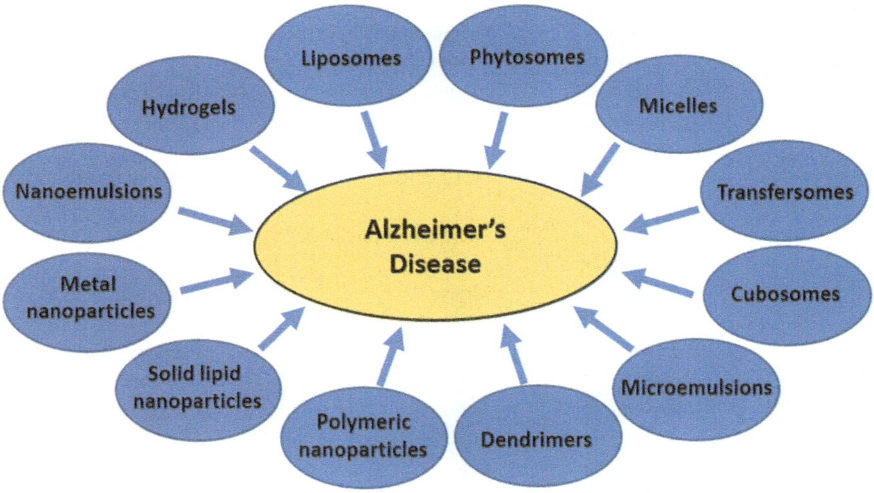

Fig. 12.1 Herbal drug delivery systems for Alzheimer's disease

phytochemical constituents and their therapeutic potential and are reported to be safe and well tolerated. They are formulated as soft/hard gelatin capsules and tablets (Harvey et al. 2010; Devi et al. 2010; Shakeri and Sahebkar 2016a).

12.1.2 Nanotechnology-Based Systems

In recent years, a tremendous progress has been achieved in nanotechnology for drug delivery. Nanosystems have been reported to achieve improved solubility of phytochemicals, reduced susceptibility to degradation in the gastrointestinal tract, increased concentrations in the bloodstream, and a lower incidence of adverse effects, particularly for water-soluble phytochemicals (Shakeri and Sahebkar 2016a, b).

12.1.3 Nanoparticles

They are sphere-like particles ranging from 1 and 100 nm in size. Numerous phytochemicals have been designed as nanoparticles (NPs) mostly by nanoprecipitation, evaporation, emulsion, lyophilization, emulsification, and homogenization procedures (Shakeri and Sahebkar 2016b; Jadhav et al. 2017). Some of the well-known nanodelivery systems for phytochemicals are liposomal formulations, nanostructured lipid carriers, polymeric nanoparticles, micelles, cubosomes, metal complexes, nanoemulsions, and nano-phytosomes. They can be administered via different drug administration routes (Devi et al. 2010; Shakeri and Sahebkar 2016a; Squillaro et al. 2018; Aw-Yong et al. 2018).

Liposomes are phospholipid vesicles that can encapsulate both hydrophilic and lipophilic molecules. Their unique structure preserves the molecule from premature degradation and dilution in the bloodstream (Squillaro et al. 2018; Hamidi et al. 2008).

Nanostructured lipid carriers (NLCs) are new generation solid liquid nanoparticles. They consist of water dispersed lipids and a surface-active compound containing aqueous solution (Squillaro et al. 2018; Aw-Yong et al. 2018). NCLs are physiological compounds which enable fast and large-scale production. They offer enhanced efficiency with targeted drug distribution to the brain. Their advantages include increased bioavailability as well as good tolerability. However, solid lipid nanocarriers have low drug-loading capacity and have stability problems(Squillaro et al. 2018; Aw-Yong et al. 2018; Ovais et al. 2018).

Polymeric nanoparticles are one of the most widely studied nanocarrier systems. They are solid colloidal particles. They are between 10 and 1000 nm in size. There are also hydrogels of polymeric nanoparticles (Hamidi et al. 2008). Typically, the drug encapsulated within the carrier is surrounded by biodegradable polymers. The unique advantages of these carriers include controlled release, stability, and degradation resistant encapsulation of the chemicals (Squillaro et al. 2018; Aw-Yong et al. 2018; Ovais et al. 2018). Chitosan, PLGA (poly (lactide-*co*-glycolide), and cyanoacrylate are polymer-based delivery systems (Ovais et al. 2018).

Surrounded by a hydrophilic shell, *micelles* are self-assembling colloidal particles with a diameter between 10 and 100 nm. The nature of the shell provides prolonged circulation time and stability of the integrated drug (Devi et al. 2010; Shakeri and Sahebkar 2016a; Squillaro et al. 2018; Ovais et al. 2018).

Metal complexes are another strategy for the improvement of new therapeutic applications. They are formed by complexation of a biologically active natural products with transition metals like copper, gadolinium, etc. The presence of the metal was shown to improve the stability, water solubility, and therapeutic effects of the phytochemicals, while reducing its side effects (Shakeri et al. 2019).

Cubosomes are cubic nanostructured liquid crystalline particles. They can encapsulate amphiphilic, hydrophobic, and hydrophilic substances. Other advantages of cubosomes are increased solubility, bioadhesion, and sustained release of the encapsulated chemical. Also, cubosomes have the potential to protect the drug from degradation in the biological environment and can be given by various routes of drug administration (Duttagupta et al. 2016).

Nanoemulsions are nano-sized ($r < 100$ nm) unmixable liquids (e.g., water and oil) balanced by convenient surfactants that can be produced in various dosage forms. They can be given via different routes of drug administration. By improving solubility, stability, permeability, and decreased metabolism in the liver, nanoemulsion formulations can increase the bioavailability and therapeutic effects of the drug while minimizing its adverse effects (Singh et al. 2017). Basically, there are two types of emulsions: nanoemulsions and microemulsions. There are differences and similarities between these emulsions, in terms of their compositions, structure, fabrication, properties. In general, microemulsions are thermodynamically more stable than nanoemulsions, an important characteristic for drug delivery (Tayeb and Sainsbury 2018).

Nano-transfersomes are novel phospholipid-containing flexible formulations developed specifically for transdermal delivery of the drug. They have better permeability characteristics compared to conventional liposomes (Rai et al. 2017).

12.2 Phytochemicals for Alzheimer's Disease: Bioavailability Problems and Novel Delivery Strategies

Phytochemicals are plant-derived, naturally occurring compounds. Polyphenols are the most abundant and widespread group of phytochemicals present as secondary metabolites of fruits, vegetables, and natural oils (Tsao 2010). Polyphenols have been shown to have preventive and/or therapeutic effects in many chronic pathological conditions including neurodegenerative disorders, cardiovascular diseases, diabetes, cancer, and obesity through their antioxidant and anti-inflammatory activities (Yao and Xue 2014). They typically have a classic phenol ring. Based upon the variations in their chemical structure, they are further divided into five major subclasses including flavonoids, phenolic acids, stilbenes, tannins, and lignans (Tsao 2010). Due to their beneficial effects, curcumin and resveratrol (RSV) are among the most extensively studied photochemicals as potential novel delivery systems for AD.

12.2.1 Curcumin

Curcumin (turmeric) is isolated from the root of *Curcuma Longa*. It is a phenolic acid derivative. There are a lot of reports suggesting its neuroprotective effects against dementia. Preventive and/or therapeutic effects of curcumin in AD involve inhibition of formation of Aβ plaques, lowering total cholesterol levels, inhibition of the apoptotic process, chelation of metal ions (esp. copper), and scavenging free radicals (Yao and Xue 2014).

For orally ingested chemicals, solubility, intestinal degradation, and hepatic metabolism are the major factors that determine their bioavailability. In humans, curcumin exhibits poor bioavailability. Insolubility in water, poor absorption, and fast metabolism are the major reasons contributing to low plasma and ineffective target tissue levels of curcumin. In an animal study, its oral bioavailability has been found to be 1% in rats. In both animals and human subjects, depending on the method of administration, curcumin was largely metabolized to the glucuronide and sulfate conjugates in the kidney, liver, and intestinal mucosa (Anand et al. 2007).

More effective new formulations have been developed and tested in order to enhance the bioavailability of curcumin. To overcome its low solubility in water, water-soluble PLGA coated-curcumin nanoparticles have been synthesized and coupled with Tet-1 peptide, to increase its affinity to neurons. It was found that encapsulated-PLGA (polyl(actide-co-glycolide)) nanoparticles are efficient in neuronal targeting, able to destroy amyloid aggregates, display similar antioxidant properties like raw curcumin and are not toxic to cells. The encapsulation does not result in changes in the original favorable properties of curcumin. Thus, this nanoparticle formulation has been suggested as a promising drug for the treatment of AD (Mathew et al. 2012). In a recent animal study, a newly designed brain tissue-targeted PLGA-PEG (poly(ethylene glycol))-coated curcumin nanoparticle conjugated with B6 peptide was found to increase curcumin cellular uptake and showed little influence on hemolysis rate and thrombo-elastography. Additionally, PLGA-PEG-B6/Cur was proven to improve memory capability and spatial learning in tested mice (Fan et al. 2018). Another study by Malvajerd et al. showed that curcumin-loaded nanostructured lipid carriers (Cur-NLCs) provided increased brain uptake of curcumin in an animal model of AD, compared to control group. In the same study, administration of Cur-NLCs effectively reduced oxidative markers, decreased Aβ deposition histopathologically, enhanced learning ability, and improved memory of rats (Sadegh Malvajerd et al. 2019). More recently, cocrystals of curcumin incorporated in micellar nanocarriers have been developed and tested to increase the bioavailability of the phytochemical. This experimental system revealed effective delivery of curcumin from nose to brain tissue and delayed clearance (Desai and Patravale 2018).

Another strategy to overcome some of the aforementioned limitations is to form metal complexes of curcumin. It is known that metal ions such as Al^{3+}, Hg^{2+}, Mn^{2+}, Cu^{2+} when found at high levels in brain may induce production of neurotoxic elements like reactive oxygen species through binding to specific sites in Aβ peptides. This is one of the factors underlying the amyloid pathology in Alzheimer's disease (Aminoff et al. 2015). Curcumin has the ability to chelate various metal ions. It was suggested that curcumin may compete with Aβ peptides for complexation with metal ions (Ovais et al. 2018). Recently, Picciano et al. reported that the Aβ protein has two binding sites, namely the metal ion-binding and curcumin-binding regions. In their study, they showed that curcumin–copper complex could bind to Cu2+ and Aβ binding sites at the same time and inhibit amyloid formation through two different binding sites (Picciano and Vaden 2013).

In a study, Kochi et al. reported that a derivative of curcumin that contains gadolinium (Gd-cur) had increased solubility and showed similar interaction with Aβ compared to curcumin. This metal derivative had also similar radical scavenging capability compared to Trolox (a vitamin E analog) (Kochi et al. 2015). Furthermore, curcumin-iron complex has been suggested to strengthen the memory. In the same study, Bicer et al. found that this complex was more effective than curcumin in terms of latent period measures (Bicer et al. 2018).

12.2.2 Resveratrol

Major dietary sources of resveratrol (RSV) include grapes, wine, blueberries, and nuts. It is a stilbenoid which is produced by herbs as a defensive response to injury or pathogens. RSV has garnered much attention owing to its healing and preventive effects in a number of chronic diseases. It has been demonstrated to reduce inflammation and oxidative damage through inhibition of Aβ synthesis and its neurotoxicity. Moreover, it has been observed to decrease insulin resistance and serum LDL cholesterol level which are associated with the development of AD (Drygalski et al. 2018). Due to its promising effects, RSV is now being evaluated in phase II clinical trials for the treatment of AD. However, the major limitation of RSV use human studies is its poor aqueous solubility, rapid metabolism, and photosensitivity in AD patients (Charytoniuk et al. 2017). Although its oral absorption has been shown to be about 75%, its half-life is only 8–14 min because of the rapid degradation into glucuronide and sulfate metabolites in hepatic cells and intestinal epithelial cells in humans. Therefore, new RSV delivery strategies addressing aforementioned issues related to its low bioavailability have been investigated (Ma et al. 2014).

The complex nature and multifactorial pathogenesis of AD led pharmaceutical manufacturers to develop multifunctional drugs with improved biological effects, namely multi-target-directed ligand (MTDL) drugs. MTDL drugs represent an innovative area in the search for novel drug candidates against neurologic diseases like AD. It is a hybrid compound which acts simultaneously on various targets (Cheng et al. 2018). Recently, Jerabek et al. reported a tacrine (an acetylcholinesterase (AChE) inhibitor)-resveratrol hybrid with good anti-AChE and anti Aβ activities. In their study, they also observed that the hybrid drug had high blood–brain permeability and low cytotoxicity in an experimental model of AD (Jeřábek et al. 2017). Similar results were reported in the literature for various hybrids of resveratrol with commercial drugs (Cheng et al. 2018; Yang et al. 2017).

In recent years, researchers focused on stability, solubility, and degradation problems of resveratrol and numerous formulations have been developed for different clinical targets (Ma et al. 2014; Amri et al. 2012). In a study by Loureiro et al., the anti-transferrin receptor monoclonal antibody (OX26 mAb) coupled with solid lipid nanoparticles (SLNs) were used to carry resveratrol and grape extract into the brain. They found that the specific OX26 mAb increased the penetration of nanoparticles across the blood–brain barrier more efficiently when compared with a non-specific LB 509Ab or solid lipid nanoparticle without any Ab. These results suggest that functionalized Ab coupled with solid lipid nanocarriers can be used for drugs that have low blood–brain barrier penetration ability (Loureiro et al. 2017). Very recently, Raiput et al. studied a nanostructured in situ gel of RSV in an AD model. They found that the new carrier system increased RSV penetration through nasal mucosa and strengthened the memory of rats better than the oral RSV suspension and in situ gel formulations of RSV. These results suggest that the novel nanostructured product is a promising approach for brain targeting of RSV in the treatment of AD (Rajput et al. 2018).

12.3 Future Perspectives and Conclusion

Phytochemical-loaded nanocarriers are promising drug delivery systems due to their higher efficacy and specificity for AD patients. However, we do not have any data on long-term treatment with phytochemicals delivered by these systems. As with many other chronic diseases, AD requires long-term, frequent administration of the phytochemicals for the desired therapeutic effectiveness. Additionally, nanotechnology is an innovative scientific field in pharmacotherapy. Despite several advantages of nanotechnological products, there are reports on the interaction of nanoparticles with endogenous molecules in the bloodstream (Subbiah et al. 2010). Once the nanoparticles enter the body, they travel more freely than larger molecules in the body and even reach the fetus. They can enter the cells easily and in an uncontrolled fashion and may cause oxidative stress, protein denaturation, organelle and cellular dysfunctions, genotoxicity, inflammation, leading to exacerbation of asthma, allergy, fibrosis, chronic inflammation, and carcinogenesis. Specifically, polymeric nanoparticles used for drug delivery were found to be toxic for macrophages and may cause concentration-dependent mortality in tested animals (Gupta and Xie 2018).

The binding of a metal ion to a biologically active compound represents a new therapeutic intervention to modify pharmacological properties of the ligand. Although there are many reports on the use of numerous metal complexes in biology including diagnostic, anti-cancer, anti-diabetic, anti-inflammatory, antimicrobial, and anti-ulcer applications, no data exists on the pharmacokinetic properties of most metal-curcumin complexes following administration into the human body (Ovais et al. 2018; Liu et al. 2018).

Nasal drug delivery is a promising area especially when the brain is targeted. It offers many advantages including escape from the blood–brain barrier (hence, faster onset of action), bypass hepatic first-pass metabolism, less systemic side effects, and non-invasive route of drug administration. However, nasal mucosa barrier, enzymatic degradation, mucociliary activity, local toxicity of the chemical, and several other drug- or patient-related factors may complicate the delivery of the formulation into the brain tissue and limit the efficacy of the therapy (Agrawal et al. 2018).

Additionally, whatever the delivery system is, increased blood–brain barrier permeability may elevate the concentration of the phytochemical in the brain tissue, potentially causing neurotoxicity.

There are still a lot of questions to be answered regarding the cost, tolerability, mechanism of action, efficiency, side effects, and long-term toxicities of both phytochemicals and new delivery systems in chronic diseases such as AD.

References

Aalinkeel R, Kutscher HL, Singh A, Cwiklinski K, Khechen N, Schwartz SA et al (2018) Neuroprotective effects of a biodegradable poly (lactic-co-glycolic acid)-ginsenoside Rg3 nanoformulation: a potential nanotherapy for Alzheimer's disease? J Drug Target 26 (2):182–193

Agrawal M, Saraf S, Saraf S, Antimisiaris SG, Chougule MB, Shoyele SA et al (2018) Nose-to-brain drug delivery: an update on clinical challenges and progress towards approval of anti-Alzheimer drugs. J Control Release 281:139–177

Akram M, Nawaz A (2017) Effects of medicinal plants on Alzheimer's disease and memory deficits. Neural Regen Res 12(4):660

Aminoff M, Greenberg D, Simon R (2015) Clinical neurology. McGraw-Hill Education, New York

Amri A, Chaumeil J, Sfar S, Charrueau C (2012) Administration of resveratrol: what formulation solutions to bioavailability limitations? J Control Release 158(2):182–193

Anand P, Kunnumakkara AB, Newman RA, Aggarwal BB (2007) Bioavailability of curcumin: problems and promises. Mol Pharm 4(6):807–818

Aw-Yong PY, Gan PH, Sasmita AO, Mak ST, Ling A (2018) Nanoparticles as carriers of phytochemicals: recent applications against lung cancer. Int J Res Biomed Biotechnol 7:1–11

Bicer N, Yildiz E, Yegani AA, Aksu F (2018) Synthesis of curcumin complexes with iron (iii) and manganese (ii), and effects of curcumin–iron (iii) on Alzheimer's disease. New J Chem 42 (10):8098–8104

Charytoniuk T, Drygalski K, Konstantynowicz-Nowicka K, Berk K, Chabowski A (2017) Alternative treatment methods attenuate the development of NAFLD: a review of resveratrol molecular mechanisms and clinical trials. Nutrition 34:108–117

Cheng G, Xu P, Zhang M, Chen J, Sheng R, Ma Y (2018) Resveratrol-maltol hybrids as multi-target-directed agents for Alzheimer's disease. Bioorg Med Chem 26(22):5759–5765

Desai PP, Patravale VB (2018) Curcumin cocrystal micelles—multifunctional nanocomposites for management of neurodegenerative ailments. J Pharm Sci 107(4):1143–1156

Devi VK, Jain N, Valli KS (2010) Importance of novel drug delivery systems in herbal medicines. Pharmacogn Rev 4(7):27

Drygalski K, Fereniec E, Koryciński K, Chomentowski A, Kiełczewska A, Odrzygóźdź C et al (2018) Resveratrol and Alzheimer's disease. From molecular pathophysiology to clinical trials. Exp Gerontol 113:36

Duttagupta AS, Chaudhary HM, Jadhav KR, Kadam VJ (2016) Cubosomes: innovative nanostructures for drug delivery. Curr Drug Deliv 13(4):482–493

Elnaggar YS, Etman SM, Abdelmonsif DA, Abdallah OY (2015a) Intranasal piperine-loaded chitosan nanoparticles as brain-targeted therapy in Alzheimer's disease: optimization, biological efficacy, and potential toxicity. J Pharm Sci 104(10):3544–3556

Elnaggar YS, Etman SM, Abdelmonsif DA, Abdallah OY (2015b) Novel piperine-loaded Tween-integrated monoolein cubosomes as brain-targeted oral nanomedicine in Alzheimer's disease: pharmaceutical, biological, and toxicological studies. Int J Nanomedicine 10:5459

Etman SM, Elnaggar YS, Abdelmonsif DA, Abdallah OY (2018) Oral brain-targeted microemulsion for enhanced piperine delivery in Alzheimer's disease therapy: in vitro appraisal, in vivo activity, and nanotoxicity. AAPS PharmSciTech 19(8):3698–3711

Fan S, Zheng Y, Liu X, Fang W, Chen X, Liao W et al (2018) Curcumin-loaded PLGA-PEG nanoparticles conjugated with B6 peptide for potential use in Alzheimer's disease. Drug Deliv 25(1):1091–1102

Goodman LS (1996) Goodman and Gilman's the pharmacological basis of therapeutics. McGraw-Hill, New York

Gupta R, Xie H (2018) Nanoparticles in daily life: applications, toxicity and regulations. J Environ Pathol Toxicol Oncol 37(3):209

Hamidi M, Azadi A, Rafiei P (2008) Hydrogel nanoparticles in drug delivery. Adv Drug Deliv Rev 60(15):1638–1649

Harvey AL, Clark RL, Mackay SP, Johnston BF (2010) Current strategies for drug discovery through natural products. Expert Opin Drug Discovery 5(6):559–568

Huegel HM (2015) Brain food for Alzheimer-free ageing: focus on herbal medicines. In: Natural compounds as therapeutic agents for amyloidogenic diseases. Springer, Cham, pp 95–116

Jadhav NR, Nadaf SJ, Lohar DA, Ghagare PS, Powar TA (2017) Phytochemicals formulated as nanoparticles: inventions, recent patents and future prospects. Recent Pat Drug Deliv Formul 11 (3):173–186

Jeřábek J, Uliassi E, Guidotti L, Korábečný J, Soukup O, Sepsova V et al (2017) Tacrine-resveratrol fused hybrids as multi-target-directed ligands against Alzheimer's disease. Eur J Med Chem 127:250–262

Kimura I (2006) Medical benefits of using natural compounds and their derivatives having multiple pharmacological actions. Yakugaku Zasshi 126(3):133–143

Kochi A, Jin Lee H, Vithanarachchi SM, Padmini V, Allen JM, Hee Lim M (2015) Inhibitory activity of curcumin derivatives towards metal-free and metal-induced amyloid-β aggregation. Curr Alzheimer Res 12(5):415–423

Kuo Y-C, Tsao C-W (2017) Neuroprotection against apoptosis of SK-N-MC cells using RMP-7- and lactoferrin-grafted liposomes carrying quercetin. Int J Nanomedicine 12:2857

Liu H, Qu Y, Wang X (2018) Amyloid β-targeted metal complexes for potential applications in Alzheimer's disease. Future Med Chem 10(6):679–701

Liu Y, Zhou H, Yin T, Gong Y, Yuan G, Chen L et al (2019) Quercetin-modified gold-palladium nanoparticles as a potential autophagy inducer for the treatment of Alzheimer's disease. J Colloid Interface Sci 552:388–400

Loureiro J, Andrade S, Duarte A, Neves A, Queiroz J, Nunes C et al (2017) Resveratrol and grape extract-loaded solid lipid nanoparticles for the treatment of Alzheimer's disease. Molecules 22 (2):277

Lu X, Ji C, Xu H, Li X, Ding H, Ye M et al (2009) Resveratrol-loaded polymeric micelles protect cells from Aβ-induced oxidative stress. Int J Pharm 375(1-2):89–96

Ma T, Tan M-S, Yu J-T, Tan L (2014) Resveratrol as a therapeutic agent for Alzheimer's disease. Biomed Res Int 2014:1

Malekpour-Galogahi F, Hatamian-Zarmi A, Ganji F, Ebrahimi-Hosseinzadeh B, Nojoki F, Sahraeian R et al (2018) Preparation and optimization of rivastigmine-loaded tocopherol succinate-based solid lipid nanoparticles. J Liposome Res 28(3):226–235

Mathew A, Fukuda T, Nagaoka Y, Hasumura T, Morimoto H, Yoshida Y et al (2012) Curcumin loaded-PLGA nanoparticles conjugated with Tet-1 peptide for potential use in Alzheimer's disease. PLoS One 7(3):e32616

Meng Q, Wang A, Hua H, Jiang Y, Wang Y, Mu H et al (2018) Intranasal delivery of Huperzine A to the brain using lactoferrin-conjugated N-trimethylated chitosan surface-modified PLGA nanoparticles for treatment of Alzheimer's disease. Int J Nanomedicine 13:705

Misra S, Chopra K, Saikia UN, Sinha VR, Sehgal R, Modi M et al (2016) Effect of mesenchymal stem cells and galantamine nanoparticles in rat model of Alzheimer's disease. Regen Med 11 (7):629–646

Mourtas S, Lazar AN, Markoutsa E, Duyckaerts C, Antimisiaris SG (2014) Multifunctional nanoliposomes with curcumin–lipid derivative and brain targeting functionality with potential applications for Alzheimer disease. Eur J Med Chem 80:175–183

Mufamadi MS, Choonara YE, Kumar P, Modi G, Naidoo D, van Vuuren S et al (2013) Ligand-functionalized nanoliposomes for targeted delivery of galantamine. Int J Pharm 448(1):267–281

Mutlu NB, Değim Z, Yılmaz Ş, Eşsiz D, Nacar A (2011) New perspective for the treatment of Alzheimer diseases: liposomal rivastigmine formulations. Drug Dev Ind Pharm 37(7):775–789

Ovais M, Zia N, Ahmad I, Khalil AT, Raza A, Ayaz M et al (2018) Phyto-therapeutic and nanomedicinal approaches to cure Alzheimer's disease: present status and future opportunities. Front Aging Neurosci 10:284

Patel PA, Patil SC, Kalaria DR, Kalia YN, Patravale VB (2013) Comparative in vitro and in vivo evaluation of lipid based nanocarriers of Huperzine A. Int J Pharm 446(1-2):16–23

Picciano AL, Vaden TD (2013) Complexation between Cu (II) and curcumin in the presence of two different segments of amyloid β. Biophys Chem 184:62–67

Rai S, Pandey V, Rai G (2017) Transfersomes as versatile and flexible nano-vesicular carriers in skin cancer therapy: the state of the art. Nano Rev Exp 8(1):1325708

Rajput A, Bariya A, Allam A, Othman S, Butani SB (2018) In situ nanostructured hydrogel of resveratrol for brain targeting: in vitro-in vivo characterization. Drug Deliv Transl Res 8 (5):1460–1470

Rishitha N, Muthuraman A (2018) Therapeutic evaluation of solid lipid nanoparticle of quercetin in pentylenetetrazole induced cognitive impairment of zebrafish. Life Sci 199:80–87

Sadegh Malvajerd S, Izadi Z, Azadi A, Kurd M, Derakhshankhah H, Sharif Zadeh M et al (2019) Neuroprotective potential of curcumin-loaded nanostructured lipid carrier in an animal model of Alzheimer's disease: behavioral and biochemical evidence. J Alzheimers Dis 69:671

Salem HF, Kharshoum RM, Abou-Taleb HA, Naguib DM (2019) Brain targeting of resveratrol through intranasal lipid vesicles labelled with gold nanoparticles: in vivo evaluation and bioaccumulation investigation using computed tomography and histopathological examination. J Drug Target 27:1127–1134

Shah BM, Misra M, Shishoo CJ, Padh H (2015) Nose to brain microemulsion-based drug delivery system of rivastigmine: formulation and ex-vivo characterization. Drug Deliv 22(7):918–930

Shakeri A, Sahebkar A (2016a) Opinion paper: phytosome: a fatty solution for efficient formulation of phytopharmaceuticals. Recent Pat Drug Deliv Formul 10(1):7–10

Shakeri A, Sahebkar A (2016b) Opinion paper: nanotechnology: a successful approach to improve oral bioavailability of phytochemicals. Recent Pat Drug Deliv Formul 10(1):4–6

Shakeri A, Panahi Y, Johnston TP, Sahebkar A (2019) Biological properties of metal complexes of curcumin. Biofactors 45(3):304–317

Shoji Y, Nakashima H (2004) Nutraceutics and delivery systems. J Drug Target 12(6):385–391

Singh N, Seedat F, Pillay V, Sweet JL, Danckwerts MP (2006) Formulation and statistical optimization of novel double-incorporated PLA-PLGA microparticles within an alginate-pectinate platform for the delivery of nicotine. J Microencapsul 23(2):153–167

Singh Y, Meher JG, Raval K, Khan FA, Chaurasia M, Jain NK et al (2017) Nanoemulsion: concepts, development and applications in drug delivery. J Control Release 252:28–49

Sood S, Jain K, Gowthamarajan K (2014) Optimization of curcumin nanoemulsion for intranasal delivery using design of experiment and its toxicity assessment. Colloids Surf B: Biointerfaces 113:330–337

Squillaro T, Cimini A, Peluso G, Giordano A, Melone MAB (2018) Nano-delivery systems for encapsulation of dietary polyphenols: an experimental approach for neurodegenerative diseases and brain tumors. Biochem Pharmacol 154:303–317

Subbiah R, Veerapandian M, Yun KS (2010) Nanoparticles: functionalization and multifunctional applications in biomedical sciences. Curr Med Chem 17(36):4559–4577

Sun D, Li N, Zhang W, Zhao Z, Mou Z, Huang D et al (2016) Design of PLGA-functionalized quercetin nanoparticles for potential use in Alzheimer's disease. Colloids Surf B: Biointerfaces 148:116–129

Tayeb HH, Sainsbury F (2018) Nanoemulsions in drug delivery: formulation to medical application. Nanomedicine 13(19):2507–2525

Tsao R (2010) Chemistry and biochemistry of dietary polyphenols. Nutrients 2(12):1231–1246

Uddin MS, Al Mamun A, Kabir MT, Jakaria M, Mathew B, Barreto GE et al (2019) Nootropic and anti-Alzheimer's actions of medicinal plants: molecular insight into therapeutic potential to alleviate Alzheimer's neuropathology. Mol Neurobiol 56(7):4925–4944

Van der Schyf CJ, Geldenhuys WJ, Youdim MB (2006) Multifunctional drugs with different CNS targets for neuropsychiatric disorders. J Neurochem 99(4):1033–1048

Wahba SM, Darwish AS, Kamal SM (2016) Ceria-containing uncoated and coated hydroxyapatite-based galantamine nanocomposites for formidable treatment of Alzheimer's disease in ovariectomized albino-rat model. Mater Sci Eng C 65:151–163

Wang T, Wu C, Fan G, Li T, Gong H, Cao F (2018) Ginkgo biloba extracts-loaded starch nanospheres: preparation, characterization, and in vitro release kinetics. Int J Biol Macromol 106:148–157

Wilson B, Samanta MK, Muthu MS, Vinothapooshan G (2011) Design and evaluation of chitosan nanoparticles as novel drug carrier for the delivery of rivastigmine to treat Alzheimer's disease. Ther Deliv 2(5):599–609

Woo FY, Basri M, Masoumi HRF, Ahmad MB, Ismail M (2015) Formulation optimization of galantamine hydrobromide loaded gel drug reservoirs in transdermal patch for Alzheimer's disease. Int J Nanomedicine 10:3879

Yang CS, Sang S, Lambert JD, Lee MJ (2008) Bioavailability issues in studying the health effects of plant polyphenolic compounds. Mol Nutr Food Res 52(S1):S139–SS51

Yang X, Qiang X, Li Y, Luo L, Xu R, Zheng Y et al (2017) Pyridoxine-resveratrol hybrids Mannich base derivatives as novel dual inhibitors of AChE and MAO-B with antioxidant and metal-chelating properties for the treatment of Alzheimer's disease. Bioorg Chem 71:305–314

Yao EC, Xue L (2014) Therapeutic effects of curcumin on Alzheimer's disease. Adv Alzheimers Dis 3(04):145

Yusuf M, Khan M, Khan RA, Ahmed B (2013) Preparation, characterization, in vivo and biochemical evaluation of brain targeted Piperine solid lipid nanoparticles in an experimentally induced Alzheimer's disease model. J Drug Target 21(3):300–311

Conclusion

13

Grazia D'Onofrio and Seyed M. Nabavi

Abstract

The main topic is the promotion of active and healthy ageing (AHA) through the use of appropriate nutrients or diets. AHA is one of the societal challenges we will be facing in the upcoming decades with an impact on society and economy.

Nutrients and nutraceuticals may be used to improve health, supporting the functions of the body, improving the cognitive and physical performance, reducing the risk factors for chronic diseases, increasing the neuronal cell survival, and increasing the life expectancy. The dietary constituents have anti-cholinergic, antioxidant, and anti-inflammatory features. The nutritional approach has the advantage of being safe and devoid of adverse and side effects, cost effective, and easy to be acceptable for the patients.

In this chapter, the paper collection of the book is summarized. The papers target an audience of practicing researchers, academics, and scientists from Iran, Italy, Pakistan, India, Saudi Arabia, and Turkey. The book contents were written by multiple authors and edited by experts in clinical and research fields in order to increase the knowledge on nutrients and nutraceuticals considering age-related and multidimensional approaches to the purpose of appropriate and personalized treatment.

Keywords

Ageing · Nutraceuticals · Anti-oxidant properties · Anti-inflammatory activity · Anti-cholinergic activity

G. D'Onofrio
Clinical Psychology Service, Health Department, Fondazione IRCCS Casa Sollievo della Sofferenza, San Giovanni Rotondo, Italy

S. M. Nabavi (✉)
Baqiyatallah University of Medical Sciences, Tehran, Iran

S. M. Nabavi et al. (eds.), *Nutrients and Nutraceuticals for Active & Healthy Ageing*,
https://doi.org/10.1007/978-981-15-3552-9_13

Age is a major risk factor in several diseases of developed countries (Niccoli et al. 2012). The process of aging involves a matched program of changes. Then, positive changes to factors such as diet and lifestyle can promote active and healthy aging (Mak and Caldeira 2014) (AHA), and help to delay the onset of the negative age-related diseases. AHA is one of the societal challenges we will be facing in the upcoming decades with an impacts on society and economy (Smart Specialisation Platform. European Commission n.d.). AHA is the process of developing and maintaining the functional ability that enables wellbeing in older age, and includes a person's ability to meet his/her basic needs, to learn, to grow and make decisions, to be mobile, to build and maintain relationships, and to contribute to the society (World Health Organization n.d.).

The main topic of the book was the promotion of AHA through the use of appropriate nutrients or diets.

Nutrients and nutraceuticals may be used to improve health, supporting the functions of the body, improving the cognitive and physical performance, reducing risk factors for chronic diseases, increasing neuronal cell survival, and increasing life expectancy. The dietary constituents have anti-cholinergic, anti-oxidant, and anti-inflammatory features. The nutritional approach has the advantage of being safe and devoid of adverse and side effects, cost effective, and easy to be acceptable for the patients. The book is a collection of papers targeting an audience of practicing researchers, academics, and other scientists from Iran, Italy, Pakistan, India, Saudi Arabia, and Turkey. Its contents was written by multiple authors and edited by experts in clinical and researcher fields.

They contributed to increasing the knowledge on nutrients and nutraceuticals considering age-related and multidimensional approaches to the purpose of appropriate and personalized treatment.

In the first chapter, Della Grace Thomas Parambi and colleagues discussed about demographic and epidemiological aspects of aging confirming that health promotion schemes and health care management strategies have propelled an abrupt rise in the survival rates of elders, accompanied by the increasing need for trained personnel, specialized care, and budgetary policies to earmark funds from young taxpayers to pay for geriatric care. By the authors, the governments should formulate feasible strategies for training in specialized care, expanding insurance cover, generating supportive technologies, and creating a skilled workforce, in addition to legally empowering helpless elders who are vulnerable to exploitation.

In the successive chapter, S. Sathya and K. Pandima Devi supported to conceptualize the healthy aging phenotypes (HAP) which is at the early stage of research. The data discussed in the study helps other researchers to assess the possible utility of the tools to measure the HAP and able them to include still more tools in respective domains of HAP for their enhanced assessment. The authors concluded that currently only holistic assessment of HAP can be achieved by measuring the five domains of HAP such as physiological and metabolic health, physical capability, cognitive function, psychological and subjective wellbeing, and social wellbeing. Other types of HAP assessment are still in development.

In the third chapter, Touqeer Ahmed et al. discussed various manifestation of aging, highlighting that the social and mental stressors are also a major cause of early aging with consequent issues of isolation, cognitive decline, and depression. The authors have described the main contributors of aging associated diseases.

In the Chap. 4, Giuseppe Annunziata and colleagues described the nutritional status assessment as a crucial item in management of elderly people and a fundamental tool to improve life span. The authors eviscerated all clinical outcomes, including evaluation of clinical signs, blood parameters, nutritional status, anthropometric parameters, and body composition. Therefore, Annunziata et al. provided a useful and practical support for nutritional assessment in old subjects.

In the successive chapter, Mahshid Hodjat et al. discussed recent findings on the most widely studied nutrients and nutraceuticals that assist to prevent age-related diseases and decelerate aging process. The authors categorized the nutrients based on their contributions to different age-related symptoms.

In the sixth chapter, Amir Hossein Abdolghaffari and colleagues reported the recent data from preclinical and clinical studies which indicate that dietary fibers exert pervasive and prominent role in controlling the mechanisms of aging and its associated diseases.

In the Chap. 7, Testai L. and Calderone V. documented the ability of polyphenolic compounds to attenuate the effects of a wide variety of human disease models.

In the eighth chapter, Sepideh Goudarzi and Mohammad Abdollahi showed that prevention of inflammation could be beneficial at any age. Many of the anti-inflammatory nutrients and nutraceutical substances mentioned in the chapter are able to combat contrast inflammation through several pathways, preventing cancers, diabetes dyslipidemia, atherosclerosis, etc.

In the successive chapter, Abida Zulfiqar and colleagues discussed about the antioxidant nutrients and nutraceuticals in aging confirming that they can decrease the process of oxidative damage to the cells and tissues. Found in various fruits and vegetable, such as cherries, orange carrots, purple berries, etc., anti-oxidants can fight against cancer and a wide range of other diseases. The authors reported that the anti-oxidants slow the process of aging by reducing the amount of free radicals production which causes damage to our body.

In the tenth chapter, Zahra Bayramia et al. attempt to explain cognitive impairment related to aging, definition of functional foods and dietary patterns and their classifications, the role of functional foods and dietary patterns on cognitive function, and the possible mechanisms and pathways involved in the effects of dietary patterns on cognitive decline related to aging.

In the following chapter, Nida Noreen and colleagues discussed about the valuable impact of Mediterranean diet (MD) on healthy aging. By the authors, the MD diet has proven to be the best tool to counteract the degenerative processes and promote an active healthy aging.

In the final chapter, Mehtap Ozkur et al. reported the technological advances in improving bioavailability of phytochemicals for the treatment of Alzheimer's disease. The authors documented that nanotechnology is an innovative scientific field in pharmacotherapy and about its several advantages, there are reports on the interaction of nanoparticles with endogenous molecules in the bloodstream.

In light of this book, a substantial impact on the potential benefits of nutrients and nutraceuticals has been correlated with AHA. Further studies are required to fully understand the complex interactions between natural compounds and age-related diseases and to determine the exact nature of their subsequent local and systemic health benefits.

References

Mak T, Caldeira S (2014) The role of nutrition in active and healthy ageing. For prevention and treatment of age-related diseases: evidence so far. Publications Office of the European Union, Luxembourg. https://doi.org/10.2788/83557

Niccoli T, Partridge L (2012) Ageing as a risk factor for disease. Curr Biol 22(17):R741–R752

Smart Specialisation Platform. European Commission (n.d.) Active and healthy ageing. https://s3platform.jrc.ec.europa.eu/active-and-healthy-ageing

World Health Organization (n.d.) Ageing and life-course. https://www.who.int/ageing/healthy-ageing/en/

The manufacturer's authorised representative in the EU is Springer Nature Customer Service Centre GmbH, Europaplatz 3, 69115 Heidelberg, Germany. If you have any concerns regarding our products, please contact ProductSafety@springernature.com

Printed and bound by CPI Group (UK) Ltd, Croydon, CR0 4YY
30/09/2026
02225333-0010